Manuel Graña and Richard J. Duro (Eds.)

Computational Intelligence for Remote Sensing

Studies in Computational Intelligence, Volume 133

Editor-in-Chief
Prof. Janusz Kacprzyk
Systems Research Institute
Polish Academy of Sciences
ul. Newelska 6
01-447 Warsaw
Poland
E-mail: kacprzyk@ibspan.waw.pl

Further volumes of this series can be found on our homepage: springer.com

Vol. 113. Gemma Bel-Enguix, M. Dolores Jiménez-López and Carlos Martín-Vide (Eds.)
New Developments in Formal Languages and Applications, 2008
ISBN 978-3-540-78290-2

Vol. 114. Christian Blum, Maria José Blesa Aguilera, Andrea Roli and Michael Sampels (Eds.)
Hybrid Metaheuristics, 2008
ISBN 978-3-540-78294-0

Vol. 115. John Fulcher and Lakhmi C. Jain (Eds.)
Computational Intelligence: A Compendium, 2008
ISBN 978-3-540-78292-6

Vol. 116. Ying Liu, Aixin Sun, Han Tong Loh, Wen Feng Lu and Ee-Peng Lim (Eds.)
Advances of Computational Intelligence in Industrial Systems, 2008
ISBN 978-3-540-78296-4

Vol. 117. Da Ruan, Frank Hardeman and Klaas van der Meer (Eds.)
Intelligent Decision and Policy Making Support Systems, 2008
ISBN 978-3-540-78306-0

Vol. 118. Tsau Young Lin, Ying Xie, Anita Wasilewska and Churn-Jung Liau (Eds.)
Data Mining: Foundations and Practice, 2008
ISBN 978-3-540-78487-6

Vol. 119. Slawomir Wiak, Andrzej Krawczyk and Ivo Dolezel (Eds.)
Intelligent Computer Techniques in Applied Electromagnetics, 2008
ISBN 978-3-540-78489-0

Vol. 120. George A. Tsihrintzis and Lakhmi C. Jain (Eds.)
Multimedia Interactive Services in Intelligent Environments, 2008
ISBN 978-3-540-78491-3

Vol. 121. Nadia Nedjah, Leandro dos Santos Coelho and Luiza de Macedo Mourelle (Eds.)
Quantum Inspired Intelligent Systems, 2008
ISBN 978-3-540-78531-6

Vol. 122. Tomasz G. Smolinski, Mariofanna G. Milanova and Aboul-Ella Hassanien (Eds.)
Applications of Computational Intelligence in Biology, 2008
ISBN 978-3-540-78533-0

Vol. 123. Shuichi Iwata, Yukio Ohsawa, Shusaku Tsumoto, Ning Zhong, Yong Shi and Lorenzo Magnani (Eds.)
Communications and Discoveries from Multidisciplinary Data, 2008
ISBN 978-3-540-78732-7

Vol. 124. Ricardo Zavala Yoe
Modelling and Control of Dynamical Systems: Numerical Implementation in a Behavioral Framework, 2008
ISBN 978-3-540-78734-1

Vol. 125. Larry Bull, Bernadó-Mansilla Ester and John Holmes (Eds.)
Learning Classifier Systems in Data Mining, 2008
ISBN 978-3-540-78978-9

Vol. 126. Oleg Okun and Giorgio Valentini (Eds.)
Supervised and Unsupervised Ensemble Methods and their Applications, 2008
ISBN 978-3-540-78980-2

Vol. 127. Régie Gras, Einoshin Suzuki, Fabrice Guillet and Filippo Spagnolo (Eds.)
Statistical Implicative Analysis, 2008
ISBN 978-3-540-78982-6

Vol. 128. Fatos Xhafa and Ajith Abraham (Eds.)
Metaheuristics for Scheduling in Industrial and Manufacturing Applications, 2008
ISBN 978-3-540-78984-0

Vol. 129. Natalio Krasnogor, Giuseppe Nicosia, Mario Pavone and David Pelta (Eds.)
Nature Inspired Cooperative Strategies for Optimization (NICSO 2007), 2008
ISBN 978-3-540-78986-4

Vol. 130. Richi Nayak, Nikhil Ichalkaranje and Lakhmi C. Jain (Eds.)
Evolution of the Web in Artificial Intelligence Environments, 2008
ISBN 978-3-540-79140-9

Vol. 131. Roger Lee and Haeng-Kon Kim (Eds.)
Computer and Information Science, 2008
ISBN 978-3-540-79186-7

Vol. 132. Danil Prokhorov (Ed.)
Computational Intelligence in Automotive Applications, 2008
ISBN 978-3-540-79256-7

Vol. 133. Manuel Graña and Richard J. Duro (Eds.)
Computational Intelligence for Remote Sensing, 2008
ISBN 978-3-540-79352-6

Manuel Graña
Richard J. Duro
(Eds.)

Computational Intelligence for Remote Sensing

Springer

Manuel Graña
Universidad Pais Vasco
Facultad de Informática
20018 San Sebastian
Spain
e-mail: manuel.grana@ehu.es

Richard Duro
Universidad de A Coruña
Grupo Integrado de Ingeniería
Escuela Politécnica Superior
c/ Mendizábal s/n
15403 Ferrol (A Coruña)
Spain
e-mail: richard@udc.es

ISBN 978-3-540-79352-6 e-ISBN 978-3-540-79353-3

DOI 10.1007/978-3-540-79353-3

Studies in Computational Intelligence ISSN 1860-949X

Library of Congress Control Number: 2008925271

Typeset & Cover Design: Scientific Publishing Services Pvt. Ltd., Chennai, India.

Printed on acid-free paper

9 8 7 6 5 4 3 2 1

springer.com

Preface

This book is a composition of diverse points of view regarding the application of Computational Intelligence techniques and methods into Remote Sensing data and problems. It is the general consensus that classification, and related data processing, and global optimization methods are the main topics of Computational Intelligence. Global random optimization algorithms appear in this book, such as the Simulated Annealing in chapter 6 and the Genetic Algorithms proposed in chapters 3 and 9. Much of the contents of the book are devoted to image segmentation and recognition, using diverse tools from regions of Computational Intelligence, ranging from Artificial Neural Networks to Markov Random Field modelling. However, there are some fringe topics, such the parallel implementation of some algorithms or the image watermarking that make evident that the frontiers between Computational Intelligence and neighboring computational disciplines are blurred and the fences run low and full of holes in many places.

The book starts with a review of the current designs of hyperspectral sensors, more appropriately named Imaging Spectrometers. Knowing the shortcomings and advantages of the diverse designs may condition the results on some applications of Computational Intelligence algorithms to the processing and understanding of them Remote Sensing images produced by these sensors. Then the book contents moves into basic signal processing techniques such as compression and watermarking applied to remote sensing images. With the huge amount of remote sensing information and the increasing rate at which it is being produced, it seems only natural that compression techniques will leap into a prominent role in the near future, overcoming the resistances of the users against uncontrolled manipulation of "their" data. Watermarking is the way to address issues of ownership authentication in digital contents. The enormous volume of information asks also for advanced information management systems, able to provide intelligent query process, as well as to provide for cooperative manipulation of the images through autonomously provided web services, streamed through special web portals, such as the one provided by the European Space Agency (ESA).

The main contents of the book are devoted to image analysis and efficient (parallel) implementations of such analysis techniques. The processes include image segmentation, change detection, endmember extraction for spectral unmixing, and feature extraction. Diverse kinds of Artificial Neural Networks, Mathematical Morphology and Markov Random Fields are applied to these tasks. The kind of images are mostly multispectral-hyperspectral images, with some examples of processing Synthetic Aperture Radar images, whose appeal lies in its insensitivity to atmospheric conditions. Two specific applications stand out. One is forest fire detection and prevention, the other is quality inspection using hyperspectral images.

Chapter 1 provides a review of current Imaging Spectrometer designs. They focus on the spectral unit. Three main classes are identified in the literature: filtering, dispersive and interferometric. The ones in the first class only transmit a narrow spectral band to each detector pixel. In dispersive imaging spectrometers the directions of light propagation change by diffraction, material dispersion or both as a continuous function of wavelength. Interferometric imaging spectrometers divide a light beam into two, delay them and recombine them in the image plane. The spectral information is then obtained by performing a Fourier transform.

Chapter 2 reviews the state of the art in the application of Data Compression techniques to Remote Sensing images, specially in the case of Hyperspectral images. Lossless, Near-Lossless and Lossy compression techniques are reviewed and evaluated on well known benchmark images. The chapter includes summaries of pertinent materials such as Wavelet Transform, KLT, Coding and Quantization algorithms, compression quality measures, etc.

Chapter 3 formulates the watermarking of digital images as a multi-objective optimization problem and proposes a Genetic Algorithm to solve it. The two conflicting objectives are the robustness of the watermark against manipulations (attacks) of the watermarked image and the low distortion of the watermarked image. Watermarking is proposed as adding the image mark DCT coefficients to some of the watermarked image DCT coefficients. In the case of hyperspectral images the DCT is performed independently on each band image. The careful definition of the robustness and distortion fitness functions to avoid flat fitness landscapes and to obtain fast fitness evaluations is described.

Chapter 4 refers the current efforts at the European Space Agency to provide Service Support Environments (SSE) that: (1) Simplify the access to multiple sources of Earth Observation (EO) data. (2) Facilitate the extraction of information from EO data. (3) Reduce the barrier for the definition and prototyping of EO Services. The objective of the chapter is to provide an overview of the systems which can be put in place to support various kinds of user needs and to show how they relate each other, as well as how they relate with higher level user requirements. The chapter reviews several apparently un-related research topics: service oriented architecture, service publishing, service orchestration, knowledge based information mining, information and feature extraction, and content based information retrieval. The authors stress their relative roles and integration into a global web-based SSE for EO data.

Chapter 5 reviews some general ideas about Content Based Image Retrieval (CBIR) Systems emphasizing the recent developments regarding Remote Sensing image databases. The authors introduce an approach for the CBIR in collections of hyperspectral images based on the spectral information given by the set of endmembers induced from each image data. A similarity function is defined and some experimental results on a collection of synthetic images are given.

Chapter 6 considers an specific problem, that of sensor deployment when trying to build up a wireless sensor network to monitor a patch of land. The Martian exploration is the metaphorical site to illustrate the problem. They propose a formal statement of the problem in the deterministic case (all node positions can be determined). This leads to the formulation of an objective function that can be easily seen to multiple local optima, and to be discontinuous due to the connectivity constraint. Simulated Annealing is applied to obtain (good approximations to) the global optimum.

Chapters 7 and 8 are devoted to the study of the efficient parallel implementation of segmentation and classification algorithms applied to hyperspectral images. They include good reviews of the state of the art of the application of mathematical morphology to spatial-spectral analysis of hyperspectral images. Chapter 7 focuses on the parallel implementation of morphological operators and morphology derived techniques for spectral unmixing, feature extraction, unsupervised and supervised classification, etc. Chapter 8 proposes parallel implementations of Multilayer Perceptron and compares with the morphology based classification algorithms. Specific experiments designed to evaluate the influence of the sample partitioning on the training convergence were carried out by the authors.

Chapter 9 deals with the detection and spatial localization (positioning) of rather elusive but also conspicuous phenomena: the line-shaped weather systems and spiral tropical cyclones. The works are performed on radar data and satellite images and tested on real life conditions. The main search engine are Genetic Algorithms based on a parametric description of the weather system. Kalman filters are used as post-processing techniques to smooth the results of tracking.

Chaper 10 proposes a Wavelet Transform procedure performed on the HSV color space to obtain the primitive features for image mining. A systematic method for decomposition level selection based on the frequency content of each decomposition level image.

Chapter 11 reviews the application of Artificial Neural Networks to land cover classification in remote sensing images and reports results on change detection using the Elmann network trained on sequences of images and of Synthetic Aperture Radar (SAR) data.

Chapter 12 is devoted to the problem of Forest Fires management. It describes two case studies of operational and autonomous processing chains in place for supporting forest fires management in Europe, focusing on the prevention and damage assessment phases of the wildfire emergency cycle, showing how computational intelligence can be effectively used for: Fire risk estimation and Burn scars mapping. The first fusing risk information and in-situ monitoring. The second based on automatic change detection with medium resolution multispectral satellite data.

Chapter 13 focus on the application of image spectrometers to quality control applications. Contrary to remote sensing settings, the imaging device is near the imaged object and the illumination can be somehow controlled. The spectral mixing problem takes also another shape, because aggregations of pixels may be needed to form an appropriate spectrum of a material. The recognition is performed applying Gaussian Synapse Neural Networks. 14 extends the application of Gaussian Synapse Neural Networks to endmember extraction.

Chapter 15 is devoted to change detection in Synthetic Aperture Radar (SAR) data. Two automatic unsupervised methods are proposed. One based on the semi-supervised Expectation Maximization (EM) algorithm and the Fisher transform. The second follows a data-fusion approach based on Markov Random Field (MRF) modeling.

Spain

Manuel Graña
Richard Duro

Acknowledgments

This book project has been supported partially by the spanish MEC grants TSI2007-30447-E, DPI2006-15346-C03-03 and VIMS-2003-20088-c04-04.

Contents

1 Optical Configurations for Imaging Spectrometers

X. Prieto-Blanco, C. Montero-Orille, B. Couce, and R. de la Fuente

Universidade de Santiago de Compostela, Departamento de Física Aplicada, Escola Universitaria de Optica e Optometria, E-15782 Santiago de Compostela, Spain
faloraul@usc.es

1.1 Introduction

Imaging spectrometers or hyperspectral imagers are radiation sensors that provide a continuous collection of spectral images of an inhomogeneous scene, allowing to obtain a spectral signature of each point in the scene. They are applied to perform many different tasks such as accurate mapping of wide areas, object identification and recognition, target detection, process monitoring and control, clinical diagnosis imaging and environment assessment and management. Application areas include forestry, geology, agriculture, medicine, security, manufacturing, colorimetry, oceanography, ecology and others.[1]

The simplest way to obtain information of a scene by optical means is taking a photograph. With a black and white photograph we get an irradiance map of the radiation reflected by each point in the scene. A colour photograph allows to get some spectral information corresponding to the superposition of the radiation reflected in three broad band of the spectrum, typically blue, green and red bands. Much more information can be obtained if the photographs are taken in tens or hundreds of different spectral bands. Imaging spectrometers do this work. They take separated images at narrow and contiguous spectral bands. This collection of images is known as a hyperspectral image. Therefore a hyperspectral image is a data cube with two spatial dimensions and one spectral or wavelength dimension. Each point in this data cube holds the irradiance data of each object point for each spectral band. As three-dimensional detectors do not exist, an imaging spectrometer has to take a stack of images to build the data cube. Different methods and devices have been developed for this purpose. They can be classified in four types: framing, whiskbroom, pushbroom and windowing, according to the data content in a single frame[2]. In Fig. 1.1 we show the cut of the data cube for each type. In the framing or staring class a conventional image is taken but only some portions of its spectrum reach the detector. For example, a collection of images for different spectral bands are obtained using a large set of filters or a tunable one[3]. The other techniques are based on the scanning of the object and are more advisable for large scene or moving objects[4]. They differ in the number of spatial dimensions of the field of view in a single frame.

M. Graña and R.J. Duro (Eds.): Comput. Intel. for Remote Sensing, SCI 133, pp. 1–25, 2008.
springerlink.com

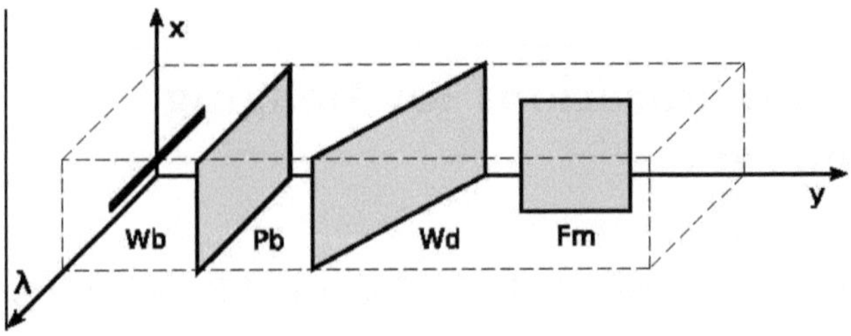

Fig. 1.1. Different ways of cutting the data cube (hyperspectral image) for whiskbroom, pushbroom, windowing and framing. Note that for FTIS λ-axis is actually a time delay (see Section 4).

In whiskbroom imaging spectrometers the spectral image of one object point (zero spatial dimension) is processed, for example, by dispersing the spectrum on a linear detector array. Scanning has to be performed in two spatial dimensions (across track scanning) to get the whole scene. In most frequently used pushbrooms systems, a stripe of the sample (one spatial dimension) is imaged onto the slit of a spectrometer. The dispersion occurs across the slit and a 2D detector array is used. Doing so the spectral dimension and one spatial dimension is acquired. The other spatial dimension has to be scanned to complete the data cube. In airborne or space-borne systems the platform movement itself provides the scanning (along track scanning). Alternatively a scan mirror can be used. Pushbroom spectrometers are usually preferred to their whiskbroom counterparts because they are generally lighter and less expensive, and can gather more light because they look at a particular area for a longer time. However one drawback of pushbroom sensors is the varying sensitivity of the individual detectors. Finally, a windowing instrument also scans the scene in one dimension as the pushbroom does; however it acquires a 2D spatial image in a single acquisition as in framing instruments. Unlike framing class, different portions of the spectrum reach different stripes on the detector. A typical windowing imager is very simple: it employs a conventional camera with a wedged or linear variable filter placed just in front of the 2D detector array. The wavelength of light transmitted by the filter varies with position providing for the spectral dimension. Both framing and windowing instruments build the spectrum of a given point from a lot of images taken at successive times. So lighting fluctuations can result in spectral artifacts. On the other hand, whiskbroom and pushbroom imaging spectrometers need a care control of the scanning process to reproduce spatial information with high fidelity.

Imaging spectrometers can also be classified attending to the method they use for spectral processing[2]. Thus they belong to filtering, dispersive or interferometric class. The first ones only transmit a narrow spectral band to each detector pixel. In dispersive imaging spectrometers the directions of light propagation change by diffraction, material dispersion or both as a continuous function of wavelength. Spectral processing in interferometric imaging spectrometers is a little cumbersome. They divide a beam into two, delay them and recombine them in the image plane. The spectral information is obtained by performing a Fourier transform.

In the following sections we discuss some representative imaging spectrometers attending to this last classification. In general, a spectrometer comprises three units: imaging, spectral and detector unit. The spectral unit determines to a large extent most of the characteristics of the spectrometer as spatial and spectral resolution, spectral range, field of view, detection threshold, and so on. In our analysis we will concentrate our attention in this unit, its principles of operation and its optical configurations.

1.2 Filtering Imaging Spectrometers

In filtering imaging spectrometers the spectral unit transmits selected spectral bands. In spectrometers that use tunable filters, the same spectral band is transmitted to all the pixels of the detector. The whole spectrum is scanned sequentially, so they are of framing class. We discuss them in section 1.2.1. In section 1.2.2 we consider wedge filters spectrometers. In this case, different spectral bands are transmitted to different stripes on the detector unit. Wedge filters spectrometers belong to the windowing class.

1.2.1 Electronically Tunable Filter Spectrometers

As discussed in the introduction hyperspectral images can be built by taken successive photographs in tens or hundreds of different spectral bands. These photographs can be obtained in practice with a monochrome detector and a set of spectral filters with a narrow band pass. A rotating wheel can be used for changing the filter. A more elegant and practical solution is some kind of tunable filter with spectral tuning performed electronically without any movable parts. In this section we discuss two types of electronically tunable filters that are widely used in the field of spectral imaging: the acousto-optic filter[5, 6, 7, 8, 9] and the birefringent filter[10, 11, 12].

Acousto-Optic Filters

An acousto-optic device works on diffraction of an optical beam by sound waves in a medium with a suitable elastooptic coefficient. The sound wave can be generated by applying a radio frequency (RF) signal to a piezoelectric transducer bounded to the medium. This wave modulates periodically the index of refraction in the region where it propagates. The modulation of the refractive index acts as a phase volume grating that can diffract an electromagnetic wave incident on the medium.

A unique property of diffraction in a volume grating is its selectivity. Only waves that verify Bragg's or phase matching condition are diffracted efficiently. This condition is given by the relation:

$$\mathbf{k}_d = \mathbf{k}_i \pm \mathbf{k}_s \tag{1.1}$$

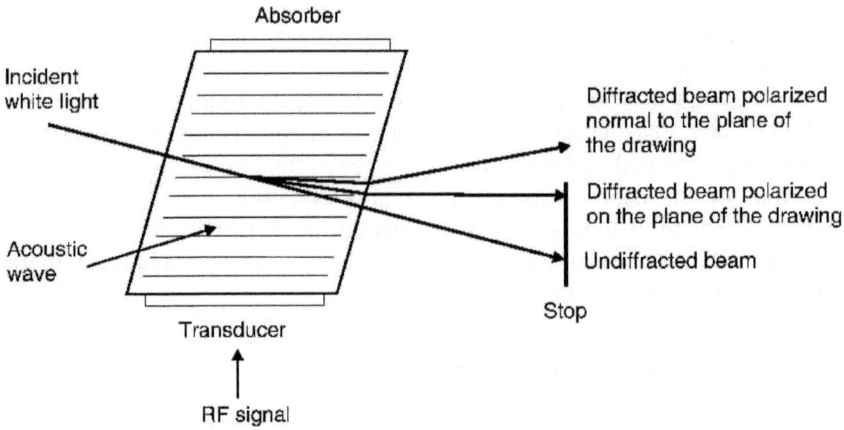

Fig. 1.2. Acousto-optic filter in a birefringent material with a non collinear geometry

where $\mathbf{k}_d$, $\mathbf{k}_i$ are the wave vector of diffracted and incident wave, ($| \mathbf{k}_d |= \frac{2\pi}{\lambda} n_d$, $| \mathbf{k}_i |= \frac{2\pi}{\lambda} n_i$, with n_d, n_i the respective refractive index and λ the optical wavelength), and $\mathbf{k}_s$ is the acoustic wave vector ($| \mathbf{k}_s |= \frac{2\pi f}{v}$, with f the RF and v the acoustic phase velocity). This selective property explains why an acousto-optic device can acts as a filter. For incident white light only the spectral components near the one that verifies exactly the Bragg condition are diffracted efficiently. When used in a non collinear geometry the direction of the diffracted beam is different from the incident beam (see Fig. 1.2). We can then use a stop to block the last one and so the device transmits only a selected portion of the spectrum. Furthermore, an acousto-optic filter typically uses a birefringent material such as $LiNbO_3$ or TeO_2. In this case the incident and diffracted beams have different states of polarization[13, 14]. For example, an ordinary beam incident in a uniaxial crystal is diffracted into an extraordinary beam and viceversa. For an arbitrary polarization (a mixture of ordinary and extraordinary waves), there are two diffracted beams in different polarization states and also in different directions. These properties can be used to stop the undesired beams by disposing the acousto-optical device between two cross polarizers. This is particularly useful when the acousto-optical device is used in a collinear geometry.

The two main parameters that characterize a spectral filter are the tuning range and the spectral bandwidth of the transmitted light. As Eq. 1.1 shows the wavelength of the diffracted light is (nearly) inversely proportional to the acoustic frequency. So the acousto-optic filter is tuned by varying the RF signal. They are easily tuned in an octave; that is, the extreme wavelengths double one each other. A larger range can be achieved by using several transducers. Besides tuning is very fast (tens of microseconds) and requires small powers (some watts or less). Furthermore, for all the tuning range a large field of view is obtained by using a non critical phase matching geometry[14]. Under this

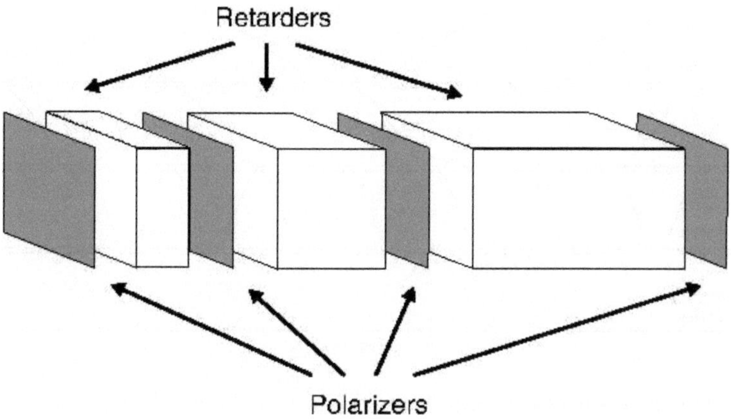

Fig. 1.3. Three stage Lyot-Ohman filter

geometry the tangents to the incident and diffracted light wave vectors surfaces are parallel. This means that the same $\mathbf{k}_s$ matches a wide range of $\mathbf{k}_i$ and $\mathbf{k}_d$ directions. Typical fields of view are lesser than 10°. Overlapping with the undiffracted beam limits this value. The bandwidth is mainly limited by two factors. First, the sound wave diverges on the medium so there is a spread of acoustic wave vectors that also spreads the range of transmitted wavelength. The bandwidth due to the angular acoustic spread is nearly proportional to the square wavelength[14]. Second, when a cone of light is incident on the device, rays incident on different directions will be diffracted with a small difference in wavelength. This also causes spectral broadening, this time proportional to the wavelength[13].

Finally, it is to note that there is an unwanted spatial shift between successive spectral images. This is due to the fact that the deflection angle can vary one degree along the tuning range.

Birefringent Filters

The most popular birefringent filter is the Lyot-Ohman filter[15]. This filter (see Fig. 1.3) is made up of a set of birefringent retarders and linear polarizers disposed in an alternate sequence. All the retarders have their fast (and low) axis pointing in the same direction which is rotated 45° from the polarizers axes (that are also parallel). Furthermore, each retarder doubles its thickness with respect to the previous one.

The Lyot filter can be considered as a series of stages formed by a birefringent retarder inserted between two polarizers with the last polarizer being the input for the next stage. In each stage the birefringent crystal changes the polarization state of light because the relative retardation between the components which are polarized along the birefringence axes. This retardation is proportional to the crystal thickness (for normal incidence) and to the frequency of light. The

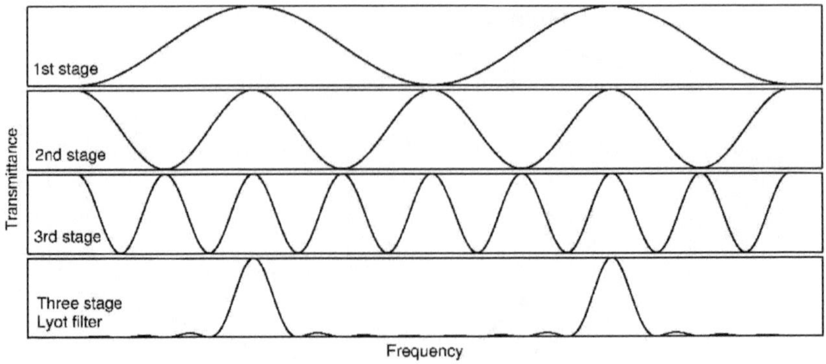

Fig. 1.4. Transmittance of the first, second and third stage of a Lyot filter as well as its total transmittance

second polarizer transmits some fraction of the incident radiation depending on its state of polarization. If ϕ is the phase difference induced in the crystal, the transmittance of the stage is

$$T = \cos^2\left(\frac{\phi}{2}\right) = \cos^2\left(\frac{\pi\Delta}{\lambda}\right) = \cos^2\left(\frac{\pi d \delta n}{\lambda}\right),$$

where Δ is the retardation, δn the index difference along the axes of the crystal, d its thickness and λ the wavelength. This transmittance has a maximum (equals to one) for a retardation multiple of λ and a minimum (equals to zero) for a retardation odd multiple of $\lambda/2$.

The transmittance of the complete Lyot filter is the product of the transmittance at each stage. As each retarder thickness doubles the previous we have

$$T = T_1 \cdot T_2 \cdot T_3 ... T_N = \cos^2\frac{\phi}{2}\cos^2\frac{2\phi}{2}\cos^2\frac{4\phi}{2}...\cos^2\frac{2^{N-1}\phi}{2} = \left(\frac{\sin(2^N\phi/2)}{2^N\sin(\phi/2)}\right)^2.$$

In Fig. 1.4 we show the transmittance as a function of frequency of single Lyot stages of thickness d, 2d and 4d as well as the transmittance of a Lyot filter made of three stages. We observe that the maxima of the first stage are also present in the other stages and the minima between these two maxima are located in different positions for each stage. Thus the filter transmittance shows some narrow isolated peaks surrounded by secondary maxima and minima. In general the distance between peaks is determined by the first stage and the peak width by the last stage. So the peak width decreases with the number of stages.

One problem related to these filters is the reduction of their transmittance because of the reflections at surfaces of each element. To suppress reflections the crystals are immersed in a match index liquid. Further problems come from the angular divergence of radiation. For non normal incidence upon the crystals, radiation crosses large thickness. So retardation depends on the incidence angle as well as transmitted wavelength. To overtake this difference in retardation a

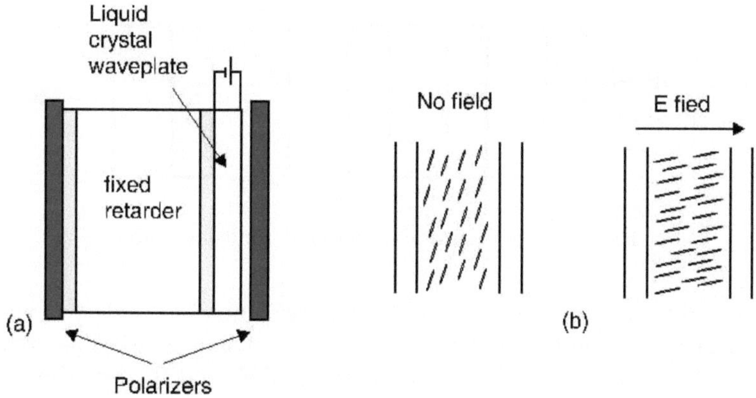

Fig. 1.5. (a) Liquid crystal tunable Lyot stage. (b) Schematic action of a liquid crystal waveplate.

retarder that is composed of two or more elements can be designed, in such a way that the retardation difference introduced by an element is compensated by another element. One choice is to use crystals of different materials; another is to use two identical crystals with its axes crossed and a half-wave plate between them. In both cases the effect is a retardation difference of opposite sign in each element[15].

As described above, the Lyot filter works for one predetermined wavelength. Several proposals have been made for the filter to tune. For example, a pair of edge crystals to vary the retarder thickness or to rotate the polarizers. More elegant and practical it is to use electronically shutters as liquid crystal retarders. A liquid crystal retarder is made up of a nematic liquid crystal between two transparent electrodes[10]. Such a liquid crystal is a uniaxial birefringent material made up of chains of molecules aligned along its optical axes. The orientation of this axis can be made parallel to the crystal surfaces. When a voltage difference between the electrodes is applied the electric field induced makes the molecules to undergo a torque. This torque reorients the molecules along the field direction. The optical effect is to change the orientation of the optical axis so retardation is reduced.

In a tuned birefringent filter each stage is composed by a thick fixed retarder and a thin liquid crystal plate. The complete tuning range is achieved by varying 2π radians the phase shift induced by the liquid crystal plate of the first stage. The effect is to shift the transmission curve of the fixed retarder from one maximum to the previous one. Note that the liquid crystal retardation has to follow the binary sequence of the Lyot filter, that is, at each stage the variable retardation has to double the one obtained in the previous range.

One of the major disadvantages of the liquid crystal Lyot filter is the large amount of radiation absorbed by the polarizers. The Solc filter has not this drawback[16]. This filter also comprises a stack of birefringent retarders but only two polarizers disposed at booth ends. In the so-called folder configuration,

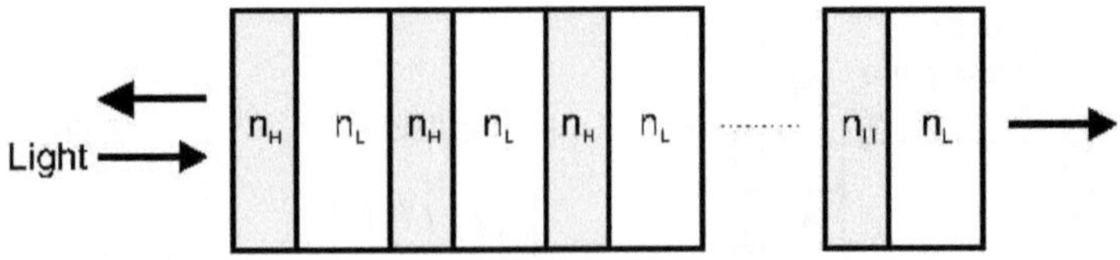

Fig. 1.6. Structure of a dielectric multilayer film

N identical retarders have their axes alternately rotated an angle $\theta = \pi/4N$ and -θ with respect to the first polarizer axis. The second polarizer is crossed. In the fan configuration, the polarizer axes are parallel, and the retarder rotations increase consecutively an angle 2θ, being θ the rotation of the first retarder. It has been shown[16] that a Solc filter with 2^N retarders is comparable to a Lyot filter made up of N stages; the thin stage having the same thickness as one retarder of the Solc filter. Nevertheless, this filter presents more parasite light than the Lyot filter.

Another tunable filter can be design with a thin plate of liquid crystal between two high reflection mirrors (a Fabry-Perot etalon)[17]. The tuning is performed by varying the index of refraction of the crystal. To make so, a polarizer with its axis parallel to the optical axis of the crystal is located before the etalon. The refractive index variation is induced by the reorientation of molecules when an electric field is applied, as in the case of the tunable Lyot filter. The tuning range of this filter is limited by the maximum refractive index variation that can be achieved: $\Delta\lambda = \lambda\Delta n/n$, where n and λ are the mean refractive index and the mean wavelength, respectively. For example, taking a wavelength in the visible range $\lambda = 600$ nm, a mean refractive index $n = 1.5$ and $\Delta n = 0.1$ we get $\Delta\lambda =$ 40 nm. This tuning range is quite moderate but this type of filter can be used for high resolution spectral imaging.

1.2.2 Multilayer Thin Films and Wedge Filter Spectrometers

A multilayer thin film is an alternated sequence of layers of transparent dielectric materials with high (n_H) and low (n_L) refractive index deposited on a transparent substrate[18]. Incident light on the film undergoes reflections and refractions on the layer interfaces. Light of a given wavelength is partially reflected and partially transmitted by the whole multilayer film as a result of the interference of the reflected waves in the dielectric interfaces, on the one hand, and the transmitted waves, on the other hand.

One example of such a system is a stack of quarter-wave layers, that is, layers in which $n_H d_H = n_L d_L = \lambda/4$ for a design wavelength λ, being $d_{H(L)}$ the thickness of the high (low) index layer. This stack works as a broadband reflector centered at the design wavelength. The peak reflectance increases with the number of layers and with the ratio between high and low refractive index. The width of the high reflectance spectral band also increases with this ratio.

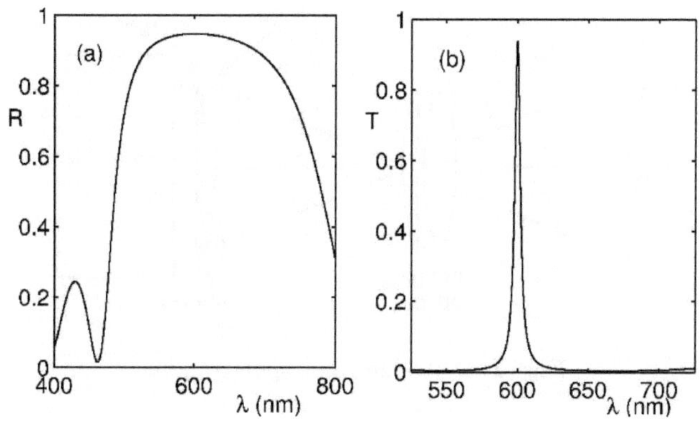

Fig. 1.7. (a) Typical reflectance of a dielectric broadband reflector HLHLHL. (b) Transmittance of a dielectric interference filter [HL]HH[LH].

This sequence of layers is typically coded as $HLHL...HL \equiv (HL)^k$, with k the number of high and low refractive index pairs of layers. A similar behaviour has the symmetric sequence $HLHL...HLHLH \equiv (HL)^k H$ where a layer of high refractive index is added. This type of multilayer films works as metallic mirrors with the advantage of a null absorption. Nevertheless, the dielectric layers have to be of high quality to avoid scattering losses. Furthermore, the reflection band in a dielectric reflector is narrower than in a metallic mirror. Two (or more) multilayer stacks with adjacent reflection bands can be added to get a broader band.

Multilayer dielectric thin films can also be used to design an interference filter or Fabry-Perot etalon. These are periodic structures of the form $(HL)^k(LH)^k$ or $(HL)^k HH(LH)^k$. They can be seen as two stacks of quarter-wave layers (the dielectric reflectors of the etalon) surrounding a half-wave layer. These multilayers have a high transmission at the design wavelength. Their transmission bandwidth is proportional to the peak wavelength and decreases with the number of layers[19]. The characteristics of these interference filters can be changed by several ways. One of them consists in using reflectors with layers of different thickness. A second way is to use several Fabry-Perot etalons in a series. For example the sequence $R_1LLR_2LLR_3$, where R stands for a multilayer reflector.

The last dielectric structure only transmits spectral components of light on a narrow band centered in a selected wavelength. To be used in an imaging spectrometer we have to tune the filter. Alternatively, we can design a filter with multilayer thin films of continuous varying thickness. This type of filters transmits a different wavelength depending on the point of incidence of light on the filter; this is a wedge dielectric interference filter. We can build a very simple, compact and rugged spectrometer by attaching a wedge filter to an array detector so that a given scene can be imaged both on the filter and the detector (see Fig. 1.8)[20]. In this spectrometer the spectral axis on the detector is parallel to the tapered direction on the filter. The detected wavelength varies along this

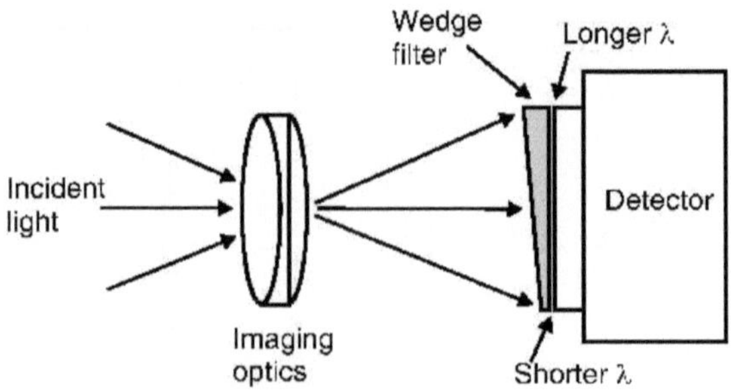

Fig. 1.8. Wedge filter spectrometer

axis. As each pixel on the detector gets radiation from a different point on the scene we obtain an image with varying colour. That is, in a single frame, we image each stripe of the scene with a different wavelength. To obtain the whole spectrum from each point in the scene, the target has to be scanned along the spectral direction in a windowing mode. Typically, additional blocking filters are added to stop out of band transmission. Furthermore, the thickness of the tapered region can follow many desired functions to adapt for the spectrometer requirements. For example, a constant bandwidth ratio $\Delta\lambda/\lambda$ is achieved with a logarithmic dependence of wavelength with position[21].

Finally, it is worthwhile to mention that high resolution spectral imaging can also be performed with a dielectric etalon composed of broadband reflectors of constant thickness separated by a wedge tapered layer[22]. In this case, the reflectors limit the spectral range of the instrument whereas the wedge spacer induces spectral filter along its tapered direction.

1.3 Dispersive Imaging Spectrometers

The spectral information in a spectrometer is normally obtained by using a dispersiveelement. This can be a prism, a transmission diffraction grating, a reflection diffraction grating or a combination of them. Furthermore the spectrometer has to incorporate some optics for selecting the region of interest of the object and for collecting the spectral information after wavelength separation. One typical dispersive imaging spectrometer can be divided in three parts: the fore-optics, the dispersive unit, and the sensor unit. In Fig. 1.9 we sketch one of the most simple configuration: the Czerny-Turner mount[23]. Its operation is quite easy to understand. The fore-optics focuses the light from a given scene onto the entrance slit (S) of the dispersive unit. Thus only a stripe of the sample is selected. The fore-optics is chosen to satisfy specific requirements: f-number, field of view, resolution, etc. Some appropriate designs are the modified double Gauss, the modified Ritchey-Chretien Cassegrains, the modified Huygens eyepiece, the three mirror anastigmat or the Schwarzschild objective[24]. Following the fore-optics the dispersive unit separates the spectral components of light coming from each point on the slit. In the example of Fig. 1.9 a first spherical

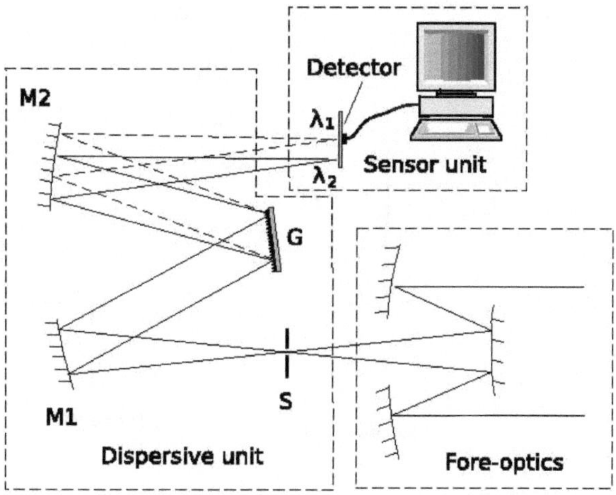

Fig. 1.9. Czerny-Turner configuration of an imaging spectrometer with a Cassegrain Ritchey-Chretien objective as the fore-optics

concave mirror (M1) collimates the light from the slit. That is, rays from each point on the slit leaves the mirror in a given direction. Next a plane diffraction grating (G) disperses the light, i.e. different wavelengths (λ_1, λ_2) leave the grating at different angles. A second spherical concave mirror (M2) focuses each wavelength onto the detector at different positions. Thus we get a collection of spectral images of the slit. One advantage of this configuration is that mirrors do not introduce any chromatic aberration. However they do introduce other aberrations which limit the resolution of the instrument. If the source is a two dimensional scene some techniques that scan the object have to be used as was mentioned in the introduction.

The Czerny-Turner configuration with some variants is usually chosen for conventional spectrometers and even for some whiskbroom imaging spectrometers because of its simplicity, low cost, and compactness. However for common imaging spectrometers Czerny-Turner configuration does not give enough optical quality. In particular, Czerny-Turner systems have high astigmatism and field curvature as well as high distortion. By varying the groove density across the grating aperture or replacing the spherical surfaces of mirrors with toroidal surfaces, some degree of aberration correction can be obtained, but would still be insufficient for the current applications. An aberration-corrected concave grating in a Rowland mount[25] is a second improvement but also falls short of the 4 to 25 micron pixel pitch of large science grade detectors. So another optical configurations are desired for imaging spectrometers that include dispersive elements. In this section we describe two of the most representative mounts[26]: one based on reflection gratings (Offner configuration) and the other one based on refractive dispersive elements (prism-grating-prism configuration). Both configurations are thought for pushbroom spectrometers. Even though these kind of spectrometers

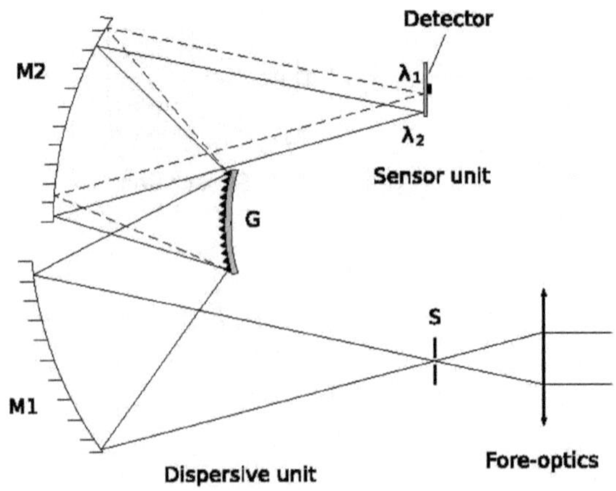

Fig. 1.10. Offner configuration of an imaging spectrometer

are the most popular, some non pushbroom configurations for dispersive imaging spectrometers have also been proposed. We present in subsection 1.3.3 two of the most representative.

1.3.1 Offner Imaging Spectrometers

In Fig. 1.10 a typical Offner spectrometer[27] is sketched. As before the fore-optics might be some kind of optics that satisfies the requirements needed. This spectrometer comprises three concentric spherical surfaces and a slit. Its configuration and operating principle is similar to the Czerny-Turner counterpart. The main difference is the reflective grating which in the Offner is of convex spherical type. Since this element is the aperture stop and it is near to the image focus of M1, the entrance pupil of the dispersive unit is at infinity. So the fore-optics should be telecentric to avoid vignetting. In addition the convex grating is the less standard element of the instrument. Until the 1990's the lack of an appropriate technology for the fabrication of efficient convex gratings made the practical realization of these designs difficult, limiting the interest in Offner spectrometers. However, progress in electron-beam lithography techniques has recently permitted the fabrication of the required high-performance convex gratings. So the Offner spectrometer can satisfy the stringent requirements of imaging spectrometry over a broad spectral band from UV to thermal IR, reducing the spectral and spatial distortion to a small fraction ($\sim$1%) of a pixel of the sensor[28]. The advantages of the Offner spectrometer are: a) its configuration allows to cancel coma and astigmatism aberrations[29, 30], b) it has very small smile (change of dispersion angle with field position) and keystone (change of magnification with spectral channel) distortions as well as a small field curvature, c) it operates with a relatively low f-number ($\geq$f/2), d) it accepts a long slit while maintaining compact size, and e) its concentric spherical surfaces facilitate the alignment during spectrometer assembly. In table 1.1 we show the main characteristics of

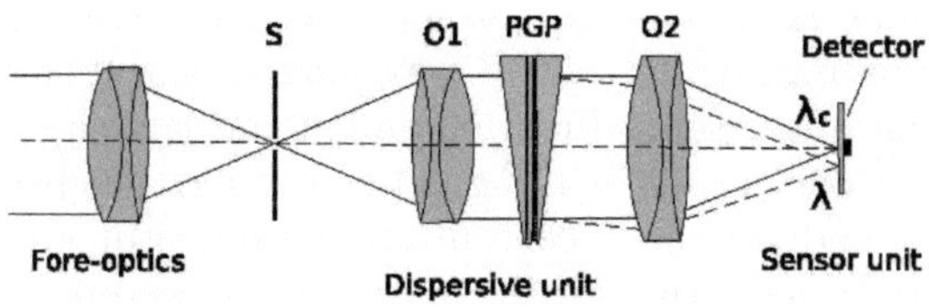

Fig. 1.11. PGP configuration of an imaging spectrometer

Table 1.1. Specifications as declared by the manufacturers of Hyperspec VS-25 and Imspector V10E imaging spectrometers

Description	Hyperspec VS-25	Imspector V10E
Size (L×W×H)	127 × 152.4 × 101.6 mm	175 × 60 × 60 mm
Spectral Range	400 - 1000 nm	400 - 1000 nm
f-number	f/2	f/2.4
Slit height	18 mm × 12 μm	14.3 mm × 30μm
Spectral resolution	∼2 nm	2.8 nm
Smile distortion	¡ 0.1%	¡ 1.5μm
Keystone distortion	¡ 0.1%	¡ 1μm
RMS spot size	¡ 24 μm	¡ 9 μm

a commercially available Offner imaging spectrometer (Hyperspec VS-25^{TM}of Headwall photonics[31]). We must stress that this instrument owns a custom made convex grating which has been specifically corrected for aberrations.

An alternative which avoids the convex grating is the design proposed by Lobb and patented by Sira Technology Ltd.[32]. It is based on the Offner relay concept but uses two curved prisms as dispersing elements. This design has been used in the spaceborne imaging spectrometer CHRIS for ESA's PROBA satellite launched late in 2000. The CHRIS spectrometer provides 10 nm spectral resolution over a range from 415 nm to 1050 nm[33, 34]. However the addition of prisms to the Offner configuration is a considerable complication as compared to a simple grating construction.

1.3.2 Prism-Grating-Prism Imaging Spectrometers

Another representative mount is the prism-grating-prism (PGP)[35, 36]. In Fig. 1.11 we show a configuration of this class which only uses refractive optics.The comparison of this system with respect to the above two shows that lenses O1 and O2 play the role of mirrors M1 and M2, respectively. O1 collimates the light beam and O2 focuses it onto the detector. The key component now is the direct vision dispersing PGP. It is made up of a volume transmission grating cemented between two identical prisms. The operating principle of this component is as follows. A light beam from the collimating optics O1 is refracted by the first prism. The grating disperses this refracted light in the symmetric direction. Finally the second prism refracts the dispersed light in such a way that light corresponding to the central wavelength of the spectrum (λ_C) exits the PGP component in the same direction and at the same height as the incident beam. Moreover the component is designed to refract the spectrum so that the light spots of both ends of the spectrum on the detector are at the same distance from the optical axis, that is, the spectrum is centered. The direct vision property, a straight optical axis, is the most important advance of the PGP. It allows to have a straight device, an easy mounting and a small size. In table 1.1 we show the main characteristics of a commercially available PGP imaging spectrometer (Imspector V10E^{TM}of Specim[37]).

1.3.3 Non Pushbroom Dispersive Imaging Spectrometers

Although dispersive imaging spectrometers used to be of pushbroom type, some configuration of non pushbroom type has been also proposed. One of them corresponds to a windowing dispersive imaging spectrometer[38]. Such a spectrometer is an easy modification of a pushbroom one. It is obtained by including another dispersive element just before the fore-optics. This dispersive element mixes 2D spatial and spectral information coming from the scene into the entrance slit. Such one-dimensional signal is further on decomposed in the dispersive unit so data are displayed in a two dimensional array. The resulting image holds all the spectral bands but each one corresponds to a different stripe of the scene like in wedge filter spectrometers. However scanning is needed to obtain the full information of the scene like in pushbroom spectrometers.

Scanning can be avoided by a computed tomography imaging spectrometer (CTIS)[39]. This instrument is based on a computer-generated hologram (CGH) disperser which allows to collect multiplexed spatial and spectral data simultaneously. This special element provides projections of the data cube on the detector array. From this projections reconstruction of the data cube is achieved by deconvolution techniques. CTIS has been used in both the visible (450-760 nm)[39] and the middle wave infrared (3000-4600 nm)[40], and has demonstrated its utility in the fluorescence microscopy field[41]. Recent advances in electron-beam lithography have allowed to fabricate the CGH element on a reflective convex substrate[42]. So this spectrometer has the advantages of Offner systems as well as those of CTIS. Anyway, this technique has limitations in terms of both the need for large enough area detector arrays to achieve higher spatial and spectral resolution and the long time needed for processing the data cube; but it continues to evolve.

1.4 Interferometric or Fourier Transform Imaging Spectrometers

Fourier transform imaging spectrometers (FTIS) also called interference imaging spectrometers can be considered as a recent evolution of Fourier transform spectrometers (FTS). Instead of obtaining directly the spectrum, these instruments

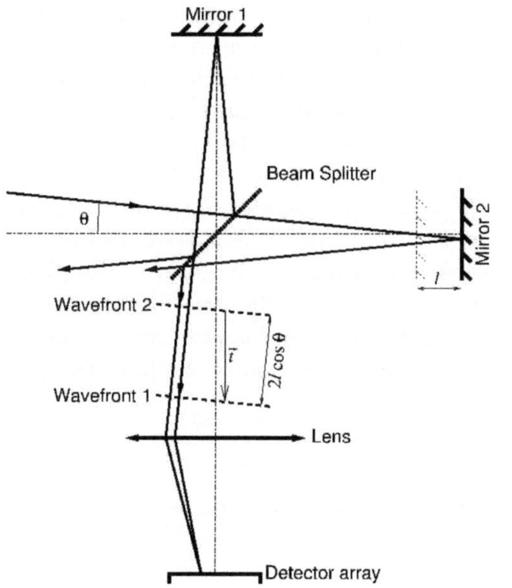

Fig. 1.12. FTIS based on classical Michelson interferometer

provide the Fourier Transform (FT) of the spectrum by two beam interference; the spectrum is retrieved by numerical computation. In this section we describe some representative configurations.

1.4.1 FTISs Based on the Michelson Interferometer with Plane Mirrors

From a conceptual point of view, the simplest FTIS consists of a light detector array located at the image plane of a lens which images a far object seen through a Michelson interferometer (Fig. 1.12). The typical Michelson interferometer comprises two perpendicular plane mirrors and a beam splitter (BS) at 45 degrees with the mirrors. One mirror is fixed and the other can be translated without any rotation. Incoming light is divided in two beams by the BS, and comes back to the BS after mirror reflections. Here both beams mix again in such a way that a fraction of the light goes to the detector array through the lens. The rest returns to the object. For instance, when both mirrors are symmetrically located with respect to the BS, the two beams follow the same optical path regardless the angle of incidence on the BS. Moreover each beam reaching the output port has just undergone a transmission and a reflection at the BS. Therefore they interfere constructively whereas waves going back to the object interfere destructively.[1] This situation is called null optical path difference (null

[1] For simplicity, we are assuming that the reflections on both sides of the BS introduce the same phase.

OPD). Now, let us consider only light propagating along the optical axis, that is, light incident on the BS at 45 degrees and normal to the mirrors. If the mobile mirror translates a distance l along the optical axis from the previous position, the optical path of the corresponding arm increases by $2l$. If $2l$ is a multiple of the light wavelength λ, the phase difference between output beams is 2π. So light whose wavelength fulfills $2l = m\lambda$, $m \in \mathbb{Z}$ interferes constructively. In contrast, the condition $2l = (m + 1/2)\lambda$, $m \in \mathbb{Z}$ means destructive interference (the relative phase is now π). Note that constructive and destructive conditions alternate each other along the spectrum. In general, the phase difference between beams is $2\pi\frac{2l}{\lambda}$ so the interference attenuates the spectral irradiance in a factor: $(1 + \cos\frac{4\pi l}{\lambda})/2$ [43]. Hence the interferometer behaves like a spectral filter with a sinusoidal transmittance as a function of the light frequency ν:

$$\frac{1 + \cos 2\pi\nu\tau}{2}, \tag{1.2}$$

being $\tau = 2l/c$ the time delay between both beams and c the speed of light. The signal $S(l)$ at the detector is:

$$S(\tau) \propto \int_0^\infty \frac{1 + \cos 2\pi\nu\tau}{2} E(\nu)\eta(\nu)\,\mathrm{d}\nu,$$

where $E(\nu)$ is the object spectral irradiance and $\eta(\nu)$ is the spectral sensitivity of the detector. Now we define $E(\nu)$ and $\eta(\nu)$ for negative frequencies as a symmetric extension of positive frequencies, that is $E(-\nu) \equiv E(\nu)$, $\eta(-\nu) \equiv \eta(\nu)$ for $\nu \in (0, \infty)$. So the previous expression becomes:

$$S(\tau) \propto \int_{-\infty}^{\infty} \exp(-\mathrm{i}2\pi\nu\tau)\, E(\nu)\eta(\nu)\,\mathrm{d}\nu + \int_{-\infty}^{\infty} E(\nu)\eta(\nu)\,\mathrm{d}\nu.$$

The first integral is the FT of $E(\nu)\eta(\nu)$ whose conjugate variable τ is proportional to l, and the second integral is a constant. Thereby, we can sample the FT of $E(\nu)\eta(\nu)$ by translating the mirror sequentially and acquiring the corresponding value of S. We retrieve $E(\nu)\eta(\nu)$ by computing the inverse FT and $E(\nu)$ from the detector calibration. In other words, as the mirror moves, a given wavelength originates a sinusoidal contribution to the detected signal whose period is specific and different from any other wavelength contribution. The inverse FT translates this sinusoid to a point at the spectrum.

The previous analysis applies to rays normal to the mirrors. Now, let us consider a collimated input beam making an angle θ with the optical axis (Fig. 1.12). Both output beams also make an angle θ with the axis, hence they are parallel. Moreover the wavefront from the longest arm is a distance $2l$ *along the optical axis* behind the other wavefront. However, the OPD is the distance between both wavefronts *along the propagation direction*. That is the OPD slightly depends on the angle of incidence as $\mathbf{tu} = 2l\cos\theta$, where $\mathbf{t}$ is the translation vector between both wavefronts (a vector of length $2l$ along the optical axis) and $\mathbf{u}$ is a unitary vector normal to these wavefronts. This dependence results in a little change of the time delay for peripheral pixels of the image that (if excessive) could be easily corrected when the inverse FT of these pixels is computed. It is worth to note that no aberrations are introduced in the interferometer because only reflections on plane surfaces occur; thus the spatial information is fully preserved. In short, images obtained by the sensor array have the same appearance as no interferometer were present but they are spectrally filtered, so this instrument is of a framing kind.

The fact that the spectrum is recovered through a FT has important consequences because range in a domain is related to resolution in the conjugate domain. So, the highest light frequency that the instrument can see determines the largest admissible step of the mobile mirror between contiguous frames. Thus, if the highest frequency corresponds to a wavelength λ_-, the largest step that avoids aliasing is $\lambda_-/4$ (note that the light path is folded). That means 100 nm for the visible range. Such a short distance could be a handicap as we will see later. On the other hand, only the mirror travel determines the spectral resolution. This allows a very good spectral resolution independent from the spatial one; a relevant advantage compared to other imaging spectrometers. Moreover, there is no fundamental restrictions to retrieve low frequencies: the minimum detectable frequency is again the spectral resolution. The theoretical longer wavelength (λ_+) is of the order of magnitude of the mirror travel. Actually BS, mirrors and detector array can limit the minimum frequency. Anyway wavelength ranges of more than one order of magnitude can be achieved in the far infrared. Other imaging spectrometers usually present a bothersome superposition of diffraction orders for spectral ranges higher than one octave. Note that non linearities of FTIS detectors (such as a saturation) also introduce harmonics into the detected signal.

The typical irradiance at the detector array is a half of the irradiance obtained without interferometer (mean value of Eq. (1.2)). This is about 500 times greater than the irradiance of filtering or dispersive imaging spectrometers, because the former reject all the spectrum except a narrow band whereas the later all the scene except a narrow stripe. Hence the detector array for FTISs can be less sensitive. It also has different consequences in the signal to noise ratio (SNR) whether the detector is ideal or not[44]. On the one hand if the detector is ideal, only photonic noise (which is proportional to the square root of the signal) is present. So the row data from FTISs -$S(\tau)$- has a better SNR by more than one order of magnitude than dispersive and filtering counterparts. However, $S(\tau)$ must be processed to obtain $E(\nu)$ which also divides its SNR by the square root of the number of images. This roughly balances the comparison between different kind of instruments. Other parameters such as the spatial/spectral point ratio, the f-number attainable by each particular design and so on, must be taken into account for a precise comparison. On the other hand, if an infrared detector is not cooled enough, thermal noise adds to photonic noise and the detector is no longer ideal. In this case, FTIS are less sensitive to the detector noise than

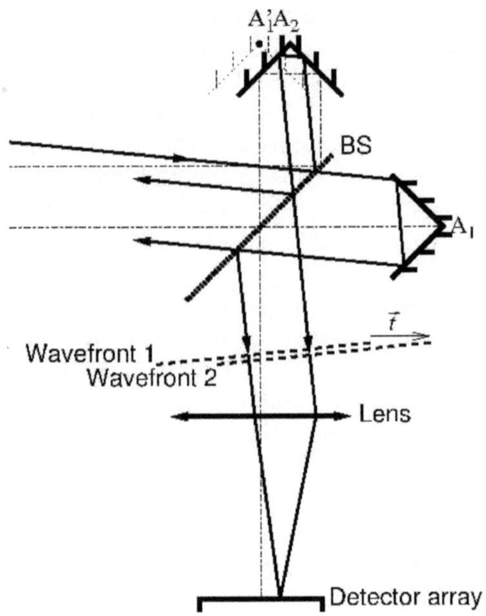

Fig. 1.13. FTIS based on a Michelson interferometer with dihedrons

filtering and dispersive ones. For instance, if the thermal noise is the 10% of the photonic noise, the SNR of a FTIS worsens by a 25% with respect to the ideal detector while the SNR of filtering and dispersive ones does by 80%[44].

In spite of the interesting features of the FTISs, this Michelson configuration is unpractical for remote sensing from aircrafts. Any vibration or acceleration will cause a larger deformation in the mechanical arrangement than the mirror step (100 nm for visible light). Even laboratory FTISs based on this configuration can need an auxiliary laser to monitorize the actual OPD. This characteristic is not a trouble for other applications; FTIS based on Mach-Zenhder interferometer are used in astronomy[45, 46] although they have a similar sensitivity to vibrations.

The following schemes are more suitable for remote sensing because they reduce the stability requirements while retain the above advantages.

1.4.2 FTISs Based on the Michelson Interferometer with Dihedrons

Let us consider the interferometer of the Fig. 1.13 which is of Michelson type but each plane mirror is replaced with two orthogonal mirrors (dihedron)[47]. Note that they could be also replaced with hollow corner cubes. The dihedrons have two relevant optical properties: first, they reflect any ray contained in the plane of the figure in the same direction than the incident one; and second, the optical path is not affected by rotations of the element around its edge. Therefore the OPD of the interferometer only depends on the position of the dihedron edges (A_1 and A_2). Specifically the OPD is null when both edges are placed symmetrically respect to the BS, that is, when the image of A_1 through the BS (A'_1) meets with A_2. If A'_1 is anywhere else the wavefront of arm 2 is moved a distance $\mathbf{t} = 2\overrightarrow{A'_1A_2}$ from the wavefront of arm 1. In general the OPD depends on the ray direction as $\mathbf{t}\,\mathbf{u} = 2|A'_1A_2|\cos\alpha$, being α the angle that each ray makes with $\mathbf{t}$ (see discussion in the previous section). A translation of a dihedron along the optical axis has the same effect than in a conventional Michelson interferometer. Then such a device is of a framing type[48]. A normal translation to the optical axis in the plane of Fig. 1.13 causes no OPD to rays propagating along the optical axis ($\alpha = \pi/2$), but it does to other rays. For such a translation, constructive interference occurs for all wavelengths at the vertical central line of the image (located at the image focal plane), but a range of OPD is obtained from left to right edge of the image. The direction set with a constant OPD makes a cone whose axis is parallel to $\mathbf{t}$. The image points with same OPD make hyperbolas resulting from the intersection of the image plane with these cones whose common vertex is at the lens center. Nevertheless, these hyperbolas can be approximated to quasi-parallel vertical lines if the angular field of the lens is small ($\cos\alpha \simeq \pi/2 - \alpha$, for angles α near $\pi/2$). The spatial information remains unaltered through this interferometer as in conventional Michelson interferometer. Hence, we obtain a conventional image that seems to be modulated with a vertical pattern of interference fringes; actually each object point is modulated with a different function that depends on its spectrum. As the object tracking is performed along the horizontal direction, a given object point crosses the field of view. A different OPD occurs to this object in each frame, which allows to capture successive points of the FT of its spectrum. Once this object point goes out from the field of view, its spectrum can be recovered by performing the inverse FT to this signal. So, this instrument is a windowing FTIS with no mobile internal parts. This is its main advantage against the Michelson interferometer with plane mirrors. The only movement is the tracking, which is also necessary in framing schemes for objects larger than the instrument field of view; thus the instrument is more robust and easier to construct. Moreover, a dihedron vibration transverse to the light does not cause a strong loss of interference, only change the fringe scaling through $|A'_1A_2|$. Unfortunately longitudinal vibrations blur the interferences because the direction of $\mathbf{t}$ changes. The sensitivity against vibrations or deformations can be greatly reduced by using either a monolithic interferometer[49] or a common path interferometer. In the later kind of instruments, both beams overlap and are reflected or refracted in the same elements, so a movement of any element changes the optical path of both beams in the same amount keeping the OPD unaltered. The following instruments comprise common path interferometers.

1.4.3 FTISs Based on the Sagnac Interferometer

Sagnac interferometers have been used as a component of both windowing and pushbroom FTISs. In a windowing configuration the interferometer is placed before the imaging lens as in the previous subsection[51]. The simplest Sagnac

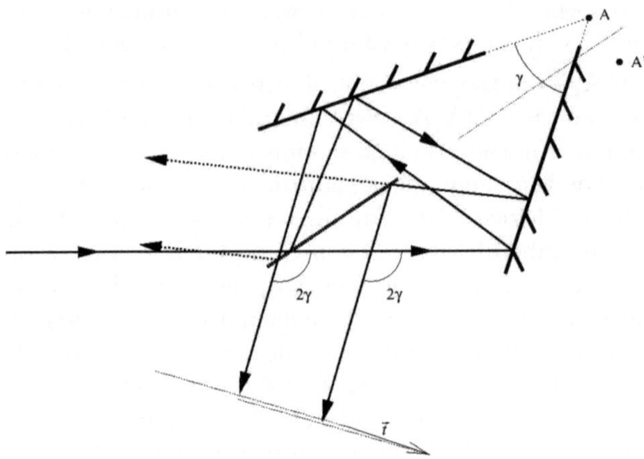

Fig. 1.14. Sagnac interferometer

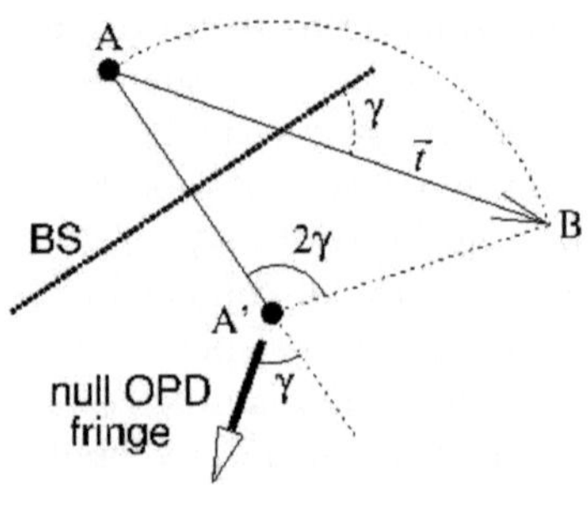

Fig. 1.15. Drawing showing the translation vector **t**. A' is the image of A through the BS. B is the image of A by rotation an angle 2γ around A'.

interferometer comprises a plane BS and two mirrors making an angle γ. Light leaving the BS is reflected in both mirrors and comes back to the BS from the opposite side (see Fig. 1.14). Thus we have two beams describing a loop, one does it counterclockwise and the other clockwise. Consequently one beam is transmitted twice through the BS to reach the output port; the other beam is reflected twice. The former beam behaves as if there were no BS, it is simply reflected in the mirrors. This is equivalent to a rotation of the beam by an angle 2γ around the line where both mirrors meet (A). The later beam suffers a rotation around the image of A in the BS (A$'$) by an angle 2γ as well. As both rays are rotated the same angle, they always exit the interferometer parallel regardless the input direction, the relative location of the elements and their orientations. If A$'$ meets A, that is if A is in the plane defined by the BS surface, both rays rotate around the same point and the OPD is null for all incidence angles. If A' does not meet A, the wavefronts are separated one from the other a distance $|\mathbf{t}| = 2|\mathrm{AA}'|\sin\gamma$ in a direction that makes an angle γ with the BS

(Fig. 1.15). Therefore the OPD is $\mathbf{t}\,\mathbf{u} = 2|\mathrm{AA}'|\sin\gamma\cos\alpha$, where α is the angle that the ray makes with $\mathbf{t}$. If the optical axis of the imaging lens is perpendicular to $\mathbf{t}$, the fringes superimposed to the image are quasi-linear as in the previous subsection. In this case, the optical axis makes an angle γ with the normal to the BS and the null OPD fringe is in the center of the image.

The advantage of the Sagnac interferometer is the stability of its central fringe whose position only depends on the orientation of the BS and the angle between both mirrors; that is, it is immune to translations of any element. Its handicap lies in the vignetting between BS and mirrors. Vignetting leads to a lower field of view compared to that of Michelson interferometer with dihedrons. The optimal throughput (which is proportional to the electromagnetic flux) of the Michelson interferometer is more than five times greater than that of the Sagnac interferometer for the same BS diameter[47, 50].

Sagnac interferometers have been also used in pushbroom FTISs [52, 53, 54, 55]. The first element of these instruments is a lens that focuses the image on a slit to select a strip of the object as in pushbroom dispersive imaging spectrometers. The interferometer provides two mutually coherent virtual images of the slit. A so-called Fourier lens placed after the interferometer gives a range of OPD in its focal plane. Unfortunately we loose the spatial information along the slit because the slit images are fully blurred in this plane. Therefore a cylindrical lens is added to focus the image along the slit but keeping the image blurred across to preserve the interference pattern. In this pushbroom configuration the interferometer limits the f-number rather than the field of view. The best aperture is obtained for a monolithic square -three mirrors- Sagnac interferometer which allows $f/\#¿5/n$, with n its refractive index[50]. Furthermore pushbroom FTISs have a poorer SNR than dispersive counterparts because FT computing rises the noise[44].

1.4.4 FTISs Based on Polarization

This kind of instruments makes interference between two linear polarization states of plane waves by using birefringent prisms or plates located between linear polarizers. A conventional optics focuses the transmitted waves on the detector array. For instance, a windowing instrument is obtained with a Savart polariscope. This birefringent element is formed by two identical plane-parallel uniaxial crystals cut to make their optical axes an angle of 45° with their faces[56, 57]. Both plates are joined and rotated 90° around the normal to the faces. Therefore an ordinary wave in the first plate becomes extraordinary in the second one and viceversa. For normal incidence (see Fig. 1.16), incident, ordinary and emerging rays are parallel and both extraordinary rays are deflected. So the emerging rays are parallel to the incident but laterally displaced in orthogonal directions. Moreover, both cross the same optical path inside the plates so the translation vector $\mathbf{t}$ is parallel to the plate faces. For an imaging lens parallel to the plate faces this instrument is equivalent to a windowing Sagnac interferometer, but it has not such strong field limitations. Moreover it is very compact.

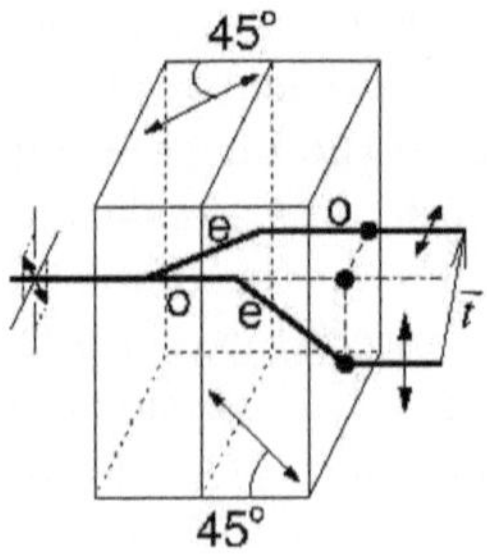

Fig. 1.16. Savart polariscope

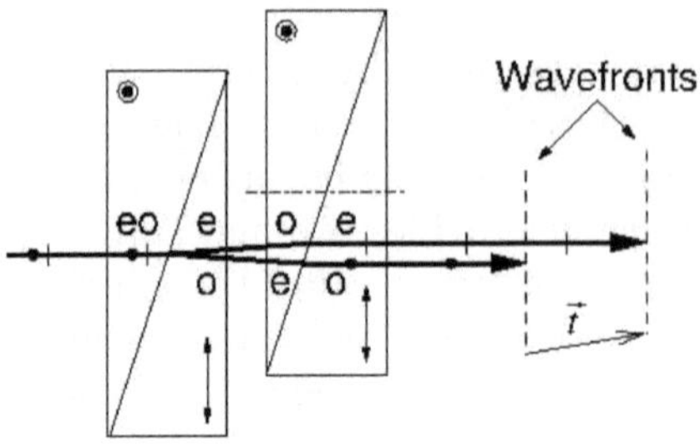

Fig. 1.17. Wollaston prisms for a framing FTIS

A framing birefringent FTIS based on two identical Wollaston prisms instead of a Savart polariscope has been also proposed [58]. The prism edges are parallel and make an angle of 45° with the transmission axes of the polarizers. The first prism is fixed and the second one can be translated transversely to the optical axis. Light with one of the polarization states propagates along a larger distance as ordinary wave than it does as extraordinary wave. The opposite happens to orthogonally polarized light. Only when the central lines of both prisms meet each other, these thicknesses match for both polarization states. That is, a wavefront is overtaken and the other one is slowed down as the second prism translates. Actually this explanation is very simplified because the vector **t** has also a small transverse component that depends on the incident direction. This results in hyperbolic fringes in the detector plane, but essentially this instrument works like a Michelson interferometer with plane mirrors. The mobile prism introduces an OPD about 200 times smaller than the mirror of a Michelson interferometer does, so accuracy requirements for the scanning arrangement are relaxed accordingly.

The disadvantages of polarization FTISs are the dependence of the signal on the polarization state of the incident light, a loss of intensity by 50% for natural polarization with respect to previous schemes, and the limitation of the spectral range by the crystal material. Anyway these instruments are very attractive.

References

1. Kerekes, J.: Imaging spectrometers go commercial, pp. 63–68. Laser Focus World (July 2006)
2. Sellar, R.G., Boreman, G.D.: Classification of imaging spectrometers for remote sensing applications. Opt. Eng. 44, 1–3 (2005)
3. Gat, N.: Imaging spectroscopy using tunable filters: a review. In: Proc. SPIE, vol. 4056, pp. 50–64 (2000)
4. Nieke, J., Schwarzer, H., Neumann, A., Zimmermann, G.: Imaging spaceborne and airborne sensor systems in the beginning of the next century. In: Proc. SPIE, vol. 3221, pp. 581–592 (1997)
5. Gao, G.H., Lin, Z.: Acousto-optic supermultispectral imaging. Appl. Opt. 32, 3081–3086 (1993)
6. Gupta, N.: Remote sensing using hyperspectral and polarization images. In: Proc. SPIE, vol. 4574, pp. 184–192 (2002)
7. Vo-Dinh, T., Stokes, D.L., Wabuyele, M.B., Martin, M.E., Song, J.M., Jagannathan, R., Michaud, E., Lee, R.J., Pan, X.: A hyperspectral imaging system for in vivo optical diagnostics. IEEE Eng. Med. Biol. 23, 40–49 (2004)
8. Gupta, N., Voloshinov, V.: Hyperspectral imager, from ultraviolet to visible, with a KDP acousto-optic tunable filter. Appl. Opt. 43, 2752–2759 (2004)
9. Gupta, N., Voloshinov, V.: Hyperspectral imaging performance of a TeO2 acousto-optic tunable filter in the ultraviolet region. Opt. Lett. 30, 985–987 (2005)
10. Miller, P.J., Hoyt, C.C.: Multispectral imaging with a liquid crystal tunable filter. In: Proc. SPIE, vol. 2345, pp. 354–365 (1995)
11. Koop, G.: Tunable birefringent filters using liquid crystal variable retarders. In: Proc. SPIE, vol. 2873, pp. 324–327 (1996)
12. Slawson, R.W., Ninkov, Z., Horch, E.P.: Hyperspectral imaging: wide-area spectrophotometry using a liquid crystal tunable filter. Publications of the Astronomical Society of the Pacific 111, 621–626 (1999)
13. Yano, T., Watanabe, A.: Acoustooptic TeO2 tunable filter using far-off-axis anisotropic Bragg diffraction. Appl. Opt. 15, 2250–2258 (1976)
14. Glenar, D.A., Hilman, J., Saif, B., Bergstralh, J.: Acousto-optic imaging spectropolarimetry for remote sensing. Appl. Opt. 33, 7412–7424 (1994)
15. Evans, J.W.: The birefringent filter. JOSA 39, 229–242 (1949)
16. Evans, J.W.: Solc birefringent filter. JOSA 48, 142–145 (1958)
17. Daly, J.T., Bodkin, A., Schneller, W., Kerr, R., Noto, J., Haren, R., Eismann, M., Karch, B.: Tunable narrow band filter for LWIR hyperspectral imaging. In: Proc. SPIE, vol. 3948, pp. 104–115 (2000)
18. Knittl, Z.: Optics of thin films. John Wiley & Sons, London (1976)
19. Franon, M.: Sparation des radiations par les filters optiques. Masson, Paris (1984)
20. Mika, A.M.: Linear-wedge spectrometer. In: Proc. SPIE, vol. 1298, pp. 127–131 (1990)
21. Reuter, D.C., Jennings, D.E., McCabe, G.H., Travis, J.W., Bly, V.T., La, A.T., Nguyen, T.L., Jhabyala, M.D., Shu, P.K., Endres, R.D.: Hyperspectral sensing using the linear etalon imaging spectral array. In: Proc. SPIE, vol. 2957, pp. 154–161 (1997)
22. Wahl, J.A., Van Delden, J.S., Tiwary, S.: Tapered Fabry-Prot Filters. IEEE Phot. Tech. Lett. 16, 1873–1875 (2004)
23. Czerny, M., Turner, A.F.: ber den Astigmatismus bei Spiegelspektrometern. Zeit. Phys. 61, 792–797 (1930)

24. Smith, W.J.: Modern lens design. McGraw-Hill Professional, New York (2004)
25. Hutley, M.C.: Diffraction Gratings. Academic Press, London (1982)
26. Fisher, J., Baumback, M., Bowles, J., Grossmann, J., Antoniades, J.: Comparison of low-cost hyperspectral sensors. In: Proc. SPIE, vol. 3438, pp. 23–30 (1998)
27. Offner, A.: Annular field systems and the future of optical microlithography. Opt. Eng. 26, 294–299 (1987)
28. Mouroulis, P., Green, R.O., Chrien, T.G.: Design of pushbroom imaging spectrometers for optimum recovery of spectroscopic and spatial information. Appl. Opt. 39, 2210–2220 (2000)
29. Chrisp, M.P.: Convex diffraction grating imaging spectrometer, US Patent 5,880,834 (1999)
30. Prieto-Blanco, X., Montero-Orille, C., Couce, B., de la Fuente, R.: Analytical design of an Offner imaging spectrometer. Opt. Express 14, 9156–9168 (2006)
31. `http://www.headwallphotonics.com`
32. Lobb, D.R.: Imaging spectrometer, WO Patent WO9837389 (1998)
33. Cutter, M.A., Lobb, D.R., Williams, T.L., Renton, R.E.: Integration & testing of the compact high-resolution imaging spectrometer (CHRIS). In: Proc. SPIE, vol. 3753, pp. 180–191 (1999)
34. Cutter, M.A.: Compact high-resolution imaging spectrometer (CHRIS) design and performance. In: Proc. SPIE, vol. 5546, pp. 126–131 (2004)
35. Aikio, M.: Optical component comprising prisms and a grating, WO Patent WO9321548 (1993)
36. Aikio, M.: Hyperspectral prism-grating-prism imaging spectrograph. VTT Publications, Technical research centre of Finland, Espoo (2001)
37. `http://www.specim.fi`
38. Couce, B., Prieto-Blanco, X., Montero-Orille, C., de la Fuente, R.: A windowing/pushbroom hyperspectral imager. In: Gabrys, B., Howlett, R.J., Jain, L.C. (eds.) KES 2006. LNCS (LNAI), vol. 4253, pp. 300–306. Springer, Heidelberg (2006)
39. Descour, M.R., Volin, C.E., Dereniak, E.L., Gleeson, T.M., Hopkins, M.F., Wilson, D.W., Maker, P.D.: Demonstration of a computed-tomography imaging spectrometer using a computer-generated hologram disperser. Appl. Opt. 36, 3694–3698 (1997)
40. Volin, C.E., Garcia, J.P., Dereniak, E.L., Descour, M.R., Sass, D.T., Simi, C.G.: MWIR computed tomography imaging spectrometer: calibration and imaging experiments. In: Proc. SPIE, vol. 3753, pp. 192–202 (1999)
41. Ford, B.K., Volin, C.E., Murphy, S.M., Lynch, R.M., Descour, M.R.: Computed tomography-based spectral imaging for fluorescence microscopy. Biophys. J. 80, 986–993 (2001)
42. Johnson, W.R., Wilson, D.W., Bearman, G.: All-reflective snapshot hyperspectral imager for ultraviolet and infrared applications. Opt. Lett. 30, 1464–1466 (2005)
43. Born, M., Wolf, E.: Principles of optics. Cambridge University Press, Cambridge (1998)
44. Sellar, R.G., Boreman, G.D.: Comparison of relative signal-to-noise ratios of different classes of imaging spectrometer. Appl. Opt. 44, 1614–1624 (2005)
45. Swinyard, B.M., Dohlen, K., Ferand, D., Baluteau, J.P., Pouliquen, D., Dargent, P., Michel, G., Martignac, J., Ade, P., Hargrave, P.C., Griffin, M.J., Jennings, D.E., Caldwell, M.E.: Imaging FTS for Herschel SPIRE. In: Proc. SPIE, vol. 4850, pp. 698–709 (2003)

46. Naylor, D.A., Gom, B.G., Schofield, I., Tompkins, G., Davis, G.R.: Mach-Zenhder Fourier transform spectrometer for astronomical spectroscopy at submilimeter wavelengths. In: Proc. SPIE, vol. 4855, pp. 540–551 (2003)
47. Ferrec, Y., Taboury, J., Sauer, H., Chavel, P.: Optimal geometry for Sagnac and Michelson interferometers used as spectral imagers. Opt. Eng. 45, 115601–115606 (2006)
48. Farley, V., Belzile, C., Chamberland, M., Legault, J.F., Schwantes, K.: Development and testing of a hyper-spectral imaging instrument for field spectroscopy. In: Proc. SPIE, vol. 5546, pp. 29–36 (2004)
49. Harnisch, B., Posselt, W., Holota, K., Otto Tittel, H., Rost, M.: Compact Fourier-transform imaging spectrometer for small satellite missions. Acta Astronautica 52, 803–811 (2003)
50. Sellar, R.G., Boreman, G.D.: Limiting aspect ratios of Sagnac interferometers. Opt. Eng. 42, 3320–3325 (2003)
51. Barducci, A., Marcoionni, P., Pippi, I.: Spectral measurement with a new Fourier transform imaging spectrometer (FTIS). In: Geoscience and Remote Sensing Symposium (IGARSS 2003) Proc. IEEE Int., vol. 3, pp. 2023–2026 (2003)
52. Rafert, J.B., Sellar, R.G., Blatt, J.H.: Monolitic Fourier-transform imaging spectrometer. Appl. Opt. 34, 7228–7230 (1995)
53. Sellar, R.G., Rafert, J.B.: Fourier-transform imaging spectrometer with a single toroidal optic. Appl. Opt. 34, 2931–2933 (1995)
54. Sellar, R.G., Rafert, J.B.: Effects of aberrations on spatially modulated Fourier transform spectrometers. Opt. Eng. 42, 3087–3092 (1994)
55. Tyo, J.S., Turner Jr., T.S.: Variable-retardance, Fourier-transform imaging spectropolarimeters for visible spectrum remote sensing. Appl. Opt. 40, 1450–1458 (2001)
56. Zhan, G., Oka, K., Ishigaki, T., Baba, N.: Birefringent imaging spectrometer. Appl. Opt. 41, 734–738 (2002)
57. Zhang, C., Xiangli, B., Zhao, B., Yuan, X.: A static polarization imaging spectrometer based on a Savart polariscope. Opt. Commun. 203, 21–26 (2002)
58. Harvey, A.R., Fletcher-Holmes, D.W.: Birefringent Fourier-transform imaging spectrometer. Opt. Express 12, 5368–5374 (2004)

2

Remote Sensing Data Compression

Joan Serra-Sagristà and Francesc Aulí-Llinàs

Group on Interactive Coding of Images
Department of Information and Communications Engineering
Universitat Autònoma de Barcelona
E-08290 Cerdanyola del Vallès (Barcelona), Spain
joan.serra@uab.es, fauli@deic.uab.es

Summary. The interest in remote sensing images is growing at an enormous pace in the last years. However, transmission and storage of remote sensing images pose a special challenge, and multiple efficient image compression systems have appeared. This chapter contributes an overview of several techniques for image coding systems, focusing on lossy approaches.

2.1 Introduction

Remote sensing is the capability of extracting information about an object without being in physical contact with it [15]. Materials comprising the various objects in a scene reflect, absorb, and emit, electromagnetic radiation depending on their molecular composition and shape. By measuring the radiation over a broad spectral range, the spectrum may be used to identify the materials.

In the last years, several sensors have come into play, allowing applications of remote sensing images such as target detection, military surveillance, farming practices assessment, fire detection, weather forecasting,... [37]. In addition, modern sensors have improved their acquisition capabilities, so that radiometric resolution, spatial resolution, spectral resolution, and time resolution have increased at least one order of magnitude, enabling better results in the aforementioned applications. Nevertheless, these improved capabilities come also at the price of an increase in the data volume size.

Remote sensing sensors are either airborne or spaceborne. For instance, the NASA Jet Propulsion Laboratory (JPL) *Airborne Visible/Infrared Imaging Spectrometer* [50] records the visible and the near-infrared spectrum of the reflected light, and is capable of producing images of size 2048 rows $\times$ 614 columns $\times$ 224 bands $\times$ 2 bytes per pixel for a total of 563.347.456 bytes (over 537 Megabytes) per flight. Also, the NASA JPL *Atmospheric Infrared Sounder* [51], which is a standard ultraspectral sounder data, records thousands of bands from the infrared spectrum, produces 240 granules per day, each of size 135 rows $\times$ 90 columns $\times$ 2107 bands $\times$ 2 bytes per pixel, for a total of 12.288.024.000 bytes (over 12 Gigabytes) data daily. Finally, NASA's project Earth Observing System satellite [49] will generate data at an unprecedented rate, estimated in over a

M. Graña and R.J. Duro (Eds.): Comput. Intel. for Remote Sensing, SCI 133, pp. 27–61, 2008.
springerlink.com

Terabyte of data every day. These enormous data volumes ask for the aid of data compression.

Data compression may be beneficial to both data storage and to data transmission. Data compression may be broadly divided in three classes: *lossless compression* refers to the process whereby the decoded image after coding and decoding is exactly the same as the input image, and achieves compression factors of at most 4:1; *near-lossless compression* refers to the process whereby the decoded image after coding and decoding is perceptually equivalent to the input image, and achieves compression factors of about 30:1; *lossy compression* refers to the process whereby some information is discarded, leading to much higher compression ratios. The data that may be discarded depends on the final application, so that in some scenarios, lossy-to-lossless compression is desired.

Data compression for remote sensing images is effective because, as natural images, they exhibit large spatial correlation. In addition, they exhibit large spectral correlation; nevertheless, since the spectrum is usually the most important feature that must be preserved for the final application, and since the dynamic range in the spectral domain is much higher than in natural images, a simple extension of 2D dimensional coding systems is not appropriate. Therefore, new methods exploiting both the spatial and the spectral redundancy have recently appeared.

Considering the final remote sensing applications, the compression ratio might not be the single parameter to evaluate the coding performance of a coding system, and it has been shown that neither the variance of the input data, nor the Signal-to-Noise Ratio (SNR), and not even the common Mean Squared Error (MSE), are often significant enough [63] for a posteriori tasks such as classification, thus other distortion measures have been proposed to help the evaluation and even to guide the compression approach.

This chapter begins with a discussion of some measures proposed to assess the performance of the coding system. Then an overview of several techniques used in recent coding systems, both for lossless compression and for lossy compression is provided. Next, some experimental results for hyperspectral and ultraspectral images are presented. Finally, a conclusions section ends the chapter.

2.2 Measures for Evaluating Compression

The performance of a coding system is established as a trade-off between the rate achieved by the coding process, and the quality of the recovered image after decoding.

For 2D dimensional images, let $x(i,j)$ denote a pixel in the input image of size $N_x \cdot N_y$ pixels, the mean of all input pixels be $\bar{x} = \frac{1}{N_x N_y} \sum_i \sum_j x(i,j)$, and let $\hat{x}(i,j)$ denote a pixel in the decoded image after coding and decoding. Bit depth of input pixels is b bits per pixel (bpp).

For 3D dimensional images, let $\mathbf{x}$ denote a pixel vector, a spectral vector of length N_z, from the input volumetric image of size N_z bands (components), with $N_x \cdot N_y$ pixels in each band, and $x(i,j,k)$ be the pixel at position (i,j) in band

k. Let $\hat{x}(i, j, k)$ denote a pixel vector in the decoded image after coding and decoding. Bit depth of input pixel vectors is b bits per pixel per band (bpppb).

2.2.1 Compression Measures

Compression may be evaluated according to several measures [66].

1. Compression Ratio

$$CompressionRatio = \frac{\text{size of the compressed stream}}{\text{size of the input image}}$$

   A value of 0.6 means that the data occupies 60% of its original size after compression. Values greater than 1 mean a compressed stream bigger than the input image (negative compression).

   Sometimes the quantity [100 × (1 - compression ratio)] is used to express the quality of compression. A value of 60 means that the compressed stream occupies 40% of the input image original size (or that the compression has resulted in a savings of 60%).
2. Compression Factor

$$CompressionFactor = \frac{\text{size of the input image}}{\text{size of the compression stream}}$$

   The compression factor is the inverse of the compression ratio. Values greater than 1 mean compression, and values less than 1 imply expansion.
3. Compression Gain

$$CompressionGain = 100 \cdot log_e \frac{\text{reference size}}{\text{compressed size}}$$

   where the reference size is, either the size of the input image, or the size of the compressed stream produced by some standard lossless compression method, and the compressed size is the size of the compressed stream.
4. Bits per pixel (Bits per pixel per band)

$$Bpp\ (Bpppb) = \frac{b}{\text{Compression Factor}}$$

   where b is the bit depth of input image pixels (pixel vectors).

2.2.2 Distortion Measures

Distortion may be evaluated according to the radiometric distortion, to the spectral distortion, or to the preservation of the capability for a posteriori processing tasks such as classification [4, 23].

Radiometric Distortion

Radiometric distortion measures for remote sensing 3D images may be defined as extensions of pixel distortion measures for natural 2D dimensional images.

2D distortion measures

1. Mean Absolute Error (MAE), or L_1 (norm)
$$MAE = \frac{1}{N_x N_y} \sum_i \sum_j \mid x(i,j) - \hat{x}(i,j) \mid$$
2. Mean Squared Error (MSE), or L_2^2
$$MSE = \frac{1}{N_x N_y} \sum_i \sum_j [x(i,j) - \hat{x}(i,j)]^2$$
3. Root MSE (RMSE), or L_2
$$RMSE = \sqrt{MSE}$$
4. Signal to Noise Ratio (SNR)
Given $E(x) = \frac{1}{N_x N_y} \sum_{i,j} x(i,j)^2$ the energy of the input image,
$$SNR_{Energy} = 10 \cdot log_{10} \frac{E(x)}{MSE} \quad (dB)$$
Given $\sigma^2 = \frac{1}{N_x N_y} \sum_{i,j} (x(i,j) - \bar{x})^2$ the variance of the input image, sometimes an alternative expression for SNR is used:
$$SNR_{Variance} = 10 \cdot log_{10} \frac{\sigma^2}{MSE} \quad (dB)$$
5. Peak Signal to Noise Ratio (PSNR)
$$PSNR = 10 \cdot log_{10} \frac{(2^b - 1)^2}{MSE} \quad (dB)$$
For images whose pixels may have negative values, an alternative expression may be used:
$$PSNR_{Negative} = 10 \cdot log_{10} \frac{\max_{i,j} \mid x(i,j) \mid}{MSE} \quad (dB)$$
6. Maximum Absolute Distortion (MAD), or Peak Absolute Error (PAE), or L_∞
$$MAD = \max_{i,j} \{ \mid x(i,j) - \hat{x}(i,j) \mid \}$$
7. Percentage Maximum Absolute Distortion (PMAD)
$$PMAD = \max_{i,j} \{ \frac{\mid x(i,j) - \hat{x}(i,j) \mid}{x(i,j)} \} \cdot 100$$
8. Mean Error (ME)
Although not strictly a measure, the Mean Error is also sometimes considered:
$$ME = \frac{1}{N_x N_y} \sum_i \sum_j [x(i,j) - \hat{x}(i,j)]$$

3D distortion measures

1. Mean Absolute Error (MAE), or L_1 (norm)

$$MAE = \frac{1}{N_x N_y N_z} \sum_i \sum_j \sum_k \mid x(i,j,k) - \hat{x}(i,j,k) \mid$$

2. Mean Squared Error (MSE), or L_2^2

$$MSE = \frac{1}{N_x N_y N_z} \sum_i \sum_j \sum_k [x(i,j,k) - \hat{x}(i,j,k)]^2$$

3. Average Vector RMSE

$$VRMSE_{\mathrm{avg}} = \frac{1}{N_x N_y N_z} \sum_{i,j} \sqrt{\sum_k [x(i,j,k) - \hat{x}(i,j,k)]^2}$$

4. Peak Vector RMSE

$$VRMSE_{\mathrm{max}} = \max_{i,j} \sqrt{\sum_k [x(i,j,k) - \hat{x}(i,j,k)]^2}$$

5. Signal to Noise Ratio (SNR)
   Given $E(x) = \frac{1}{N_x N_y N_z} \sum_{i,j,k} x(i,j,k)^2$ the energy of the input image,

$$SNR_{Energy} = 10 \cdot log_{10} \frac{E(x)}{MSE} \quad (dB)$$

   Given $\sigma^2 = \frac{1}{N_x N_y N_z} \sum_{i,j,k} (x(i,j,k) - \bar{x})^2$ the variance of the input image, sometimes an alternative expression for SNR is used:

$$SNR_{Variance} = 10 \cdot log_{10} \frac{\sigma^2}{MSE} \quad (dB)$$

6. Peak Signal to Noise Ratio (PSNR)

$$PSNR = 10 \cdot log_{10} \frac{(2^b - 1)^2}{MSE} \quad (dB)$$

   This measure only has a meaningful sense when all the bands (components) of the volume input image have the same bit per pixel per band bit depth b.
7. Maximum Absolute Distortion (MAD)

$$MAD = \max_{i,j,k} \{\mid x(i,j,k) - \hat{x}(i,j,k) \mid\}$$

8. Percentage Maximum Absolute Distortion (PMAD)

$$PMAD = \max_{i,j,k} \{ \frac{\mid x(i,j,k) - \hat{x}(i,j,k) \mid}{x(i,j,k)} \} \cdot 100$$

Spectral Distortion

Let $\mathbf{x}$ denote a pixel vector, a spectral vector of length N_z, from the input volumetric image of size N_z bands, with $N_x \cdot N_y$ pixels in each band, and let $\hat{\mathbf{x}}$ denote a pixel vector in the decoded image after coding and decoding.

1. Spectral Angle Mapper (SAM)

$$SAM(\mathbf{x}, \hat{\mathbf{x}}) \doteq \arccos\left(\frac{<\mathbf{x}, \hat{\mathbf{x}}>}{\|\mathbf{x}\|_2 \cdot \|\hat{\mathbf{x}}\|_2}\right)$$

SAM denotes the absolute value of the spectral angle between the couple of vectors, where $< \cdot, \cdot >$ stands for scalar product, and can be measured in either degrees or radians.

2. Spectral Information Divergence (SID) [14]

$$SID(\mathbf{x}, \hat{\mathbf{x}}) = D(\mathbf{x}\|\hat{\mathbf{x}}) + D(\mathbf{x}\|\hat{\mathbf{x}})$$

where $D(\mathbf{x}\|\hat{\mathbf{x}})$ is the Kullback-Leibler distance (KLD), also known as entropy divergence or discrimination, and is defined as

$$D(\mathbf{x}\|\hat{\mathbf{x}}) \doteq \sum_{k=1}^{N_z} p_k \log\left(\frac{p_k}{q_k}\right)$$

where $p_k \doteq \frac{x_k}{\|\mathbf{x}\|_1}$ and $q_k \doteq \frac{\hat{x}_k}{\|\hat{\mathbf{x}}\|_1}$. In practice, SID is equal to the symmetric KLD. Its measure unit depends on the base of the logarithm: for logarithm in base two, *bit / vector*.

3. Spectral Unmixing
Because the spatial resolution of some sensors may not have the needed granularity, it is expected that pixels in a spectral band contain the radiance or reflectance of several materials, so that each pixel is a mixture of *endmembers e*.

$$\mathbf{x} = \sum_l \omega_l e_l + r$$

where ω_l is the weight of endmember e_l and r is a random noise. Several methods have appeared to find the endmembers [8, 58, 71].

Preservation of Classification

The performance of the final application depends heavily on the quality of the decoded images, so that the final application dictates the quality of the coding system. As examples of easily implemented final applications, unsupervised classification methods as Isodata [6] or k-means [43] have been considered. Then the classification of the input image is compared to the classification of the decoded image, and the **Preservation of Classification** accounts for the percentage of pixel vectors that do not change their classification in the decoded image taking the classification of the input image as the ground-truth.

Considerations about Distortion Measures

Radiometric distortion does not necessarily imply spectral distortion, and it is known that radiometric distortion may not be appropriate as a measure for image interpretation [13]. In contrast, spectral distortion implies radiometric distortion: when a couple of pixel vectors have either the same Euclidean distance (L_2 norm) for *SAM*, or the same absolute value distance (L_1 norm) for *SID*, the radiometric distortion is minimal. Also, it seems that radiometric distortion is somehow tied to the preservation of classification [23].

2.3 Lossless / Near-Lossless Compression

Remote sensing data is sometimes very sensitive to data quality, so that lossless or near-lossless compression is in order.

With no intention of being exhaustive, this section provides a brief enumeration of several recent proposals for lossless and near-lossless coding systems. It is apparent that the best performance is achieved when the coding process incorporates a prediction stage [47]. This prediction takes usually the form of a simple decorrelation among adjacent pixels within a spectral band, or among adjacent coordinates along the spectral dimension.

2.3.1 Lossless Compression

In [28] the authors evaluate several coding systems for lossless compression of ultraspectral sounder data. Since this data presents a strong correlation in disjoint spectral regions affected by the same type of absorbing gases, a Bias-Adjusted Reordering (BAR) is devised, applicable to any 2D compression method. When spatial and spectral BAR is carried out, the compression factor may grow by no less than 0.4 for such data.

Other band orderings have been considered [47] for AVIRIS data. For such data, band orderings schemes based on Prediction Mean Squared Error (PMSE) and on Compressed File Size (CFS) may increase the compression factor by no less than 0.9.

Lossless coding techniques may be classified into transform-based, prediction-based, and clustering-based methods, although several hybrid approaches have also been presented.

- Transform-based methods perform a transform such as a Discrete Cosine Transform (DCT) [2], a Discrete Wavelet Transform (DWT) [62], or a Karhunen Loève Transform (KLT) [53] (also known as Principal Component Analysis (PCA), a well-known technique in multivariate data analysis), as the first step of the coding process, and then a bit-plane coding technique like SPIHT [65] or JPEG2000 [32] is performed.

- Prediction-based methods include coding techniques like JPEG-LS [31], CALIC [73], JPEG-7 and Correlation-based Conditional Average Prediction (CCAP) [72].
- Clustering-based methods are related to Vector Quantization [48, 59].

For the ultraspectral sounder data, the best reported results are achieved by a Predictive Partitioned Vector Quantization (PPVQ) [28], a clustering-based method, yielding, on average, a compression factor of 3.31, and by an Adaptive Vector Quantization with Linear Prediction (AVQLP) [29], achieving a compression factor of 3.33.

For AVIRIS hyperspectral data, the best reported results are achieved by a clustered prediction method [46], a mixture of a clustering-based and a prediction-based method, yielding, on average, a compression factor of 3.43, by a crisp and fuzzy adaptive spectral predictor [3], yielding, on average, a compression factor of 3.45, and by a method using Look-Up Tables [45] along with a predictor selection mechanism [30], yielding, on average, a compression factor of 3.47.

2.3.2 Near-Lossless Compression

Most of the above-mentioned lossless coding techniques may be also devised for near-lossless compression, with higher compression factors, but still preserving a high quality for the decoded volume data, usually with bounded pixel (pixel vector) error.

For the transform-based methods, using a integer wavelet transform [7] allows to provide lossy to lossless compression: SPIHT [74], JPEG2000 [52].

For the prediction-based methods, a differential pulse code modulation (DPCM) compression for AVIRIS hyperspectral data is presented in [4]. The crisp and fuzzy adaptive spectral predictor [3] may be also applied for near-lossless compression.

For the clustering-based methods, a near-lossless coding technique that may be applied onboard a hyperspectral satellite is presented in [60].

2.4 Lossy Compression

When a certain degree of loss does not alter significantly the a posteriori processing task, lossy compression of remote sensing images may be helpful to reduce transmission time and requested storage capability.

Most modern lossy coding techniques rely on a two- or three-stage approach: first, a transform is applied to decorrelate the input signal; second, a quantization, and a bit-plane coding followed by an arithmetic coding is performed; third, a rate-distortion optimization may be applied to select the best coding segments (mostly according to a MSE distortion).

This section focuses on the latest image coding system standard as related to remote sensing compression, namely JPEG2000 multicomponent as contemplated in Part 2 of the JPEG2000 Standard [33]. First, a review of several coding systems, providing competitive performance is presented.

2.4.1 Prominent Coding Systems

In the sequel, a single or few references are provided for each reviewed system. The interested reader is referred to the references in each paper for a more extensive bibliography.

Coding systems

Image coding systems based on Set Partitioning, as SPIHT or Set Partitioned Embedded bloCK (SPECK) [67], have also been extended to 3D applications. They rely on a DCT, a DWT, or a KLT, as the first stage of the coding process, next they generate an embedded bit stream by analyzing the significance of each transformed coefficient, and then successively refining its magnitude. SPIHT and SPECK are scalable coding techniques, in a purely rate-distortion sense and in a resolution sense, thus allowing progressive transmission. Headers may also be embodied to produce code-streams with the random access property (scalability by spatial location). Coding performance of SPIHT and SPECK may be considered state-of-the-art, and their low computational complexity make these two techniques highly competitive.

Moreover, since the coding performance of these two techniques do not heavily depend on the entropy coding –usually a computationally demanding arithmetic encoding subsystem–, hardware implementations for onboard space applications have also been proposed [17].

The Consultative Committee for Space Data Systems (CCSDS) has proposed the "Image Data Compression" (IDC) Recommended Standard [22] applicable to digital data from payload instruments and devised for near-lossless and lossy compression. Coding performance of CCSDS IDC Recommendation is slightly inferior to JPEG2000, but it has also been designed for onboard implementation [76, 75] achieving competitive results.

Onboard applications

Remote sensing images captured onboard satellites are also prone to compression. The image coding system onboard satellites has several strong memory and computation requirements. Alternative coding techniques to JPEG2000 –seemingly more complex–, have been devised.

Successive approximation multistage vector quantization [61], hierarchical self-organizing cluster vector quantization [60], a fast vector quantization based on the full search generalized Lloyd algorithm [24, 59], and a class-based coder [10] –also based on vector quantization– have been proposed, reporting gains larger than 10 dB over JPEG2000.

2.4.2 JPEG2000

JPEG2000 is the latest international standard for image and video compression and transmission, designed to behave as an image processing system and not only as an input-output image compression filter.

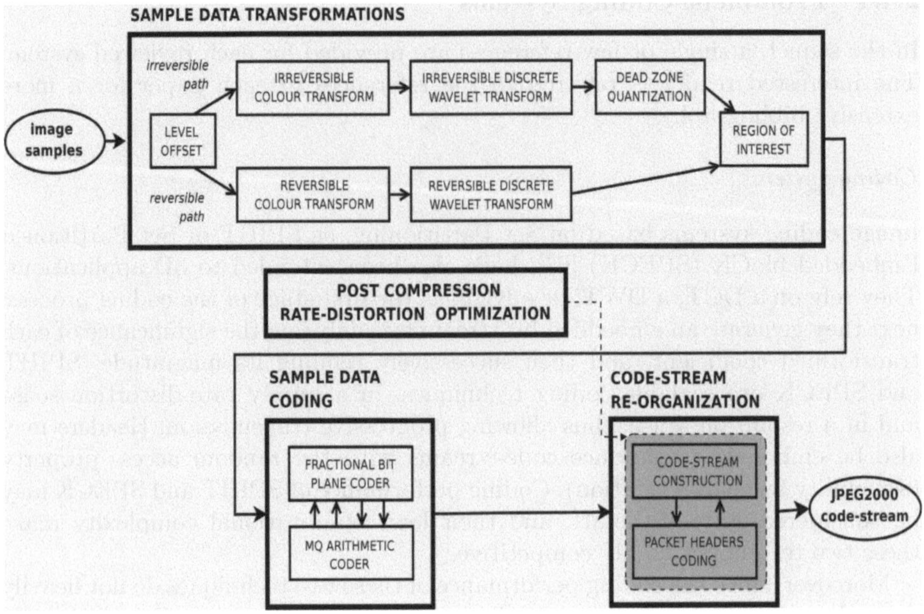

Fig. 2.1. JPEG2000 core coding system: main stages and operations

The standard is structured in several parts. The parts more closely related to remote sensing image compression are Part 1 [32], Part 2 [33], Part9 [34], and Part 10 [35]. In the following, we lay down the basics of the core coding system, Part 1, and guide the interested reader to [70] and the other published Parts for a larger description.

Figure 2.1 illustrates the four main stages of the whole coding system of a tile-component (see below): sample data transformations, sample data coding, rate distortion optimization, and code-stream organization, as well as some of the operations within each stage. Some of these operations are now briefly described.

Tiling

The input 2D dimensional image may be logically divided into non-overlapping rectangular regions called tiles; each tile may be as large as $2^{32} - 1$ pixels in either dimension, and a single tile may include the whole image.

For volumetric images, all pixels from a component lying within a rectangular spatial tile form a tile-component.

Each tile is processed independently of the others, easing low-memory implementations, but, at the same time, possibly introducing ringing and discontinuities artifacts, mainly at low bit rates.

Component Transform

As said before, most remote sensing images are three dimensional in nature. Several approaches for dealing with the third dimension correlation have appeared, furnishing maximum efficiency and random accessibility in different applications as video [18, 42] or medical imagery [42, 39]. These approaches have been exploited for remote sensing within the framework of JPEG2000 standard [64, 23]. JPEG2000 allows up to 16384 components.

The standard provides two means for the multi-component transform, either a wavelet-based or an array-based [41]. Wavelet transform is delayed until the next subsection. Array-based transform in Part 2 are further classified as a dependency transform, or a decorrelation transform.

Array-based dependency transform may be thought of as an extension to prediction methods for 2D dimensional images, it tries to predict next input component based on already known components, similar to DPCM.

Array-based decorrelation transform, notably the Karhunen Loève Transform, has been lately recognized as the most promising approach to account for the correlation in the spectral domain [54, 20]. KLT has been considered too complex for the spatial domain, so that DCT and DWT are widespread elections for transform coding of 2D dimensional image coding. Nevertheless, it has lately been recognized that the KLT (or PCA) provides superior performance for decorrelating the spectral information in remote sensing images [16].

The KLT is the optimal block-based transform, because it approximates a signal in the transform domain using the smallest number of coefficients, minimizing the MSE. The covariance matrix $C_{\mathbf{x}}$ or a random column vector $\mathbf{x}$ with mean value $\bar{x}$ is defined as

$$C_{\mathbf{x}} = E[(\mathbf{x} - \bar{x})(\mathbf{x} - \bar{x})^T].$$

Then the KLT transform matrix V is obtained by aligning column-wise the eigenvectors $C_{\mathbf{x}}$. It can be shown that the transformed random vector $\mathbf{y} = V^T(\mathbf{x} - \bar{x})$ has uncorrelated components, i.e. $C_{\mathbf{y}} = V^T C_{\mathbf{x}} V$ is a diagonal matrix.

Employing KLT as the array-based transform for spectral decorrelation has brought over 7 dB compared to a 3D rectangular wavelet transform, but it has not been widely used because of its expensive computational complexity.

The best and most recent results are obtained not discarding some of the least significant KLT transform components, but keeping all components. In this case, to decrease the high computational complexity of the KLT, a low complexity KLT [56] is carried on a subset of spectral vectors selected at random. The coding performance of the low complexity KLT is very close to the full-complexity transform, with a maximum loss of 0.27 dB at high bit-rates.

A low complexity KLT has also been recently integrated in the framework of JPEG2000 multicomponent [21], where the down-sampling is applied to the spatial pixels, and the correlation matrix instead of the covariance matrix is used. About 1% of the total number of pixels seem to provide competitive enough results, improving the coding performance at very low bit rates, and introducing very small degradation for low to medium bit rates.

Without needing to reduce the dimensionality of the data-set, a novel approach where the KLT inverse matrix is also encapsulated in JPEG2000 headers, helps to improve the ratio of anomalous pixels detection without degrading the coding performance [55].

Wavelet Transform

The Discrete Wavelet Transform [5] is the choice of modern 2D image coders, because it decomposes the input signal into low-frequency coefficients, that contain the majority of the signal energy, and high-frequency coefficients, that account for the details of the image. In addition, low-frequency components retain much of the spatial structure of the input image, so that this decomposition may be iterated in the low-frequency component by means of a filter bank scheme. Figure 2.2 illustrates two common representations of the dyadic decomposition of an image in three DWT decomposition levels (four resolution levels).

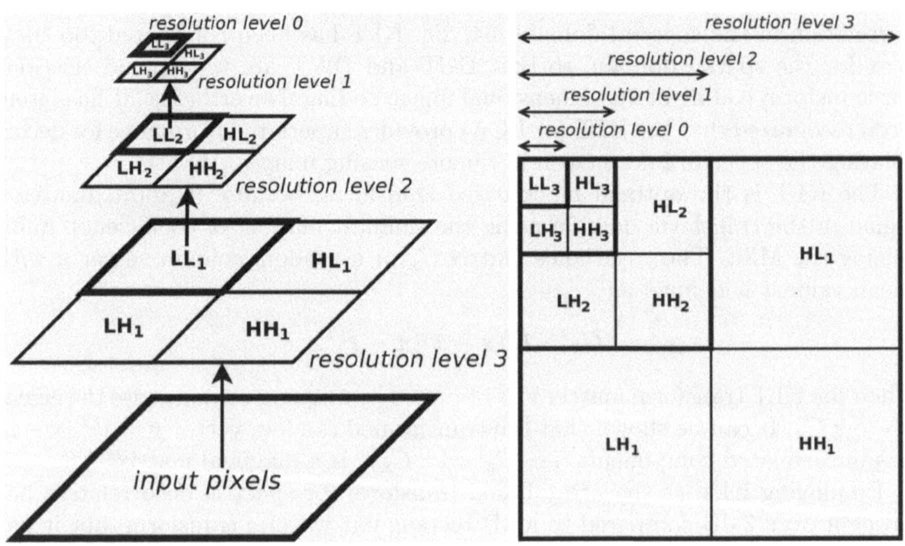

Fig. 2.2. Two common representations of the dyadic DWT decomposition

Extension of 2D DWT to 3D DWT may be of three types, referred to as square, rectangular, and hybrid transform respectively.

- 3D square transform is obtained by first computing one decomposition level in all dimensions, and then iterating this decomposition on the low-frequency cube (LLL); square DWT transform is usually known as dyadic transform.
- 3D rectangular transform is obtained applying first the complete transform along the first dimension, then along the second dimension, and last along the third dimension.

- 3D hybrid transform is obtained by first applying a complete transform in one dimension, and then a 2D square transform in the other two dimensions.

The 3D hybrid rectangular (spectral) / square (spatial) has been found to provide better coding performance than 3D square (dyadic) transform for hyperspectral images. This approach is sometimes called "wavelet packet" transform in the literature, but it is not related to discrete wavelet packet transform (DWPT), which is a generalization of DWT where all subbands (low-frequency and high-frequency components) are allowed to be further splitted into approximation and detail. When all low-frequency and high-frequency components are always splitted, the DWPT is named a complete wavelet packet transform.

In some particular applications, object-based methods have begun to appear for compression of multispectral images [9]. Of course, shape adaptive spatial transforms and encoding techniques are needed. From the experimental results, conclusions seem to indicate that there is a rough equivalence between object-based and ordinary techniques in terms of rate-distortion performance, as long as the optimal allocation is properly performed [11]. This shape adaptive wavelet transform also helps to improve the coding performance of class-based coding methods [12].

As has been previously reported, when the wavelet-based image coding system is to be employed for lossless compression scenarios, a reversible integer wavelet transform [7] is applied, furnishing several advantages, notably, that a fully embedded (rate or quality scalability) code-stream may be generated.

When using the DWT for hyperspectral image data, caution has to be taken to avoid or to overcome Spectral Ringing Artifacts [38]. Adaptive directional lifting-based wavelet transform [19] may be appropriate in such case.

Sample Data Partitioning

Wavelet transformed data is logically partitioned into geometric structures, following the scheme proposed in EBCOT [68]. The smallest geometric structure is the codeblock, a rectangular region within a wavelet subband. The dimension of codeblocks is the same for all wavelet resolution levels and subbands (but for boundaries), and each dimension must be a power of two.

A second geometric structure is the precinct, a rectangular region within a wavelet resolution level. The dimension of precincts may change across resolution levels, and each dimension must be a power of two. The precinct size determines the minimum accessible region size when decoding image areas of that resolution level.

Figure 2.3 illustrates the logical partitioning into geometrical structures.

Quantization

After the input pixels x have been wavelet transformed, precision of the wavelet coefficients y may be modified through a quantization, in order to ease the next coding step. Part 1 of the standard proposes a dead-zone scalar quantization Q,

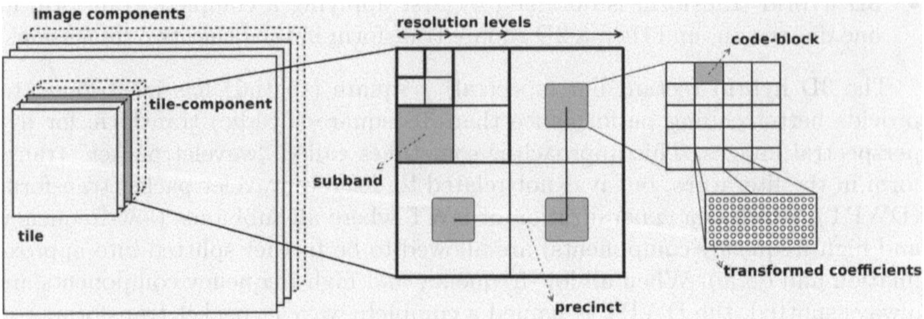

Fig. 2.3. JPEG2000 logical partitions into geometrical structures

where a step size parameter Δ_s is defined for each wavelet subband s. Then, the quantized index q is a signed integer computed as

$$q = Q(y) = sign(y) \left\lfloor \frac{|\, y \,|}{\Delta_s} \right\rfloor .$$

Figure 2.4 illustrates a dead-zone quantizer with a step size of Δ_s.

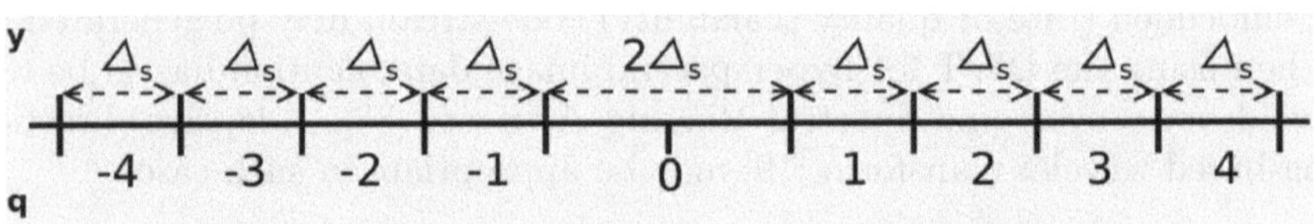

Fig. 2.4. Dead-zone quantizer with a step size of Δ_s

Trellis-Coded Quantization

Trellis-Coded Quantization (TCQ) [1] has been proposed in JPEG2000 Part 2 as a substitute to the classical dead-zone quantization. TCQ has a higher computational complexity than simple scalar quantization, but it provides an outstanding coding performance, both quantitative and subjective. Several variations have been considered: fixed-rate TCQ, entropy-constrained TCQ, entropy-constrained predictive TCQ, and channel-optimized TCQ. Additionally, robustness against transmission errors in noisy channels is easily incorporated, supplying error resilience.

Fractional Bit-Plane Coder

The quantized indices q of a codeblock are entropy coded by means of a fractional bit-plane coder followed by an adaptive binary contextual arithmetic coder. Codeblocks are coded independently of others, thus allowing parallelization at encoding time, and also independent processing at decoding time.

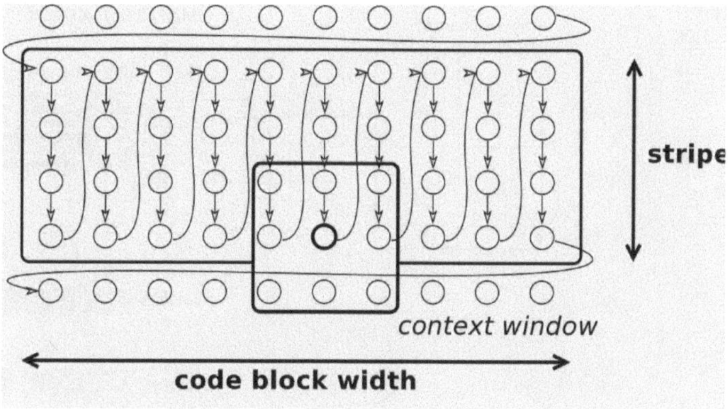

Fig. 2.5. JPEG2000 code-stream organization

A binary magnitude bitplane of a codeblock is defined as an array of bits, one bit from each quantized index in the codeblock, all with the same bit-depth. Each bitplane is encoded in three passes, named coding passes: the significance propagation pass, the magnitude refinement pass, and the clean-up pass.

Figure 2.5 illustrates the scanning order of the fractional bit-plane coder. In each coding pass, all quantized indices are visited, but each quantized index will be coded in only one coding pass, depending on the significance of the current and neighboring coefficients. The aim of the fractional bitplane coding is to encode first the information contributing more largely to decrease the image distortion. In addition, the fractional bitplane coding produces an embedded code-stream with a large collection of truncation points, which are later considered by the rate distortion optimization.

Rate Distortion Optimization

In JPEG2000 Part 1, the first proposed method for performing the rate distortion optimization was applied after the coding segments corresponding to the coding passes were formed, and is, thus, referred to as Post Compression Rate Distortion (PCRD) optimization.

If a coding pass is included in the final code-stream, it contributes to reduce the distortion of the decoded image by a certain amount, ΔD, and, at the same time, it contributes a certain length at the final code-stream, ΔR; therefore, each coding pass has associated a distortion-length or distortion-rate slope $\frac{\Delta D}{\Delta R}$.

The higher the distortion-length slope of a coding pass, the more beneficial it is. PCRD only considers coding passes of codeblocks with strictly decreasing distortion-length slope and conceptually sorts coding passes of all image codeblocks in descending order, selecting those ones included in the final code-stream starting from the top of the sorted list until the targeted rate is achieved.

The PCRD optimization algorithm requires that the whole code-stream be kept in memory, which is sometimes not convenient in remote sensing applications.

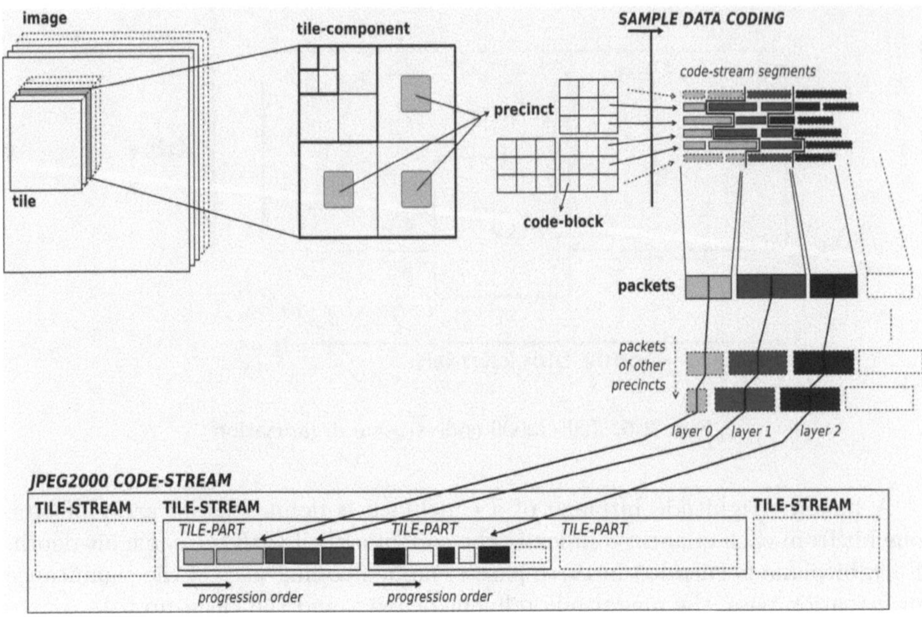

Fig. 2.6. JPEG2000 code-stream organization

Reduced memory approaches for multi-component rate allocation for JPEG2000 have been proposed [40, 41].

Code-Stream Organization

Once the encoder has selected the coding passes to be included in the code-stream, the last operation is to form the code-stream. JPEG2000 defines a flexible code-stream organization, supported by a rich syntax.

The fundamental unit for constructing the code-stream, and the smallest identifiable container, is a packet, which contains compressed data from a varying number (including zero) of coding passes from each codeblock in one precinct of one tile-component. A packet is one quality increment for one spatial region (precinct) at one resolution level of one tile-component.

Packets are then organized into layers, which is a collection of all the packets, one from every precinct, of every resolution level of every tile-component of a tile. Since a layer consists of one packet from every precinct of each resolution level of each tile-component, a layer is one quality increment for an entire tile.

Figure 2.6 illustrates the whole code-stream organization.

This organization into quality layers facilitates progression by quality (SNR scalability) decoding. The final code-stream includes information (headers) necessary for decoding the code-stream, for example, sizes of the geometric structures (image, tile, precinct, codeblock), as well as coding parameters.

Thanks to headers included in the final code-stream, and to the logical partitions, JPEG2000 offers four types of scalability: quality scalability, resolution (or size) scalability, random access (position) scalability, and component scalability. This enables several types of progression. The standard contemplates five progression orders.

2.5 Experimental Results

Experiments for both lossless and lossy coding approaches are reported in this section. Results are reported for JPEG2000 coding system.

Experiments have been conducted on a Dual Core AMD Opteron^{TM} Processor 275, CPU MHz: 2189.781, Linux version 2.6.21.1, gcc version 4.1.2, Debian 4.1.1-21.

Two independent JPEG2000 implementations have been used, Kakadu software v5.2.6 [69], and BOI software v1.2 [26]. The coding performance of these two implementations is equivalent, and results are mostly reported for BOI.

For the KLT implementation, GNU Scientific Library [25] has been employed.

All rate measurements are expressed in bits per pixel per band (bpppb).

The source data employed for evaluating the coding performance is described next.

Ultraspectral Sounder Data

The ultraspectral sounder observations have been obtained by NASA's Atmospheric Infrared Sounder (AIRS). The selected AIRS digital count data set on March 2, 2004 for compression studies has been publicly available via anonymous ftp [51].

It consists of ten granules, five daytime and five nighttime, corresponding to different geographical regions of the Earth [28].

This AIRS data includes 2378 infrared channels in the 3.74 to 15.4 μm region of the spectrum. A day's worth of AIRS data is divided into 240 granules, each of 6 minute duration. Each granule consists of 135 scan lines containing 90 cross-track footprints per scan line, thus, there are a total of $135 \times 90 = 12{,}150$ footprints per granule.

The digital count data ranges from 12-bit to 14-bit for different channels. These raw digital counts may be converted into radiances and further converted into brightness temperatures.

To make the selected data more generic to other ultraspectral sounders, 271 bad channels identified in the supplied AIRS infrared channel are excluded, assuming that they occur only in the AIRS sounder [27]. Each resulting granule is saved as a binary file, arranged as 2107 channels, 135 scan lines, and 90 pixels for each scan line, each pixel stored in signed 16 bpppb. No further pre-processing has been performed in these binary files.

Hyperspectral Data

The *Airborne Visible/Infrared Imaging Spectrometer* [50] records the visible and the near-infrared spectrum in 224 components (bands), from 400 nm to 2500 nm, of size 2048 rows × 614 columns × 2 bytes/sample.

For the results here, we crop the first scene in each data set to produce image cubes with dimensions of 512 × 512 × 224 pixels, each pixel stored in signed 16 bpppb.

Both unprocessed radiance (Cuprite, Jasper Ridge, Lunar Lake, Low Altitude) and processed reflectance (Cuprite, Jasper Ridge, Lunar Lake, Moffet Field) data are used. Processed reflectance data includes, among other processes, atmospheric calibration, and has, in general, a larger variation in reflectance spectra.

2.5.1 Lossless Experiments

Ultraspectral Sounder Data

The coding performance of JPEG2000 is compared against two general purpose lossless compression utilities, GZIP and BZIP2, and against the state-of-the-art technique for this kind of images, a clustering-based method, PPVQ [27].

For this set of experiments, two cases have been considered for JPEG2000.

1. In the case of JPEG2000-2D, each component is encoded independently of the others, and a three levels 5/3 wavelet transform is applied in the spatial domain.
2. In the case of JPEG2000-3D, first a six levels 5/3 wavelet transform is applied along the spectral domain, then a three levels 5/3 wavelet transform in the spatial domain, followed by the PCRD.

Table 2.1 presents lossless compression results obtained for the ten granules and their average. JPEG2000 obtains worse results than PPVQ, the best published technique, and JPEG2000-3D is clearly superior to BZIP2.

In the case of this corpus of images, there exists a pre-processing technique, called Bias-Adjusted Reordering (BAR [27]), that improves the results by no less than 0.4. Previously reported results for a BAR + JPEG2000-2D approach

Table 2.1. Ultraspectral images. Compression factors for the ten granules and their average.

Granule	9	16	60	82	120	126	129	151	182	193	**Avg**
Method	**Compression factor**										
JPEG2000 - 2D	1.88	1.96	1.76	2.08	1.90	1.79	2.16	1.87	1.84	1.80	1.91
JPEG2000 - 3D	2.68	2.76	2.56	2.85	2.66	2.55	2.87	2.59	2.57	2.55	2.66
PPVQ	3.35	3.36	3.30	3.39	3.31	3.29	3.38	3.26	3.22	3.27	3.31
GZIP	1.87	1.95	1.82	2.02	1.86	1.88	2.02	1.79	1.81	1.84	1.89
BZIP2	2.18	2.28	2.07	2.45	2.09	2.16	2.45	2.06	2.14	2.07	2.19

achieve an average compression factor of 2.71, which is better than the current JPEG2000-3D results. It is an open issue whether the combination of BAR with JPEG2000-3D, or of reversible KLT + JPEG2000 (as contemplated in Part 2 of the standard), would prove beneficial.

Hyperspectral Data

The coding performance of JPEG2000 for lossless compression of Hyperspectral Data is reported.

For this set of experiments, three cases have been considered for JPEG2000.

1. In the case of Individual *Band Independent Fixed Rate (BIFR)*, each component is encoded independently of the others, and a five levels 5/3 wavelet transform is applied in the spatial domain.
2. In the case of *Volume_PCRD*, a five levels 5/3 wavelet transform is applied in the spatial domain in each component, and then the PCRD is applied jointly for the 224 components. No spectral transform is carried out. Contrary to the previous case, *Volume_PCRD* provides lossy-to-lossless compression.
3. In the case of *MultiComponent*, first a five levels 5/3 wavelet transform is applied along the spectral domain, then also a five levels 5/3 wavelet transform in the spatial domain, followed by the PCRD.

Table 2.2 presents lossless compression results for these three different coding approaches. For all eight images, the performance of *BIFR* and *Volume_PCRD* is almost identical, the difference being due only to the savings in some headers.

Table 2.2. Top: AVIRIS Radiance images; Bottom: AVIRIS Reflectance images. Lossless results for three different coding approaches.

Image	JPEG2000 Method					
	Individual (BIFR)		Volume_PCRD		MultiComponent	
	Size	CF	Size	CF	Size	CF
Cuprite	51468482	2.281795	51444126	2.282875	38738972	3.031586
Jasper Ridge	56235288	2.088378	56210880	2.089284	40655602	2.888667
Low Altitude	57466996	2.043617	57442618	2.044484	43658054	2.690008
Lunar Lake	50774047	2.313003	50749681	2.314113	38902587	3.018836

Image	JPEG2000 Method					
	Individual (BIFR)		Volume_PCRD		MultiComponent	
	Size	CF	Size	CF	Size	CF
Cuprite	59052113	1.988761	59027708	1.989583	49571405	2.369118
Jasper Ridge	63001663	1.864086	62977246	1.864809	48179850	2.437544
Lunar Lake	57697575	2.035450	57673087	2.036314	48357112	2.428609
Moffet Field	64502329	1.820717	64477691	1.821413	50170109	2.340846

Also, performance of *MultiComponent* is above the other two approaches: in the case of the Radiance images, by about 0.7; in the case of Reflectance images, by about 0.4.

2.5.2 Lossy Experiments

The coding performance of JPEG2000 for lossy compression of Hyperspectral Data is reported.

Like in the previous set of experiments for lossless compression of Hyperspectral Data, for this set of experiments for lossy compression, the three previous cases have been considered for JPEG2000. The single difference is in the type of the wavelet transform used, which is now a bi-orthogonal 9/7 wavelet transform, known to yield better coding results. In addition, a fourth case is added, *KLT*, where the array-based spectral decorrelation is achieved applying a Karhunen-Loève Transform, followed by a spatial wavelet transform in each component, and the PCRD.

1. In the case of Individual *Band Independent Fixed Rate (BIFR)*, each component is encoded independently of the others, and a five levels 9/7 wavelet transform is applied in the spatial domain.
2. In the case of *Volume_PCRD*, a five levels 9/7 wavelet transform is applied in the spatial domain in each component, and then the PCRD is applied jointly for the 224 components. No spectral transform is carried out
3. In the case of *MultiComponent*, first a five levels 9/7 wavelet transform is applied along the spectral domain, then also a five levels 9/7 wavelet transform in the spatial domain, followed by the PCRD, applied jointly for the 224 components.
4. In the case of *KLT*, first a Karhunen-Loève Transform is applied along the spectral domain, then also a five levels 9/7 wavelet transform in the spatial domain, followed by the PCRD, applied jointly for the 224 components.

Coding performance is reported for three different radiometric distortion measures, namely Signal to Noise Ratio Variance ($SNR_{variance}$), Mean Absolute Error (MAE), and Peak Absolute Error (PAE) –also known as Maximum Absolute Distortion (MAD)–.

SNR_Variance Radiometric Distortion Measure

Radiance Images

Table 2.3 presents the SNR results for the AVIRIS Radiance images. The performance of *BIFR* is the lowest one, as expected. Letting the PCRD choose the best coding segments (*Volume_PCRD*) improves the coding performance from 3 to 7 dB as the bitrate grows. Applying a spectral transform (*MultiComponent*) is clearly a better option, supplying additional improvements of about 10 dB for all bitrates. Recent results [21] consistently report a further improvement of no

Table 2.3. AVIRIS Radiance images. $SNR_{Variance}$ (dB) results for four different coding approaches.

Image	JPEG2000 Method	BitRate (bpppb)				
		0.125	0.250	0.500	1.000	2.000
Cuprite	Individual (BIFR)	24.883	26.816	29.372	32.953	38.471
	Volume_PCRD	27.340	30.044	33.506	38.434	45.972
	MultiComponent	39.811	43.153	46.641	50.891	56.569
	KLT	46.044	48.214	50.776	54.422	59.401
Jasper Ridge	Individual (BIFR)	16.755	18.479	20.715	24.065	29.671
	Volume_PCRD	18.900	21.277	24.808	29.844	37.835
	MultiComponent	30.316	34.536	39.254	44.831	51.461
	KLT	39.919	43.6492	46.829	50.652	55.937
Low Altitude	Individual (BIFR)	15.707	17.563	20.012	23.517	29.314
	Volume_PCRD	18.157	20.911	24.691	30.122	38.299
	MultiComponent	28.629	32.797	37.794	43.633	50.527
	KLT	38.950	43.073	46.413	50.109	54.786
Lunar Lake	Individual (BIFR)	26.526	28.769	31.630	35.661	41.676
	Volume_PCRD	29.218	32.183	35.933	40.923	48.588
	MultiComponent	41.765	45.092	48.453	52.611	58.158
	KLT	47.009	49.230	51.754	55.495	60.527

less than 5 dB for all bitrates for a KLT + JPEG2000 approach (as contemplated in Part 2 of the standard). Results of *KLT* in this table show a coding gain of at least 5 dB for very low bit rates, and of at least 2 dB for high bit rates.

For the Cuprite Radiance image, Figure 2.7 plots the coding performance achieved by the four different coding approaches, for 5 different bitrates. The SNR is computed independently for each component, considering the component's variance. The curves show that some components have a much better coding performance than others, reflecting that different regions of the spectrum represent different spectral information.

Also for the Cuprite Radiance image, Figure 2.8 plots the coding performance achieved by each coding approach at 3 different bitrates. The SNR is computed independently for each component, considering the component's variance.

In order to assess the coding performance of each coding approach without relating it to the variance of the components, the radiometric distortion PSNR has been considered. Always for the Cuprite Radiance image, Figure 2.9 plots the PSNR coding performance achieved by the four different coding approaches, for 5 different bitrates. The PSNR is computed taking $(2^{15} - 1)^2$ as the numerator of the logarithm fraction, since images are signed 16 bpppb.

Reflectance Images

Table 2.4 presents the SNR results for the AVIRIS Reflectance images. Again, the performance of *BIFR* is the lowest one, as expected. Letting the PCRD choose

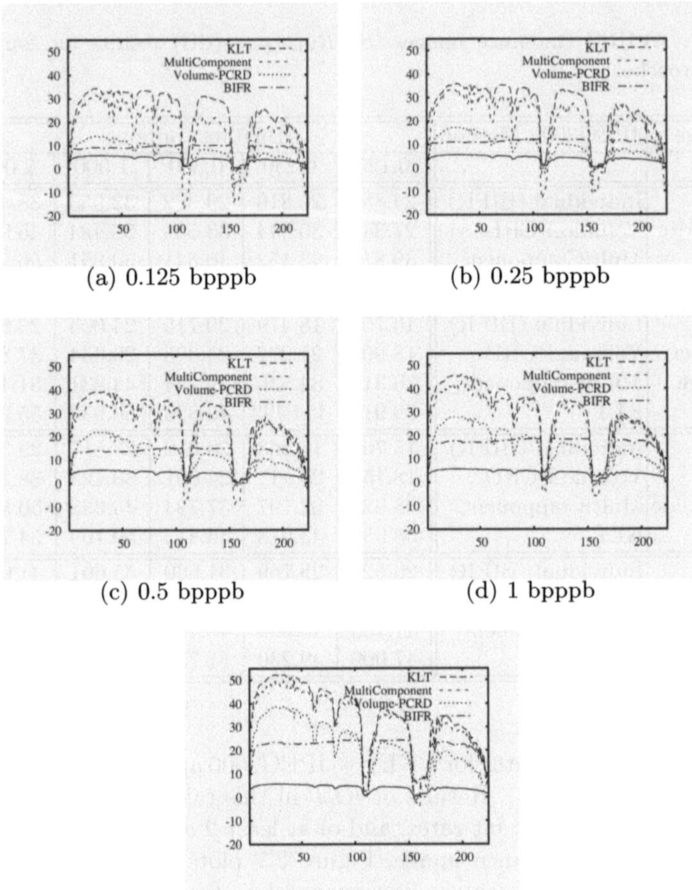

(a) 0.125 bpppb (b) 0.25 bpppb

(c) 0.5 bpppb (d) 1 bpppb

(e) 2 bpppb

Fig. 2.7. AVIRIS Cuprite Radiance. Results for four different coding approaches. X-axis: 224 components; Y-axis: $SNR_{Variance}$ (dB) per component. Bitrate shown below each figure. The logarithm of each component's variance is also drawn as a solid line (curve at the bottom).

the best coding segments (*Volume_PCRD*) improves the coding performance from 8 to 16 *dB* as the bitrate grows, except for the Jasper Ridge image, for which *BIFR* already provides good performance. Applying a spectral transform (*MultiComponent*) is usually a better option, supplying larger improvements as the bitrate grows. Nevertheless, for Cuprite Reflectance image, and for very low bitrate (very large compression factor), the *MultiComponent* approach does not seem to be as adequate as the *Volume_PCRD* approach.

Notice also that SNR results for Radiance images are much higher than for Reflectance images, about 10 *dB*, probably because reflectance images have a larger spectral variation.

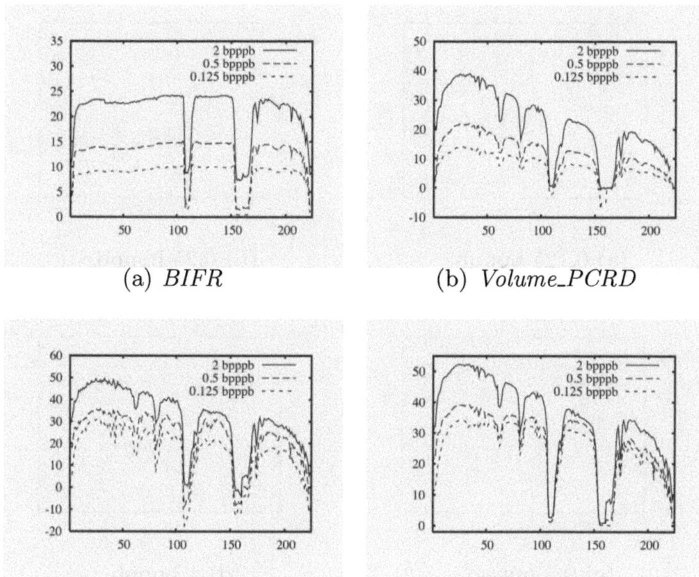

(a) *BIFR* (b) *Volume_PCRD*

(c) *MultiComponent* (d) *KLT*

Fig. 2.8. AVIRIS Cuprite Radiance. Results for the four different coding approaches. X-axis: 224 components; Y-axis: $SNR_{Variance}$ (dB) per component. Coding approach given below each figure.

Reported results [21] for reflectance images also show a further improvement of no less than 5 *dB* for all bitrates for a KLT + JPEG2000 approach (as contemplated in Part 2 of the standard). Results of *KLT* in this table are consistent with the reported results.

The process of conversion from unprocessed at-sensor radiance images to processed at-ground reflectance images provokes that several components have a null response, i.e., that all pixels in a component are 0. Exploiting this a-priory knowledge may prove useful, similar to the conclusions for lossless compression with the identification of bands that require spatial prediction, those that require spectral prediction, and those that require no prediction at all [36].

Table 2.5 reports the SNR results when the null components have been not considered in the encoding process. Cuprite Reflectance has 18 such null components, and Jasper Ridge, Lunar Lake, and Moffet Field have 20 null components.

The *KLT* coding approach has not been considered now, since applying a KLT to a volume with null components, or to a volume without null components has no impact on the final coding performance.

Table 2.6 reports, in comparison to the same coding approach, the gain (in *dB*) that is achieved when the null components are not considered in the encoding process.

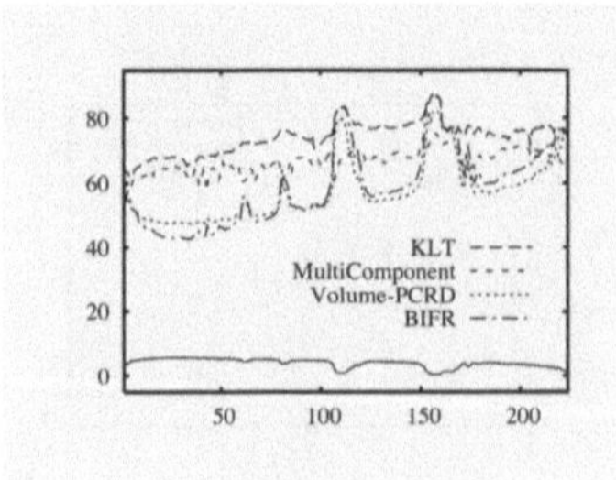

(a) 0.125 bpppb

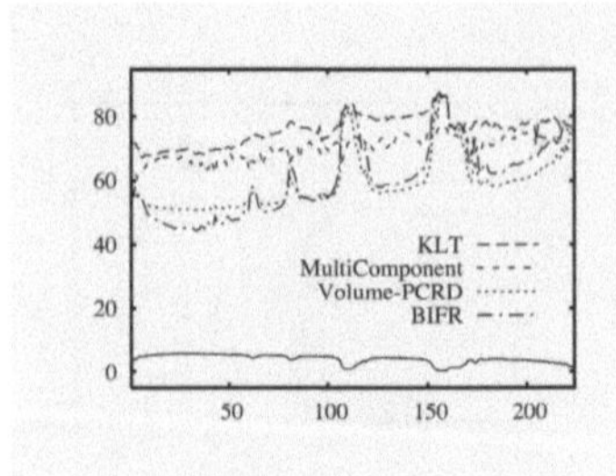

(b) 0.25 bpppb

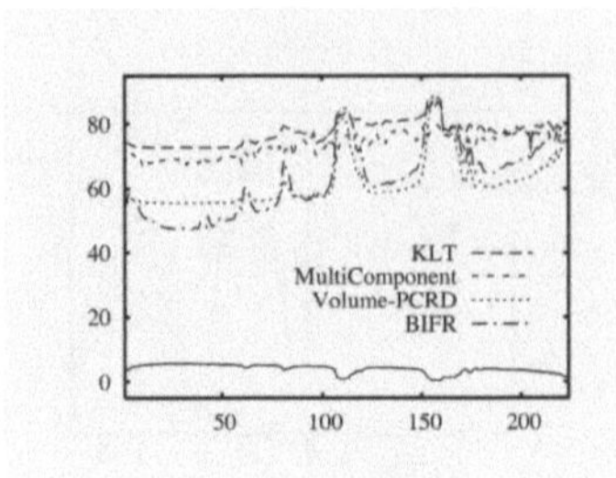

(c) 0.5 bpppb

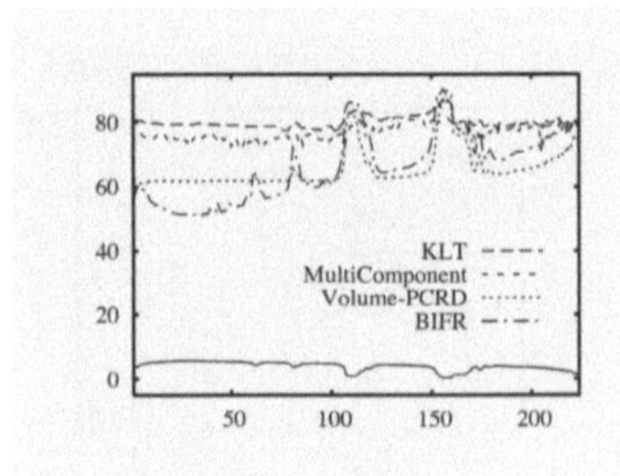

(d) 1 bpppb

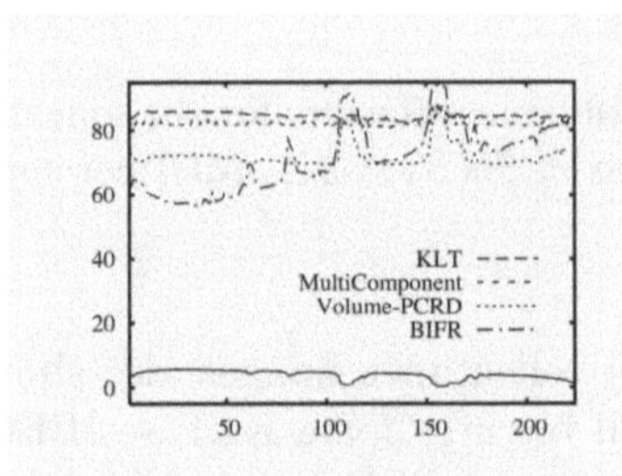

(e) 2 bpppb

Fig. 2.9. AVIRIS Cuprite Radiance. Results for four different coding approaches. X-axis: 224 components; Y-axis: $PSNR$ (dB) per component. Bitrate shown below each figure. The logarithm of each component's variance is also drawn as a solid line (curve at the bottom).

This coding gain does not depend on the signal to noise ratio distortion measure being used, either SNR_Energy, SNR_Variance, or PSNR, since it is a difference between the measure computed for the volume data without the null components (SNR^o) and the measure computed for the original reflectance image scene (SNR^i). The gain only depends on the MSE, because the numerator is always constant (supposing we consider SNR_Variance):

$$\begin{aligned} SNR^o_{Variance} - SNR^i_{Variance} &= 10 \cdot log_{10}\frac{\sigma^2}{MSE^o} - 10 \cdot log_{10}\frac{\sigma^2}{MSE^i} = \\ & 10 \cdot \left[\left(log_{10}\sigma^2 - log_{10}MSE^o\right) - \left(log_{10}\sigma^2 - log_{10}MSE^i\right)\right] = \\ & 10 \cdot \left[log_{10}MSE^i - log_{10}MSE^o\right] \end{aligned}$$

Table 2.4. AVIRIS Reflectance images. $SNR_{Variance}$ (dB) results for four different coding approaches.

Image	JPEG2000 Method	BitRate (bpppb)				
		0.125	0.250	0.500	1.000	2.000
Cuprite	Individual (BIFR)	7.536	8.051	9.096	11.331	16.066
	Volume_PCRD	18.679	21.344	24.533	28.520	34.486
	MultiComponent	15.462	20.146	26.318	35.051	43.310
	KLT	27.513	33.415	38.114	42.672	48.397
Jasper Ridge	Individual (BIFR)	12.046	13.310	15.046	17.683	22.825
	Volume_PCRD	13.430	15.385	18.086	22.312	29.106
	MultiComponent	20.382	25.180	30.883	36.993	43.756
	KLT	28.380	33.086	37.572	42.146	47.812
Lunar Lake	Individual (BIFR)	8.266	8.813	9.869	11.825	16.961
	Volume_PCRD	18.202	21.310	24.781	29.226	35.498
	MultiComponent	18.032	24.181	31.304	38.299	44.662
	KLT	27.623	33.825	38.346	42.634	48.000
Moffet Field	Individual (BIFR)	5.952	6.467	7.455	9.498	14.300
	Volume_PCRD	13.950	16.614	19.643	23.782	30.369
	MultiComponent	14.628	20.534	27.951	36.441	44.333
	KLT	24.336	32.113	38.128	43.206	49.128

Table 2.5. AVIRIS Reflectance images. Selected Components. $SNR_{Variance}$ (dB) results for four different coding approaches.

Image	JPEG2000 Method	BitRate (bpppb)				
		0.125	0.250	0.500	1.000	2.000
Cuprite	Individual (BIFR)	7.539	8.102	9.199	11.660	17.084
	Volume_PCRD	18.932	21.692	24.901	29.010	35.445
	MultiComponent	17.123	22.042	29.357	38.181	45.869
Jasper Ridge	Individual (BIFR)	12.039	13.439	15.321	18.115	23.889
	Volume_PCRD	13.631	15.651	18.575	23.081	30.434
	MultiComponent	21.326	26.991	32.987	39.269	46.380
Lunar Lake	Individual (BIFR)	8.285	8.905	10.037	12.286	18.073
	Volume_PCRD	18.596	21.732	25.281	29.953	36.721
	MultiComponent	18.927	25.992	33.821	40.626	47.105
Moffet Field	Individual (BIFR)	5.951	6.493	7.551	10.015	15.438
	Volume_PCRD	14.303	16.997	20.087	24.485	31.656
	MultiComponent	15.786	22.903	30.937	39.489	47.162

The gain is marginal for *BIFR*, being due to the savings in the headers of the null components. The gain for *Volume_PCRD* is also due to the savings in the headers of the null components, but now the PCRD is able to distribute the spared bits more efficiently. The gain for *MultiComponent* is the largest, seeming

Table 2.6. AVIRIS Reflectance images. Selected Components. Gain (dB) results for four different coding approaches.

Image	JPEG2000 Method	BitRate (bpppb)				
		0.125	0.250	0.500	1.000	2.000
Cuprite	Individual (BIFR)	0.003	0.051	0.103	0.330	1.018
	Volume_PCRD	0.252	0.348	0.368	0.490	0.960
	MultiComponent	1.660	1.896	3.039	3.130	2.559
Jasper Ridge	Individual (BIFR)	-0.008	0.129	0.275	0.432	1.065
	Volume_PCRD	0.201	0.266	0.489	0.769	1.329
	MultiComponent	0.944	1.811	2.104	2.276	2.623
Lunar Lake	Individual (BIFR)	0.019	0.092	0.168	0.461	1.112
	Volume_PCRD	0.394	0.422	0.500	0.727	1.223
	MultiComponent	0.895	1.811	2.517	2.327	2.443
Moffet Field	Individual (BIFR)	-0.001	0.026	0.097	0.517	1.139
	Volume_PCRD	0.353	0.383	0.444	0.703	1.287
	MultiComponent	1.158	2.369	2.986	3.048	2.830

to reach a maximum of 3 dB. Recall that the gain for KLT + JPEG2000 is about 5 dB for all bitrates.

The gain may be appropriate enough for applications where a low computational complexity is a compulsory requirement. As recently reported [56], the time spent by the whole coding process for encoding the Cuprite image for KLT + JPEG2000 is 816.12 seconds, 195.86 seconds for JPEG2000 *MultiComponent*. In addition, not considering the null components further reduces the computational complexity.

MAE Radiometric Distortion Measure

Tables 2.7 and 2.8 report the Mean Absolute Error achieved by the four different JPEG2000 coding approaches for, respectively, the AVIRIS Radiance, and the AVIRIS Reflectance images.

For the Cuprite Radiance image, Figure 2.10 plots the MAE coding performance achieved by the four different coding approaches, for 5 different bitrates.

Conclusions for this distortion metric also suggest that the results improve as the bitrate grows, that the best approach is, by large, *KLT*, while *BIFR* is the worst, and that performance for Radiance images is better than for Reflectance images.

PAE Radiometric Distortion Measure

Tables 2.9 and 2.10 report the Peak Absolute Error achieved by the four different JPEG2000 coding approaches for, respectively, the AVIRIS Radiance, and the AVIRIS Reflectance images.

Table 2.7. AVIRIS Radiance images. MAE results for four different coding approaches.

Image	JPEG2000 Method	BitRate (bpppb)				
		0.125	0.250	0.500	1.000	2.000
Cuprite	Individual (BIFR)	54.496	44.370	33.539	22.771	12.227
	Volume_PCRD	48.215	36.973	25.917	15.533	6.873
	MultiComponent	13.174	8.984	6.139	3.920	2.061
	KLT	6.049	4.882	3.855	2.668	1.523
Jasper Ridge	Individual (BIFR)	91.458	75.521	58.248	39.570	20.728
	Volume_PCRD	82.288	65.168	45.719	26.973	11.345
	MultiComponent	25.541	15.779	9.243	5.067	2.434
	KLT	7.845	5.391	3.902	2.659	1.476
Low Altitude	Individual (BIFR)	111.441	90.260	68.405	45.931	23.939
	Volume_PCRD	99.022	75.324	51.507	29.323	12.186
	MultiComponent	34.069	21.479	12.513	6.620	3.136
	KLT	9.705	6.548	4.743	3.279	1.928
Lunar Lake	Individual (BIFR)	51.868	41.344	30.712	20.201	10.431
	Volume_PCRD	45.525	34.460	23.671	14.175	6.242
	MultiComponent	12.466	8.653	6.010	3.904	2.084
	KLT	6.504	5.261	4.179	2.860	1.625

Table 2.8. AVIRIS Reflectance images. MAE results for four different coding approaches.

Image	JPEG2000 Method	BitRate (bpppb)				
		0.125	0.250	0.500	1.000	2.000
Cuprite	Individual (BIFR)	130.094	112.399	91.325	66.107	36.890
	Volume_PCRD	98.622	76.256	54.395	35.058	17.793
	MultiComponent	126.928	76.329	40.264	16.000	6.665
	KLT	29.995	17.011	10.757	6.769	3.573
Jasper Ridge	Individual (BIFR)	166.633	140.019	109.765	76.068	40.425
	Volume_PCRD	160.744	130.009	97.642	61.861	28.735
	MultiComponent	71.976	42.935	22.950	11.859	5.652
	KLT	27.036	16.391	10.278	6.395	3.385
Lunar Lake	Individual (BIFR)	125.707	107.669	87.415	63.894	35.254
	Volume_PCRD	99.380	73.070	51.110	32.055	15.737
	MultiComponent	94.225	51.345	23.604	11.389	5.776
	KLT	31.118	16.461	10.568	6.865	3.725
Moffet Field	Individual (BIFR)	242.984	207.107	166.352	118.346	63.827
	Volume_PCRD	197.890	148.666	107.170	67.656	32.282
	MultiComponent	163.750	93.147	39.779	16.051	6.814
	KLT	54.148	22.692	12.370	7.253	3.746

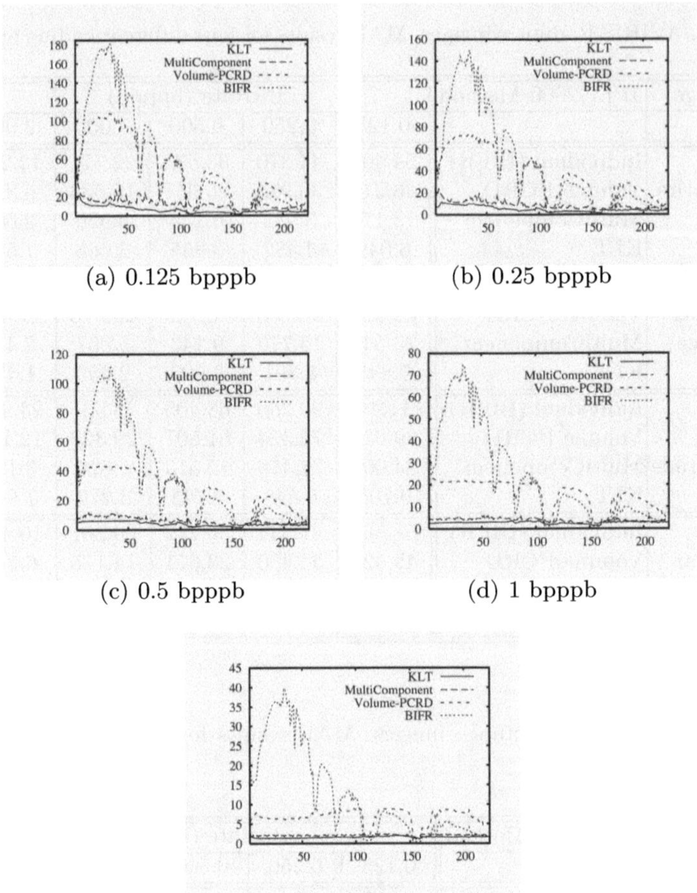

(a) 0.125 bpppb

(b) 0.25 bpppb

(c) 0.5 bpppb

(d) 1 bpppb

(e) 2 bpppb

Fig. 2.10. AVIRIS Cuprite Radiance. Results for four different coding approaches. X-axis: 224 components; Y-axis: MAE per component. Bitrate shown below each figure.

Conclusions for this distortion metric also suggest that performance for Radiance images is better than for Reflectance images.

For Radiance images, the best approach is *KLT*, while *BIFR* is the worst. Nevertheless, for Cuprite and Lunar Lake images, for very low to medium bit rates, *MultiComponent* is better than *KLT*.

For Reflectance images, for very low to low bitrate (large compression factor), the *MultiComponent* approach does not seem to be as adequate as the *Volume_PCRD* approach. For instance, for the Cuprite image, at a compression factor of 32:1, 0.5 bpppb, *Volume_PCRD* is still better than *MultiComponent*.

Also, this measure presents a bizarre behaviour for the Reflectance images for *BIFR* approach, where results do not always improve as the bitrate grows.

Table 2.9. AVIRIS Radiance images. PAE results for four different coding approaches.

Image	JPEG2000 Method	BitRate (bpppb)				
		0.125	0.250	0.500	1.000	2.000
Cuprite	Individual (BIFR)	2409	1623	1206	604	261
	Volume_PCRD	1362	726	413	217	84
	MultiComponent	358	199	111	66	29
	KLT	343	158	127	89	27
Jasper Ridge	Individual (BIFR)	3775	3228	2362	1229	707
	Volume_PCRD	2064	1321	761	370	160
	MultiComponent	631	470	217	107	35
	KLT	314	198	138	63	27
Low Altitude	Individual (BIFR)	8109	5626	3005	1851	814
	Volume_PCRD	3154	2118	1183	548	214
	MultiComponent	1121	593	332	127	55
	KLT	775	383	138	75	25
Lunar Lake	Individual (BIFR)	3037	1699	1163	560	253
	Volume_PCRD	1180	663	401	172	74
	MultiComponent	290	178	133	72	30
	KLT	428	208	144	74	21

Table 2.10. AVIRIS Reflectance images. PAE results for four different coding approaches.

Image	JPEG2000 Method	BitRate (bpppb)				
		0.125	0.250	0.500	1.000	2.000
Cuprite	Individual (BIFR)	15590	17118	15951	15828	11360
	Volume_PCRD	2236	1483	664	401	169
	MultiComponent	4262	2182	849	343	98
	KLT	1650	1045	610	151	45
Jasper Ridge	Individual (BIFR)	10913	10888	8529	8421	3605
	Volume_PCRD	3474	2363	1442	716	349
	MultiComponent	2315	1065	482	215	77
	KLT	1291	638	408	148	59
Lunar Lake	Individual (BIFR)	15474	16160	15926	14590	9774
	Volume_PCRD	3086	1357	820	344	143
	MultiComponent	3882	2024	638	328	76
	KLT	1609	850	482	217	35
Moffet Field	Individual (BIFR)	19204	17374	19543	15548	12863
	Volume_PCRD	5315	3133	1774	847	339
	MultiComponent	5383	2875	1161	337	100
	KLT	5189	946	374	199	54

2.6 Conclusions

Compression of remote sensing images should endeavour to take into account the successive intended application [44, 57]. Based on this premise, several coding systems have been devised, both supplying perfect recovery, i.e. lossless compression, or degraded recovery, i.e. lossy compression. The former attains compression factors up to 4:1; the latter may increase the compression factor by two orders of magnitude.

Regarding lossless compression, the current best results are reported for a clustered prediction method, combining band ordering, vector quantization, and prediction among neighbour pixel vectors.

Regarding lossy compression, the current best results are reported for a JPEG2000 multicomponent approach, applying a 1D KLT transform in the spectral direction, a 2D DWT in the spatial direction, and benefiting from the rate-distortion optimization inherent to JPEG2000.

Acknowledgements

The authors would like to thank Fernando García-Vílchez and Joan Bartrina-Rapesta for their help in providing the experimental results.

This work has been partially supported by the Spanish Government (MEC), by FEDER, and by the Catalan Government, under Grants MEC_BPD-2007-1040, TSI2006-14005-C02-01, and SGR2005-00319.

References

1. Abousleman, G.O.: Coding of Hyperspectral Imagery with Trellis-Coded Quantization. In: Hyperspectral Data Compression, pp. 233–272. Springer, USA (2006)
2. Abousleman, G.P., Marcellin, M.W., Hunt, B.R.: Compression of hyperspectral imagery using the 3-D DCT and hybrid DPCM/DCT. IEEE Trans. Geoscience Remote Sensing 33(1), 26–34 (1995)
3. Aiazzi, B., Alparone, L., Baronti, S., Lastri, C.: Crisp and fuzzy adaptive spectral predictions for lossless and near-lossless compression of hyperspectral imagery. IEEE Geoscience and Remote Sensing Letters 4(4), 532–536 (2007)
4. Aiazzi, B., Alparone, L., Baronti, S., Lastri, C., Santurri, L.: Near-Lossless Compression of Hyperspectral Imagery Through Crisp/Fuzzy Adaptive DPCM. In: Hyperspectral Data Compression, pp. 147–178. Springer, USA (2006)
5. Antonini, M., Barlaud, M., Mathieu, P., Daubechies, I.: Image coding using wavelet transform. IEEE Trans. Image Processing 1(2), 205–220 (1992)
6. Ball, G.H., Hall, D.I.: Some fundamental concepts and synthesis procedures for pattern recognition preprocessors. In: Proc. Int'l Conf. Microwaves, Circuit Theory, and Information Theory, September 1964, pp. 281–297 (1964)

7. Bilgin, A., Sementilli, P.J., Sheng, F., Marcellin, M.W.: Scalable image coding using reversible integer wavelet transforms. IEEE Trans. Image Processing 9(11), 1972–1977 (2000)
8. Boardman, W., Kruse, F., Green, R.: Mapping target signatures via partial unmixing of AVIRIS data. In: VI JPL airborne Earth Science Workshop, Pasadena, CA, USA (1995)
9. Cagnazzo, M., Poggi, G., Verdoliva, L.: A comparison of flat and object-based transform coding techniques for the compression of multispectral images. In: Proceeding of the IEEE International Conference on Image Processing, Genova, Italy, September 2005, vol. I, pp. 657–660 (2005)
10. Cagnazzo, M., Cicala, L., Poggi, G., Verdoliva, L.: Low-complexity compression of multispectral images based on classified transform coding. In: Elsevier Signal Processing Image Communication (August 2006), doi:10.1016/.image.2006.08.003
11. Cagnazzo, M., Parrilli, S., Poggi, G., Verdoliva, L.: Costs and advantages of object-based image coding with shape-adaptive wavelet transform. EURASIP Journal on Image and Video Processing, Article ID 78323, 13 pages (January 2007), doi:10.1155/IVP/2007/78323
12. Cagnazzo, M., Parrilli, S., Poggi, G., Verdoliva, L.: Improved class-based coding of multispectral images with shape-adaptive wavelet transform. IEEE Geoscience and Remote Sensing Letters 4(4), 566–570 (2007)
13. Chang, C.-I., Du, Q., Sun, T.S., Althouse, M.L.G.: A joint band prioritization and band decorrelation approach to band selection for hyperspectral image classification. IEEE Trans. Geoscience Remote Sensing 37(6), 2631–2641 (1999)
14. Chang, C.-I.: An information-theoretic approach to spectral variability, similarity, and discrimination for hyperspectral image analysis. IEEE Trans. Information Theory 46(5), 1927–1932 (2000)
15. Chang, C.-I. (ed.): Hyperspectral Data Exploitation. Theory and Applications. Wiley Interscience. John Wiley & Sons, USA (2007)
16. Choi, E., Choi, H., Lee, C.: Principal component analysis with pre-emphasis for compression of hyperspectral imagery. In: Proceedings of the IEEE International Geoscience and Remote Sensing Symposium, Seoul, South Korea, July 2005, vol. 2 (2005), doi:10.1109/IGARSS.2005.1525203
17. Corsonello, P., Perri, S., Staiano, G., Lanuzza, M., Cocorullo, G.: Low bit rate image compression core for onboard space applications. IEEE Trans. Circuits Syst. Video Technol. 16(1), 114–128 (2006)
18. Dagher, J.C., Bilgin, A., Marcellin, M.W.: Resource-constrained rate control for motion JPEG2000. IEEE Trans. Image Processing 12(12), 1522–1529 (2003)
19. Ding, W., Wu, F., Wu, X., Li, S., Li, H.: Adaptive directional lifting-based wavelet transform for image coding. IEEE Trans. Image Processing 16(2), 416–427 (2007)
20. Du, Q., Fowler, J.E.: Hyperspectral image compression using JPEG2000 and principal component analysis. IEEE Geoscience and Remote Sensing Letters 4(2), 201–205 (2007)
21. Du, Q., Zhu, W.: Integration of PCA and JPEG2000 for hyperspectral image compression. In: Shen, S.S., Lewis, P.E. (eds.) Proceedings of Algorithms and Technologies for Multispectral, Hyperspectral, and Ultraspectral Imagery XIII, SPIE, May 2007, vol. 6565, p. 65650L (2007)
22. Consultative Committee for Space Data Systems. Image data compression. CCSDS 122.0-B-1, Blue Book (November 2005)

23. Fowler, J.E., Rucker, J.T.: 3D wavelet-Based Compression of Hyperspectral Imagery. In: Hyperspectral Data Exploitation: Theory and Applications, pp. 379–407. John Wiley & Sons Inc., Hoboken (2007)
24. Gersho, A., Gray, R.M.: Vector Quantization and Signal Compression, Communications and Information Theory. The Kluwer International Series in Engineering and Computer Science. Kluwer Academic Publisher, Boston (1992)
25. GNU. GNU Scientific Library (GSL) – reference manual, release 1.10, `http://www.gnu.org/software/gsl/manual/html_node/`
26. Group on Interactive Coding of Images. BOI software; Open Source JPEG2000 implementation (October 2007), `http://www.gici.uab.cat/BOI`
27. Huang, B., Ahuja, A., Huang, H.-L., Schmit, T.J., Heymann, R.W.: Lossless compression of 3D hyperspectral sounding data using context-based adaptive lossless image codec with Bias-Adjusted Reorderding. SPIE Optical Engineering 43(9), 2071–2079 (2004)
28. Huang, B., Ahuja, A., Huang, H.-L.: Lossless Compression of Ultraspectral Sounder Data. In: Hyperspectral Data Compression, pp. 75–106. Springer, USA (2006)
29. Huang, B., Ahuja, A., Huang, H.-L.: Optimal compression of high spectral resolution satellite data via adaptive vector quantization with linear prediction. Journal of Atmospheric and Oceanic Technology 24(7) (to appear, 2007)
30. Huang, B., Sriraja, Y.: Lossless compression of hyperspectral imagery via lookup tables with predictor selection. In: Bruzzone, L. (ed.) Proceedings of Image and Signal Processing for Remote Sensing XII, SPIE, September 2006, vol. 6365, pp. 63650–63651 (2006)
31. JPEG-LS, Information technology 14495-1. International Organization for Standardization / International Electrotechnical Commission (ISO/IEC); International Telecommunication Union-Telecom Standardization (ITU-T) recommendation T.87 (1999)
32. JPEG 2000 image coding system - Part 1: Core coding system, Information technology 15444-1. International Organization for Standardization / International Electrotechnical Commission (ISO/IEC); International Telecommunication Union-Telecom Standardization (ITU-T) (December 2000)
33. JPEG 2000 image coding system - Part 2: Extensions, Information technology 15444-2. International Organization for Standardization / International Electrotechnical Commission (ISO/IEC); International Telecommunication Union-Telecom Standardization (ITU-T) (May 2004)
34. JPEG 2000 image coding system - Part 9: Interactivity tools, APIs, and protocols. Information technology 15444-9. International Organization for Standardization / International Electrotechnical Commission (ISO/IEC); International Telecommunication Union-Telecom Standardization (ITU-T) (December 2005)
35. JPEG 2000 image coding system - Part 10: Volumetric JPEG2000. Information technology 15444-10. International Organization for Standardization / International Electrotechnical Commission (ISO/IEC); International Telecommunication Union-Telecom Standardization (ITU-T) (March 2007), Final Draft, International Standard
36. Jain, S.K., Adjeroh, D.A.: Edge-based prediction for lossless compression of hyperspectral images. In: Storer, J.A., Marcellin, M.W. (eds.) Data Compression Conference, Snowbird, Utah, USA, March 2007, pp. 153–162. IEEE, Los Alamitos (2007)

37. Kerekes, J.P., Scott, J.R.: Hyperspectral Imaging Systems. In: Hyperspectral Data Exploitation: Theory and Applications, pp. 19–46. Wiley Interscience, John Wiley & Sons, USA (2007)
38. Klimesh, M., Kiely, A., Xie, H., Aranki, N.: Spectral Ringing Artifacts in Hyperspectral Image Data Compression. In: Hyperspectral Data Compression, pp. 379–405. Springer, USA (2006)
39. Krishnan, K., Marcellin, M.W., Bilgin, A., Nadar, M.S.: Efficient transmission of compressed data for remote volume visualization. IEEE Trans. Medical Imaging 25(9), 1189–1199 (2006)
40. Kulkarni, P., Bilgin, A., Marcellin, M.W., Dagher, J.C., Flohr, T.J., Rountree, J.C.: Reduced memory multi-layer multi-component rate allocation for JPEG2000. In: Proc. SPIE International Conference on Image and Video Communications and Processing, January 2005, pp. 139–150 (2005)
41. Kulkarni, P., Bilgin, A., Marcellin, M.W., Dagher, J.C., Kasner, J.H., Flohr, T.J., Rountree, J.C.: Compression of Earth Science Data with JPEG2000. In: Hyperspectral Data Compression, pp. 347–378. Springer, USA (2006)
42. Leung, R., Taubman, D.S.: Transform and embedded coding techniques for maximum efficiency and random accessibility in 3-D scalable compression. IEEE Transactions on Image Processing 14(10), 1632–1646 (2005)
43. MacQueen, J.: Some methods for classification and analysis of multivariate observations. In: Proceedings of the Fifth Berkeley Symposium on Math. Statistic and Probability, vol. 1, pp. 281–296 (1967)
44. Mercier, G., Lennon, M.: Joint Classification and Compression of Hyperspectral Images. In: Hyperspectral Data Compression, pp. 179–196. Springer, USA (2006)
45. Mielikainen, J.: Lossless compression of hyperspectral images using lookup tables. IEEE Signal Processing Letters 13(3), 157–160 (2006)
46. Mielikainen, J., Toivanen, P.: Lossless Hyperspectral Image Compression via Linear Prediction. In: Hyperspectral Data Compression, pp. 57–74. Springer, USA (2006)
47. Miguel, A., Lagner, R., Riskin, E., Hauck, S., Barney, D., Askey, A., Chang, A.: Predictive Coding of Hyperspectral Images. In: Hyperspectral Data Compression, pp. 197–232. Springer, USA (2006)
48. Motta, G., Rizzo, F., Storer, J.A.: Locally Optimal Partitioned Vector Quantization of Hyperspectral Data. In: Hyperspectral Data Compression, pp. 107–146. Springer, USA (2006)
49. NASA. Earth Observing System satellite (EOS), `http://eos.nasa.gov`
50. NASA. Jet Propulsion Laboratory. Airborne Visible / Infrared Imaging Spectrometer (AVIRIS), `http://aviris.jpl.nasa.gov`
51. NASA. Jet Propulsion Laboratory. Atmospheric Infrared Sounder (AIRS), `http://www-airs.jpl.nasa.gov`
52. Penna, B., Tillo, T., Magli, E., Olmo, G.: Embedded lossy to lossless compression of hyperspectral images using JPEG2000. In: Proceedings of the IEEE International Geoscience and Remote Sensing Symposium, Seoul, South Korea, July 2005, vol. I (2005), doi:10.1109/IGARSS.2005.1526124
53. Penna, B., Tillo, T., Magli, E., Olmo, G.: A new low complexity KLT for lossy hyperspectral data compression. In: Proceedings of the IEEE International Geoscience and Remote Sensing Symposium, Denver, Colorado, USA, July 2006, vol. VII, pp. 3525–3528 (2006)

54. Penna, B., Tillo, T., Magli, E., Olmo, G.: Progressive 3-D coding of hyperspectral images based on JPEG2000. IEEE Geoscience and Remote Sensing Letters 3(1), 125–129 (2006)
55. Penna, B., Tillo, T., Magli, E., Olmo, G.: Hyperspectral image compression employing a model of anomalous pixels. IEEE Geoscience and Remote Sensing Letters 4(4), 664–668 (2007)
56. Penna, B., Tillo, T., Magli, E., Olmo, G.: Transform coding techniques for lossy hyperspectral data compression. IEEE Trans. Geoscience Remote Sensing 45(5), 1408–1421 (2007)
57. Pickering, M.R., Ryan, M.J.: An Architecture for the Compression of Hyperspectral Imagery. In: Hyperspectral Data Compression, pp. 1–34. Springer, USA (2006)
58. Plaza, A., Martinez, P., Perez, R., Plaza, J.: Spatial/spectral endmember extraction by multidimensional morphological operations. IEEE Trans. Geoscience Remote Sensing 40(9), 2025–2041 (2002)
59. Qian, S.-E.: Hyperspectral data compression using a fast vector quantization algorithm. IEEE Trans. Geoscience Remote Sensing 42(8), 1791–1798 (2004)
60. Qian, S.-E., Bergeron, M., Cunningham, I., Gagnon, L., Hollinger, A.: Near lossless data compression onboard a hyperspectral satellite. IEEE Trans. Aerospace and Electronic Systems 42(3), 851–866 (2006)
61. Qian, S.-E., Bergeron, M., Serele, C., Cunningham, I., Hollinger, A.: Evaluation and comparison of JPEG2000 and vector quantization based onboard data compression algorithm for hyperspectral imagery. In: Proceedings of the IEEE International Geoscience and Remote Sensing Symposium, Toulouse, France, July 2003, vol. 3, pp. 1820–1822 (2003)
62. Rabbani, M., Joshi, R.: An overview of the JPEG2000 still image compression standard. Elsevier Signal Processing: Image Communication 17, 3–48 (2002)
63. Ramakrishna, B., Plaza, A.J., Chang, C.-I., Ren, H., Du, Q., Chang, C.-C.: Spectral/Spatial Hyperspectral Image Compression. In: Hyperspectral Data Compression, pp. 309–346. Springer, USA (2006)
64. Rucker, J.T., Fowler, J.E., Younan, N.H.: JPEG2000 coding strategies for hyperspectral data. In: Proceedings of the IEEE International Geoscience and Remote Sensing Symposium, Seoul, South Korea, July 2005, vol. 1, pp. 128–131 (2005)
65. Said, A., Pearlman, W.A.: A new, fast, and efficient image codec based on set partitioning in hierarchical trees. IEEE Trans. Circuits Syst. Video Technol. 6, 243–250 (1996)
66. Salomon, D.: Data Compression: The Complete Reference, 3rd edn. Springer, New York (2004)
67. Tang, X., Pearlman, W.A.: Three-Dimensional Wavelet-Based Compression of hyperspectral Images. In: Hyperspectral Data Compression, pp. 273–308. Springer, USA (2006)
68. Taubman, D.S.: High performance scalable image compression with EBCOT. IEEE Trans. Image Processing 9(7), 1158–1170 (2000)
69. Taubman, D.S.: Kakadu software; Commercial JPEG2000 implementation (July 2007), `http://www.kakadusoftware.com`
70. Taubman, D.S., Marcellin, M.W.: JPEG2000 Image compression fundamentals, standards and practice. Kluwer Academic Publishers, Norwell (2002)
71. Vapnick, V.N.: Statistical Learning Theory. John Wiley and Sons Inc., Chichester (1998)
72. Wang, H., Sayood, K.: Lossless Predictive Compression of Hyperspectral Images. In: Hyperspectral Data Compression, pp. 35–56. Springer, USA (2006)

73. Wu, X., Memon, N.: Context-based, adaptive, lossless image coding. IEEE Trans. Communications 45(4), 437–444 (1997)
74. Yea, S., Pearlman, W.A.: A wavelet-based two-stage near-lossless coder. IEEE Trans. Image Processing 15(11), 3488–3500 (2006)
75. Yeh, P.-S., Armbruster, P., Kiely, A., Masschelein, B., Moury, G., Schaefer, C., Thiebaut, C.: The new CCSDS image compression recommendation. In: Proceedings of the IEEE Aerospace Conference, Montana, USA, March 2005, pp. 4138–4145 (2005)
76. Yeh, P.-S., Venbrux, J.: A high performance image data compression technique for space application. In: NASA Earth Science Technology Conference 2003, NASA (2003)

3

A Multiobjective Evolutionary Algorithm for Hyperspectral Image Watermarking

D. Sal and M. Graña*

Grupo Inteligencia Computacional, UPV/EHU,
Apdo. 649, 20080 San Sebastian, Spain
manuel.grana@ehu.es

Summary. With the increasing availability of internet access to remote sensing imagery, the concern with image authentication and ownership issues is growing in the remote sensing community. Watermarking techniques help to solve the problems raised by this issue. In this paper we elaborate on the proposition of an optimal placement of the watermark image in a hyperspectral image. We propose an evolutionary algorithm for the digital semi-fragile watermaking of hyperspectral images based on the manipulation of the image discrete cosine transform (DCT) computed for each band in the image. The algorithm searches for the optimal localization in the support of an image's DCT to place the mark image. The problem is stated as a multi-objective optimization problem (MOP), that involves the simultaneous minimization of distortion and robustness criteria. We propose appropriate fitness functions that implement these conflicting criteria, and that can be efficiently evaluated. The application of an evolutionary algorithm (MOGA) to the optimal watermarking hyperspectral images is presented. Given an appropriate initialization, the algorithm can perform the search for the optimal mark placement in the order of minutes, approaching real time application restrictions.

3.1 Introduction

The hyperspectral sensor performs a fine sampling of the surface radiance in the visible and near infrared wavelength spectrum. Therefore each image pixel may be interpreted as a high dimensional vector. We are interested in the watermarking of hyperspectral images because all the new remote sensor are designed to be hyperspectral. The fact that Internet is is becoming the primary mean of communication and transport of these images, may raise authentication and ownership issues in the near future.

Watermarking is a technique for image authorship and content protection [21, 1, 15, 16, 20, 22, 13, 23]. Semi-fragile watermarking [12, 24] tries to ensure the image integrity, by means of an embedded watermark which can be recovered without modification if the image has not been manipulated. However, it is desirable that the watermark recovery is robust to operations like filtering, smoothing and lossy compression [19] which are very common while distributing

* The Spanish Ministerio de Educacion y Ciencia supports this work through grant DPI2006-15346-C03-03 and VIMS-2003-20088-c04-04.

M. Graña and R.J. Duro (Eds.): Comput. Intel. for Remote Sensing, SCI 133, pp. 63–78, 2008.
springerlink.com

images through communication networks. For instance, the JPEG lossy compression first standard deletes the image discrete cosine transform (DCT) high frequency coefficients. The JPEG 2000 standard works on the image discrete wavelet transform (DWT) coefficients, also removing high frequency ones as needed to attain the desired compression ratio. Embedding the watermark image in the image transform coefficients is the usual and most convenient approach when trying to obtain perceptually invisible watermarks. We have focused on the DCT transform for several reasons. First it is a real valued transform, so we do not need to deal with complex numbers. Second, the transform domain is continuously evolving from low to high spatial frequencies, unlike DWT which has a complex hierarchical structure in the transform domain. The definition of the fitness functions below benefits from this domain continuity. It is possible to assume some conclusions about the watermark robustness dependence on its placement. Besides being robust, we want the watermarked image must be as perceptually indistinguishable from the original one as possible, that is, the watermarking process must introduce the minimum possible visual distortion in the image.

These two requirements (robustness against filtering and minimal distortion) are the contradicting objectives of our work. The trivial watermarking approach consists in the addition or substitution of the watermark image over the high frequency image transform coefficients. That way, the distortion is perceptually minimal, because the watermark is embedded in the noisy components of the image. However, this approach is not robust against smoothing and lossy compression. The robustness can be enhanced placing the watermark in other regions of the image transform, at the cost of increased distortion. Combined optimization of the distortion and the robustness can be stated as a multi-objective optimization.

Multi-objective optimization problems (MOP) are characterized by a vector objective function. As there is no total order defined in vector spaces, the desired solution does not correspond to a single point or collection of points in the solution space with global optimal objective function value. We must consider the so-called Pareto front which is the set of non-dominated solutions. A non-dominated solution is one that is not improved in all and every one of the vector objective function components by any other solution [6]. In the problem of searching for an optimal placement of the watermark image, the trade-off between robustness and image fidelity is represented by the Pareto front discovered by the algorithm. We define an evolutive strategy that tries to provide a sample of the Pareto front preserving as much as possible the diversity of the solutions. The stated problem is not trivial and shows the combinatorial explosion of the search space: the number of possible solutions is the number of combinations of the image pixel positions over the size of the image mark to be placed.

Section 3.2 provides a review of related previous works found in the literature. Section 3.3 will review multi-objective optimization basics. Section 3.4 introduces the problem notation. Section 3.5 describes the proposed algorithm. Section 3.6

presents some empirical results and section 3.7 gives our conclusions and further work discussion.

3.2 Related Works

The growing number of papers devoted to watermarking of remote sensing images is a proof of the growing concern of this community with authentication and copyright issues. Some of the authors deal with conventional (grayscale) images [8, 5, 14], others with multispectral images (LANDSAT) [3, 4] and some of them with hyperspectral images [10, 17, 9, 18]. In [8] the watermark is applied on the coefficients of the image Hadamard transform. In [10] it is applied to a PCA dimensional reduction of the image wavelet transform coefficients. A near lossless watermarking schema is proposed in [3]. There the effect of watermarking on image classification is the measure of watermarked image quality, while in [4] the watermark placement is decided to minimize the effect on the classification of the image. In [18] two watermarking algorithms are proposed aimed to minimize the effect on target detection. The combination of watermarking and near lossless compression is reported in [5]. The exploration of semi-fragile watermarking based on the wavelet transform is reported in [14]. The watermarking of hyperspectral images performed on the redundant discrete wavelet transform of the pixel spectral signatures is proposed in [17]. The approach in [9] involves 3D wavelet transform and the watermark strength is controlled by perceptive experiments. Our approach allows for greater quantities of information to hide, and provides an variable placement to minimize the effect of the watermark measured by a correlation measure.

3.3 Multi-objective Optimization Problem

Osyczka defined the Multiobjective Optimization Problem (MOP) as "the problem of finding a vector of decision variables which satisfies constraints and optimizes a vector function whose elements represent the objective functions. These functions form a mathematical description of performance criteria which are usually in conflict with each other. Hence, the term *optimize* means finding such a solution which would give the values of all the objective functions acceptable to the decision maker" [2, 6].

The general MOP tries to find the vector $\mathbf{x}^* = [x_1^*, x_2^*, ..., x_n^*]^T$ which will satisfy m inequality constraints $g_i(\mathbf{x}) \geq 0, i = 1, 2, ..., m$, p equality constraints $h_i(\mathbf{x}) = 0, i = 1, 2, ..., p$ and will optimize the vector function $\mathbf{f}(\mathbf{x}) = [f_1(\mathbf{x}), f_2(\mathbf{x}), ..., f_k(\mathbf{x})]^T$.

A vector of decision variables $\mathbf{x}^* \in \mathcal{F}$ is Pareto optimal if it does not exist another $\mathbf{x} \in \mathcal{F}$ such that $f_i(\mathbf{x}) \leq f_i(\mathbf{x}^*)$ for all $i = 1, .., k$ and $f_j(\mathbf{x}) < f_j(\mathbf{x}^*)$ for at least one j. Here $\mathcal{F}$ denotes the region of feasible solutions that meet the inequality constraints. Each solution that carries this property, is called non-dominated solution, and the set of non-dominated solutions is called Pareto optimal set. The plot of the objective functions whose non-dominated vectors are in the Pareto optimal set is called the Pareto front.

A vector $\mathbf{u} = (u_1, ..., u_n)$ is said to dominate $\mathbf{v} = (v_1, ..., v_n)$ (denoted as $\mathbf{u} \preceq \mathbf{v}$) if and only if $\forall i \in \{1..k\}, u_i \leq v_i \wedge \exists i \in \{1, ..., k\} : u_i < v_i$.

For a given MOP $\mathbf{f}(x)$, the Pareto optimal set $\mathcal{P}^*$ is defined as: $\mathcal{P}^* := \{x \in \mathcal{F} \mid \neg\exists x' \in \mathcal{F} : \mathbf{f}(x') \preceq \mathbf{f}(x)\}$, and the Pareto front ($\mathcal{PF}^*$) is defined as: $\mathcal{PF}^* := \{\mathbf{u} = \mathbf{f} = (f_1(x), ..., f_k(x)) \mid x \in P^*\}$.

3.4 Watermarking Problem and Algorithm Notation

We have an hyperspectral image X of size m_x x n_x x $nbands$ that we want to protect. To do that, we use a mark image W of size m_w x n_w. The DCT of the image and the mark image are denoted X_t and W_t respectively. X_t is obtained by applying the bi-dimensional DCT to each band. Watermarking is performed by adding the DCT watermark image coefficients in W_t to selected DCT image coefficients in X_t. Given two coordinates k, l of the W domain, $1 \leq k \leq m_w$, $1 \leq l \leq n_w$, we denote $x(k,l)$, $y(k,l)$, $z(k,l)$ the coordinates of the X_t domain where the coefficient $W_t(k,l)$ is added in order to embed the mark.

The algorithm described below works with a population Pop of N_p individuals which are solutions to the problem. We denote O the offspring population. Let be P_s, P_m and P_c the selection, mutation and crossover probabilities, respectively.

To avoid a possible confusion between the solution vector ($\mathbf{x}$) and the original image (X), we will denote the first one as $\mathbf{s}^*$. So, the algorithm will try to find the vector $\mathbf{s}^*$ optimizing $\mathbf{f}(\mathbf{s}) = [f_1(\mathbf{s}), f_2(\mathbf{s})]$ where f_1 is the robustness fitness function and f_2 is the distortion fitness function. The algorithm returns a sampling of the Pareto optimal set $\mathcal{P}^*$ of size between 1 and N_p. The user will be able to select the solution which is better adapted to his necessities from the plotted Pareto front $\mathcal{PF}^*$.

A solution $\mathbf{s}^*$ is represented as an m_w x n_w matrix in which every position $\mathbf{s}^*(k,l)$ of the W_t domain takes three positive values: $x(k,l)$, $y(k,l)$ and $z(k,l)$. Actually, our mark is a small image or logo. The embedded information is the logo's DCT. So, the corruption of the recovered mark is detected by visual inspection, and can be measured by correlation with the original mark.

3.5 Algorithm

In this section we will start introducing the fitness functions that model the robustness and distortion of the solutions. Next we define the operators employed. The section ends with the global definition of the algorithm.

3.5.1 Multi-Objective Fitness

The fitness functions f_1 and f_2 measure the robustness and distortion of the watermark placement represented by an individual solution, respectively. Together, they compose the vector objective function ($\mathbf{f}$) to be optimized.

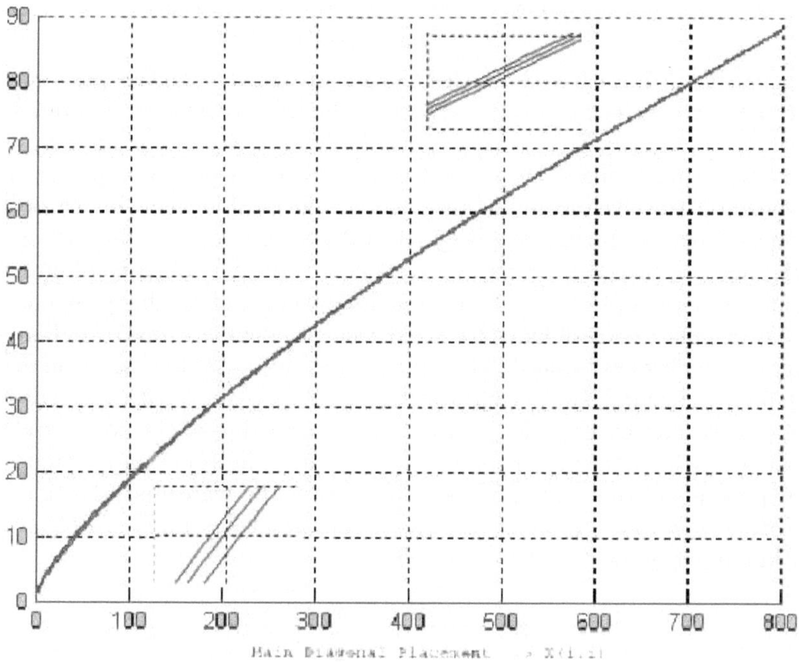

Fig. 3.1. Evolution of f_1, for $F = 4$ and $d = 3$. Insets show zooming the function in specific domains.

Robustness fitness function f_1

Watermark Robustness refers to the ability to recover the watermark image even after the watermarked image has been manipulated. We focus in obtaining robustness against lossy compression and smoothing of the watermarked image. Both transformations affect the high and preserve the low frequency image transform coefficients. Therefore the closer to the transform space origin the mark is located, the higher the robustness of the mark. As we are embedding the watermark image DCT, we note also that most of the watermark image information will be in its low frequency coefficients so.Therefore, they must have priority to be embedded in the positions that are nearer to the low frequencies of X_t. All these requirements are expressed in equations (3.1) and (3.2). Our robustness fitness function is the sum extended to all the watermark pixels of the α-root of the position norm.

$$f_1 = \sum_{k=i}^{m_w} \sum_{l=1}^{n_w} \sqrt[\alpha]{x(k,l)^2 + y(k,l)^2 + k + l} \tag{3.1}$$

where α is given by:

$$\alpha = F\frac{\sqrt{x(k,l)^2+k}+\sqrt{y(k,l)^2+l}}{x(k,l)+y(k,l)+k+l}+d \tag{3.2}$$

Equation (3.1) is based on the Euclidean distance of the position where a mark DCT coefficient $W_t(k,l)$ is placed to the DCT transform X_t domain origin : $\sqrt{x(k,l)^2+y(k,l)^2}$. The terms $k+l$ inside the root expression model the priority of the watermark image DCT low frequency coefficients to be placed on robust placements. However, this distance has an unsuitable behavior to be taken as a fitness function for minimization. Its value decreases very fast when the pixel of the mark is placed near the X_t low frequencies, but remains almost constant when the mark is placed in the low-medium frequencies. This problem is known as the big plateau problem. To avoid this problem, we try to define a fitness function which shows smooth (bounded) but non-negligible variation over all the domain of solutions. To this end we introduce the α-root, with the root exponent being controlled by equation (3.2). The higher value of the root exponent, the closer to a constant value is obtained (although the function continues to have an exponential behavior). The more important the watermark DCT coefficient, the bigger the root exponent and the lower the fitness function. Equation (3.2) is a line function on the following ratio

$$\alpha = \frac{\sqrt{x(k,l)^2}+\sqrt{y(k,l)^2}}{x(k,l)+y(k,l)}$$

which takes values between zero and one. This ratio is modulated by a factor F and a displacement d. As said before, the fitness function has to be sensible to the relative importance of k,l in the watermark image DCT $W_t(k,l)$ domain. Equation (3.2) also introduces this sensitivity by taking into account the k,l coordinates.

Figure 3.1 shows the behavior of f_1 when three different pixels of W_t are embedded in the main diagonal of X_t. The x axis of this plot represent the position in the main diagonal. The function grows smoothly and steadily without plateau effects towards the high frequency region. The insets show that the behavior of the function depends also of the watermark image DCT coefficient $W_t(k,l)$ placed (bigger the lower frequencies).

The robustness fitness does not depend on the band number, because each band DCT has been computed independently. In summary, this function possesses the following properties:

1. As the position in X_t where $S(k,l)$ is embedded is closer to the low frequency region, the function value decreases smoothly.
2. As the pixel of W_t is more important (nearest to the W_t low frequencies), the value of α increases smoothly, so the fitness function decreases smoothly.

Thus, f_1 must be minimized to maximize the robustness.

Distortion fitness function f_2

The watermarking distortion is the mean square error of the watermarked image relative to the original unmarked image. We compute it as the mean squared

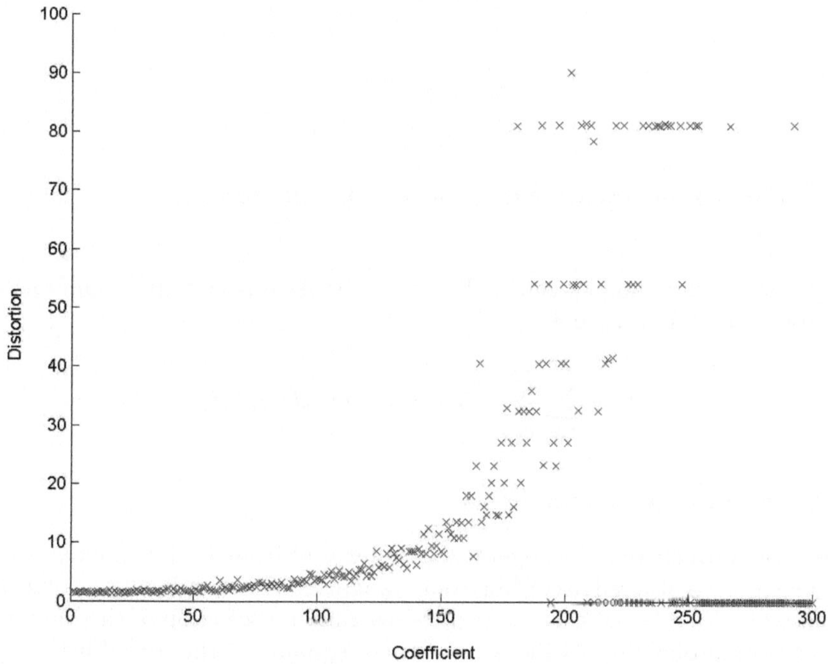

Fig. 3.2. Plot of distortion of the watermarked image versus coefficient magnitude regardless of position

difference between the original image and the inverse of the marked DCT. Minimizing the distortion would be trivially attained by placing the watermark at the higher frequency region of the DCT domain. However, this contradicts our goal of obtaining a maximally robust placement. To avoid the computational cost of the DCT inversion, we propose as the fitness function of the evolutionary algorithm an approximation that follows from the observation that the distortion introduced adding something to a DCT coefficient is proportional to the absolute value of that coefficient. An empirical validation of this assertion is shown in figure 3.2. The computational experiment consisted in repeatedly adding a constant value to single randomly selected coefficients of a test image DCT and computing the distortion of the marked image. There the x axis in the graph is the value of the affected coefficient. The ordinate axis is the distortion value respect the original image. The figure shows that modifications in coefficients with the same value generate different distortion values. This is the effect due to the coefficient placement in the transform domain. In general, the distortion decreases as the the distance to the transform domain origin increases. Nevertheless, it can appreciated that as the affected coefficient magnitude decreases the marked image distortion decreases regardless of the coefficient placement in

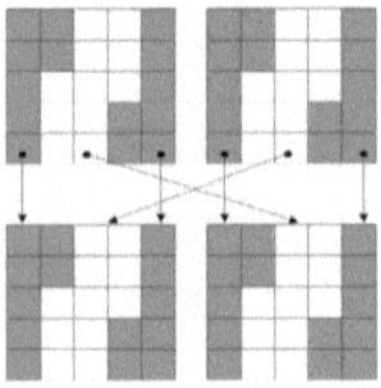

Fig. 3.3. Illustration of the Crossover Operator based on 2 cut points

the transform domain. Thus, the distortion fitness function to be minimized we propose is the following one:

$$f_2 = \sum_{k=i}^{m_w} \sum_{l=1}^{n_w} \mid X_t(x(k,l), y(k,l), z(k,l)) \mid \tag{3.3}$$

3.5.2 Evolutionary Operators

Selection Operator: This operator generates O from P. The populations has previously been ordered according to its range and distance between solutions as proposed in [7]. The selection is realized by random selection of the individuals, giving more probability to the ones at the beginning of the sorted list.

Crossover operator: This operator is applied with probability P_c and is used to recombine each couple of individuals and obtain a new one. Two points from the solution matrix are randomly selected as cut points, and the individuals are recombined as in conventional crossing operators. This operator is illustrated in 3.3.

Mutation operator: Every element of an individual solution $\mathbf{s}$ undergoes a mutation with probability P_m. The mutation of an element consists of displacing it to a position belonging to its 24-Neighborhood in the 3D DCT domain grid: given a pixel $W_t(k,l)$ located in the position $x(k,l), y(k,l), z(k,l)$ of X_t, the new placement of $\mathbf{s}(k,l) \in \{X_t(x(k,l) \pm 1, y(k,l) \pm 1, z(k,l) \pm 1)\}$. The direction of the displacement is randomly chosen. If the selected position is out of the image, or collides with another assignment, a new direction is chosen.

Reduction operator: After applying the selection, crossover and mutation operators we have two populations: parents P and offsprings O. The reduction operator determines the individuals who are going to form the next generation population. Parent and offspring populations are joined in a new one of size $2P_s$. This population is sorted according to the rank of each solution and distance between solutions[7]. This ensures an elitist selection and the diversity of the solutions through the Pareto front. The new population P is composed of the best P_s individuals according to this sorting.

3.5.3 Algorithm

The first step of the GA is the generation of an initial population P and the evaluation of each individual's fitness. The rank and distance of each individual is calculated [7] and is computed to sort P. Once done this, the genetic iteration begins: An offspring population O is calculated by means of the selection, crossover, and mutation operators. The new individuals are evaluated before joining them to the population P. Finally, after computing the reduction operator over the new rank and distance of each individual, we obtain the population P for the next iteration.

Since the GA works with many non-dominated solutions, the stopping criterion compares the actual population with the best generation, individual to individual, by means of the *crowded_comparison*() [7]. If no individual, or a number of individuals below a threshold, improves the best solution in n consecutive iterations, the process is finished. A pseudo-code for de GA is shown in figure 3.4.

```
Pob = Generate_Initial_Population();
Fitness_Function_Evaluation(Pob);
Range=fast_non_dominated_sort(Pob);
Distance=crowding_distance_assignment(Pob);
Stop = false;
While Stop == false
    Couples = Selection_Operator(Pob);
    Of = Merge_Operator(Couples);
    Of = Mutation_Operator(Of);
    Fitness_Function_Evaluation(Of);
    Pob = Join(Pob,Of);
    Range = fast_non_dominated_sort(Pob);
    Distance = crowding_distance_assignment(Pob);
    Pob = ordering(Pob, Range, Distance);
    Pob = Reduction_Operator(Pob);
    Evaluate_Stop_Criterium():
end while;
plot(Pareto-Front);
```

Fig. 3.4. Pseudo-code for the proposed GA

The Pareto front is formed by the set of solutions with rank = 1. Once finished the process and chosen a solution, the mark is embedded adding its coefficients to the coefficients of X_t according to the corresponding value of $\mathbf{s}^*$. Before the coefficients are added, they are multiplied by a small value.

3.6 Results

This results section contains an example application to a conventional gray level scale, that could correspond to a panchromatic remote sensing image. We show

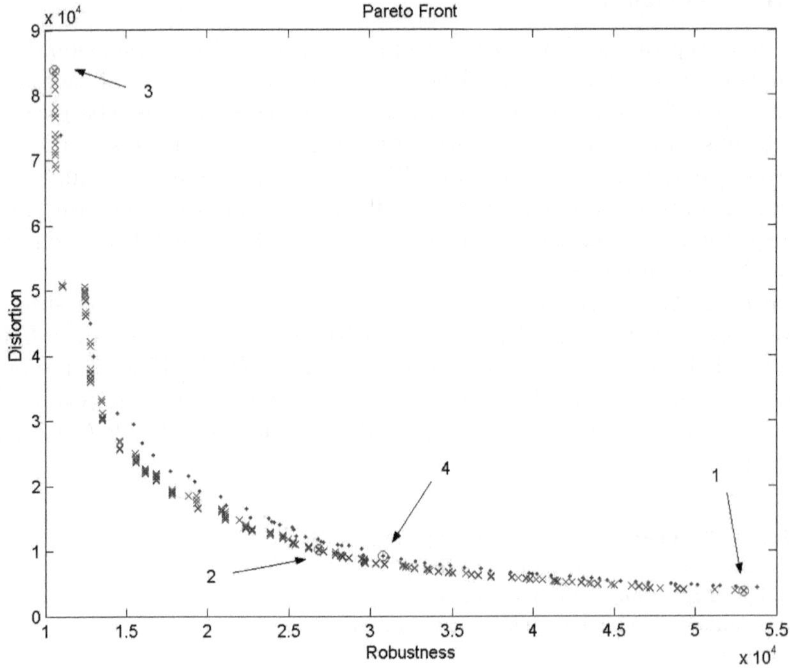

Fig. 3.5. Pareto fronts found by GA and local search, identified by 'x' and '.' respectively

Fig. 3.6. From left to right: Original Image; Images watermarked using the placement denoted in figure 3.5 as solution 3, as solution 2 and as solution 1

that proposed algorithm finds robust and low distortion watermark placements, therefore the proposed fitness functions can be assumed to model appropriately the desired watermark properties. Then we extend the results to a well known benchmark hyperspectral image.

3.6.1 Results on a Conventional Gray Level Image

The results presented in this section concern the application of the algorithm over an image of size 400 x 500. The image DCT X_t has been divided in 676

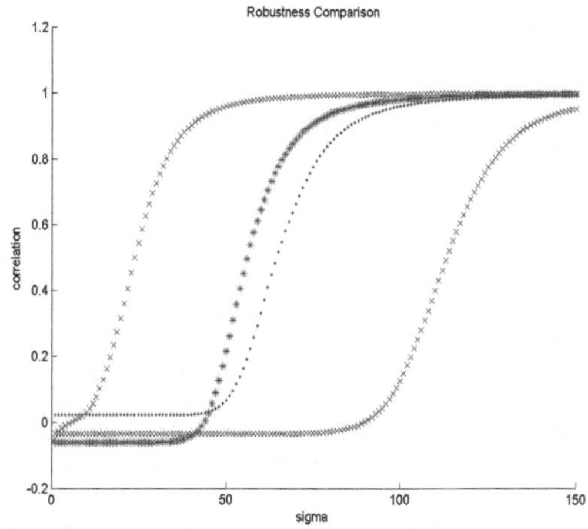

Fig. 3.7. Robustness measured as the correlation coefficient of the recovered image mark versus the radius of the gaussian smoothing filter. Each curve corresponds to a placement solution identified in in figure 3.5. Solutions 1 and 3 are represented by 'x'. Solution 2 by '*' and Solution 4 by '.'.

Fig. 3.8. Watermark logo: Original and recovered from the image watermarked using placement solution 2 in figure 3.5 after it has been low-pass filtered with *sigma* = 50, 60, 70, 80, 90, 100

overlapping and regular image blocks of size 100 x 100. The initial population is formed by 672 individuals each one placed randomly in a different quadrant. As the watermark image we have used an image of size 32 x 32. The GA was executed with $P_s = 20, P_m = 0.05$ and $P_c = 0.9$. We set the robustness fitness f_1 parameters to $F = 4$ and $d = 3$.

For comparison purposes the problem has been solved by means of a random local search starting from the same random initial conditions. This local search consist only of proposing a new placement by a random perturbation computed like the mutations above. This new placement is accepted if it does improve the current solution. The local search stops when a number of proposed placements are rejected, assuming that the algorithm is stuck in a local optimum. Figure 3.5 shows the Pareto-Front found with both algorithms. The GA has found 329 non-dominated solutions while the local search only found 62. Besides the GA solutions dominate all the solutions found by the local search.

We have pointed out and numbered in figure 3.5 some very specific solutions. The solution denoted as 1 corresponds to the solution with the lowest fitness

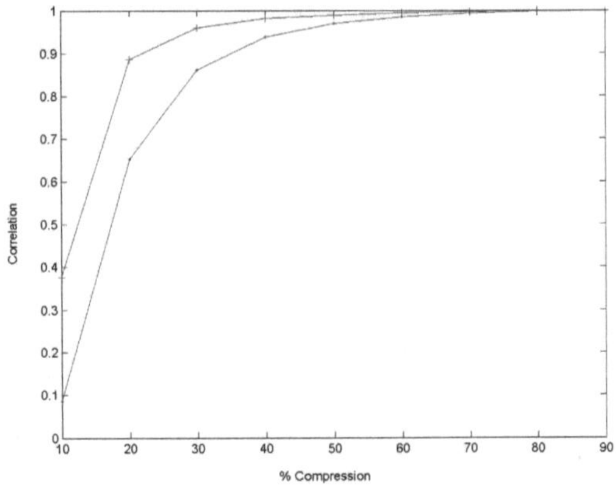

Fig. 3.9. Robustness of watermark placement solutions 2 ('.') and 4 ('+') in in figure 3.5 to JPEG compression. Correlation of the recovered mark image versus compression quality.

distortion value, regardless of the robustness fitness value (which is very high). The solution signaled as 3 corresponds to the solution with lowest robustness fitness, regardless of the fitness distortion value (again very high). These solutions correspond to optima of the objective criteria taken in isolation. We consider also compromise solutions 2 and 4 that correspond to the best robustness for a set upper limit of the distortion, taken from the Pareto fronts found by the GA (solution 2) and the local search (solution 4). Figure 3.6 shows the experimental image (left) and the visual results of GA generated watermark placement solution. The distortion is almost no perceptible, but for the image corresponding to solution 3 in figure 3.5.

To asses the robustness of the watermark placements found, we compute the correlation coefficient between the original watermark and the watermark recovered from the watermarked image after it has been smoothed by a low-pass gaussian filter applied in the Fourier transform domain. The figure 3.7 plots the correlation coefficients versus the increasing filter radius *sigma* for each of the selected watermark placement solutions selected in figure 3.5 . This plot shows that the watermark placement solution 2 obtains a good correlation coefficient for lower values of *sigma* than solution 1 (note that in figure 3.6 there are no perceptual differences between both images). That means that the GA found a solution that is much more robust than the one with minimal distortion while preserving much of the distortion quality. It can be appreciated also in figure 3.7 that the robustness is higher in the solution 2 (GA) than in the solution 4 (Local Search) . Figure 3.8 shows the visual results of the recuperation of the mark image after smoothing the image watermarked using the placement from solution 2.

The second class of attacks we are considering are the lossy compression. We apply the standard jpeg compression with increasing quality factor to the watermarked image, and we recover the watermark image from the decompressed image. Figure 3.9 shows the correlation of the recovered mark image relative to the original mark image versus compression quality, for the local search and GA watermark placement solutions identified as 4 and 2 in in figure 3.5 . It can be appreciated that the GA solution recovers much better than the local search solution from strong lossy compression.

3.6.2 Results on an Hyperspectral Image

The results presented in this section concern the application of the algorithm over the well known AVIRIS Indian Pines hyperspectral image of size 145 x 145 x 220. The image DCT transform X_t has been divided in 1452 overlapping quadrants of size 45 x 45 x 110. The initial population is formed by 1452 individuals each one placed randomly in a different quadrant. The watermark is an image of size

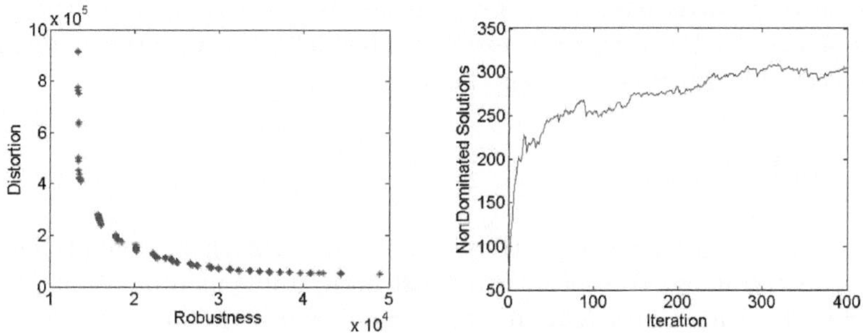

(a) Pareto-Front (b) NonDominated Evolution

Fig. 3.10. a) Pareto front found by GA. b)Evolution of the number of non-dominated solution found by the GA.

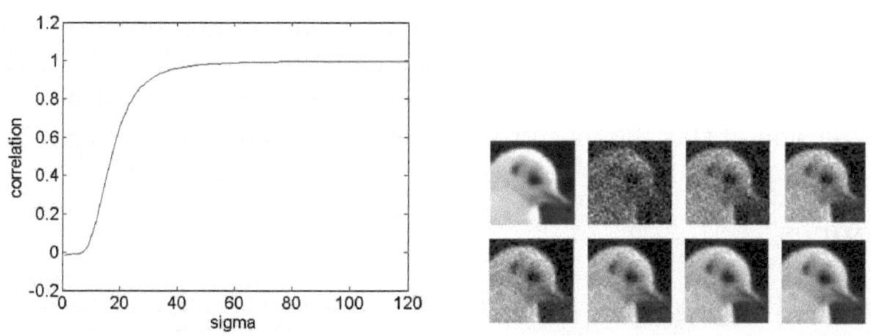

(a) Filtering (b) Recovered watermark

Fig. 3.11. a) Robustness level by means of the correlation coefficient of the recovered image mark versus the radius of the smoothing convolution kernel. b) Original watermark and watermark recovered after low pass filtering with sigma = 10, 20, 30, 40, 50, 60 and 70 respectively.

50 x 50. The GA was executed with $P_s = 20, P_m = 0.05$ and $P_c = 0.9$. We fit the response of the robustness fitness f_1 with $F = 4$ and $d = 3$.

Figure 10(a) shows the Pareto front consisting of 303 non-dominated solutions, found by the algorithm, following the evolution shown in Figure 10(b). Figure 11(a) plots the correlation coefficient between the original watermark and the watermark recovered after each band image of the watermarked image has been smoothed by a low-pass gaussian filter with increasing filter radius applied in the Fourier transform domain. Figure 11(b) shows the visual results of the recuperation of the mark image after smoothing the watermarked image.

Studying each pixel spectrum, experts can know which material form the area represented by this pixel. Automated classification systems can be constructed [11] to perform this task. This is the main objective of hyperspectral imaging, so, it is critical that the watermarking process doesn't disturb the spectral content of the pixels. For the noisiest of the solutions shown in Figure 10(a) we computed the correlation of each pixel spectrum with the corresponding one in the original image. The worst value obtained was 0.999. Therefore, this watermarking process is not expected to influence further classification processes.

3.7 Conclusions

We present an evolutionary algorithm to find a watermark's image placement in an hyperspectral image to protect it against undesirable manipulations. It is desirable that the watermark remains recognizable when the image is compressed or low-pass filtered. We state the problem as a multiobjective optimization problem, having two fitness functions to be minimized. The algorithm tries to obtain the Pareto front to find the best trade-off between distortion of the original image in the embedding process and robustness of the mark. The solutions found by the GA provide strong robustness against smoothing manipulations of the image. Because the algorithm works with the entire image DCT, it can be used to hide bigger images or data chunks than other similar approaches. Also it will be more robust than approaches based on small block embedding, experimental verification is on the way to prove this intuition. Furher work must be addressed to the extension of this approach to wavelet transforms of the images.

References

1. Augot, D., Boucqueau, J.M., Delaigle, J.F., Fontaine, C., Goray, E.: Secure delivery of images over open networks. Proceedings of the IEEE 87(7), 1251–1266 (1999)
2. Back, T., Fogel, D.B., Michalewicz, T.: Evolutionary Computation1. Basic Algorithms and Operators. Board Addison-Wesley Publishing Company, Reading (2000)
3. Barni, M., Bartolini, F., Cappellini, V., Magli, E., Olmo, G.: Near-lossless digital watermarking for copyright protection of remote sensing images. In: IGARSS 2002, vol. 3, pp. 1447–1449. IEEE Press, Los Alamitos (2002)

4. Barni, M., Magli, E., Troia, R.: Minimum-impact-on-classifier (mic) watermarking for protection of remote sensing imagery. In: IGARSS 2004, vol. 7, pp. 4436–4439 (2004)
5. Caldelli, R., Macaluso, G., Barni, M., Magli, E.: Joint near-lossless watermarking and compression for the authentication of remote sensing images. In: IGARSS 2004, vol. 1, p. 300 (2004)
6. Coello Coello, C.A., Toscano Pulido, G., Mezura Montes, E.: Current and future research trends in evolutionary multiobjective optimization. In: Graña, M., Duro, R., d'Anjou, A., Wang, P.P. (eds.) Information Processing with Evolutionary Algorithms, pp. 213–232. Springer, New York (2004)
7. Deb, K., Pratap, A., Agarwal, S., Meyarivan, T.: A fast and elitist muliobjective genetic algorithm: Nsga-ii. IEEE transactions on evolutionary computation 6(2), 182–197 (2002)
8. Ho, A.T.S., Jun, S., Hie, T.S., Kot, A.C.: Digital image-in-image watermarking for copyright protection of satellite images using the fast hadamard transform. In: IGARSS 2002, vol. 6, pp. 3311–3313 (2002)
9. Kaarna, A., Parkkinen, J.: Multiwavelets in watermarking spectral images. In: IGARSS 2004, vol. 5, pp. 3225–3228 (2004)
10. Kaarna, A., Toivanen, P.: Digital watermarking of spectral images in pca/wavelet-transform domain. In: IGARSS 2003, vol. 6, pp. 3564–3567 (2003)
11. Landgrebe, D.A.: Signal Theory Methods in Multispectral Remote Sensing Signal Theory Methods in Multispectral Remote Sensing. Wiley-Interscience, Chichester (2003)
12. Maeno, K., Qibin, S., Shih-Fu, C., Suto, M.: New semi-fragile image authentication watermarking techniques using random bias and nonuniform quantization. Multimedia, IEEE Transactions 8(1), 32–45 (2006)
13. Nikolaidis, A., Pitas, I.: Region-based image watermarking. IEEE Transactions on image processing 10(11), 1726–1740 (2001)
14. Qiming, Q., Wenjun, W., Sijin, C., Dezhi, C., Wei, F.: Research of digital semi-fragile watermarking of remote sensing image based on wavelet analysis. In: IGARSS 2004, vol. 4, pp. 2542–2545 (2004)
15. Schneck, P.B.: Persistent access control to prevent piracy of digital information. Proceedings of the IEEE 87(7), 1239–1250 (1999)
16. Young, K.T., Hyuk, C., Kiryung, L., Taejeong, K.: An asymmetric watermarking system with many embedding watermarks corresponding to one detection watermark. Signal Processing Letters 11(3), 375–377 (2004)
17. Tamhankar, H., Bruce, L.M., Younan, N.: Watermarking of hyperspectral data. In: IGARSS 2003, vol. 6, pp. 3574–3576 (2003)
18. Tamhankar, H., Mathur, A., Bruce, L.M.: Effects of watermarking on feature efficacy in remotely sensed data. In: IGARSS 2004, vol. 1, p. 280 (2004)
19. Tang, X., Pearlman, W.A., Modestino, J.W.: Hyperspectral image compression using three-dimensional wavelet coding. In: Image and Video Communications and Processing 2003, Proceedings of SPIE, vol. 5022, pp. 1037–1047. SPIE Press (2003)
20. Vleeschouwer, C., Delaigle, J.F., Macq, B.: Invisibility and application functionalities on perceptual watermarking- an overview. Proceedings of the IEEE 90(1), 64–77 (2002)
21. Voyatzis, G., Pitas, I.: The use of watermarks in the protection of digital multimedia products. Proceedings of the IEEE 87(7), 1197–1207 (1999)

22. Wolfgang, R.B., Podilchuk, C.I., Delp, E.J.: Perceptual watermarking for digital images and video. Proceedings of the IEEE 87(7), 1108–1126 (1999)
23. Yuan, H., Zhang, X.P.: Multiscale fragile watermarking based on the gaussian mixture model. Image Processing, IEEE Transactions 15(10), 3189–3200 (2006)
24. Zou, Y.Q., Shi, D., Ni, Z., Su, W.: A semi-fragile lossless digital watermarking scheme based on integer wavelet transform. Circuits and Systems for Video Technology, IEEE Transactions 16(10), 1294–1300 (2006)

4

Architecture and Services for Computational Intelligence in Remote Sensing

Sergio D'Elia[1], Pier Giorgio Marchetti[1], Yves Coene[2], Steven Smolders[3], Andrea Colapicchioni[4], and Claudio Rosati[4]

[1] ESA-European Space Agency
[2] SPACEBEL
[3] GIM-Geographic Information Management
[4] Advanced Computer Systems ACS S.p.A.

4.1 Introduction

4.1.1 Background

The Earth is facing unprecedented climatic and environmental changes [1], which require a global monitoring [2] and large scale actions which are addressed by European and world wide programmes of similar wide scale. The overall objective of a sustainable growth demands for a cleaner, safer and healthier global environment. The 7th Framework Programme (FP7) of the European Commission (EC) assumes that the Information and Communication Technologies (ICT) may play a role in combating the unsustainable trends that risk undermining the future economic growth and impact on the quality of life in Europe:

- environmental degradation and unsustainable use of depleting natural resources;
- pollution and waste generation;
- increasing exposure and risk to man made and natural disasters.

Furthermore other initiatives are addressing other challenges which if uncoordinated may undermine the objective of a single information space in Europe, which the EC has set as one of the targets for the so-called Lisbon Agenda (i2010):

- INSPIRE - infrastructure for spatial information in Europe. The INSPIRE directive addresses the European fragmentation of datasets and sources, gaps in availability, lack of harmonisation between datasets at different geographic scales and duplication of information collection. The initiative intends to trigger the creation of a European spatial information infrastructure that delivers to the users integrated spatial information services. These services should allow the policy makers, planners and managers at European, national and local level, the citizens and their organisations to identify and access spatial or geographical information from a wide range of sources.
- GMES - Global Monitoring for Environment and Security is a joint EC - ESA programme with the objective to establish a European capacity for

M. Graña and R.J. Duro (Eds.): Comput. Intel. for Remote Sensing, SCI 133, pp. 79–123, 2008.
springerlink.com

global monitoring by 2008. Work is in progress on the development of pilot and fast track services (land, marine, emergency response), on multi-mission facilities and on the integration of national and European missions to guarantee continuity of data and services, as well as on the development of the space component (the Sentinels missions) in order to ensure the necessary space data capacity by 2012.
- GEO / GEOSS - The Group of Earth Observations (GEO) is an organisation dedicated to developing and instituting a Global Earth Observation System of Systems (GEOSS).

These initiatives demand data from Earth Observation (EO) satellites, operating as single missions or constellations, which increase revisit time. Sensors on board of these satellites acquire data in increasing variety, amount and resolution, resulting in huge amounts of data to be handled. On the other side, users, service industry and the above initiatives require timely delivery of focused information. Therefore EO satellites' data, which are not immediately usable, require chained transformations before becoming the needed "information product" (easily understandable and ready-to-use without further manipulations) offered as on-demand or systematic services.

Different actors perform these transformations using own processes, which require specific knowledge, experience and possibly also data or information from domains other than EO. Today, a number of specialised companies, operating independently in specific application domains, tend or are forced to build and keep full control of the entire process, with efforts going beyond their core interest and expertise. This often results in an inefficient system, where parts of the processes are redeveloped many times without benefits for the end user (no competition increase) or the service provider (deploying resources on non-core activities).

Additional cost efficiency could come from the reduction of the information extraction time through the automation of such processes, by using systems which can learn and / or apply knowledge. Automatic or semi-automatic Image Information Mining (IIM or I2M) techniques would permit to quickly identify the relevant subset among the large quantity of images, as well as to support and integrate the expert's interpretation. This approach would also permit to apply these processes to a bigger portion of the petabytes of archived or new data, which currently are systematically processed only in limited quantities.

The development of related systems must be harmonised and aligned to latest technologies and standards in order to foster cooperation by all partners and feed the "single information space" for GEOSS, GMES and INSPIRE. Given the variety and diversity of initiatives, this "single information space" will most likely consist of more than one architecture due to the considerable investments already made by the various actors. Although Service Oriented Architectures (SOA) are being adopted as the architectural paradigm in many European Commission (EC) and ESA supported ICT infrastructure projects for managing environmental risks and GMES [3], the standards and approaches differ considerably and need further convergence and standardisation.

The SOA approach is well suited to support the resulting "system of systems". For example interoperability between the SOA approaches chosen by EC WIN, EC ORCHESTRA, and ESA SSE projects was recently demonstrated [4]. This chapter describes in detail the architecture of two ESA environments, instances of which may become components of this hierarchical "system of systems" for GMES and GEO. They are in the field of Service Provisioning Support and Image Information Mining. A view on the selected underlying architecture precedes the descriptions of these two environments. The descriptions are provided according to two different methodologies, an engineering one and a user / history focused one.

4.1.2 Service Oriented Architectures

The benefits of the proposed Service Oriented Architecture are;

- Loosely-coupling and independence
  - Increases organisational agility; allows service providers to easily design, assemble, and modify business processes in response to market requirements;
  - Provides a competitive advantage by offering greater flexibility in the way computer systems can be used to support the business;
  - Reduces time to market, fostering re-use of available services;
  - Lowers implementation costs by increasing reusability;
- Independence
  - Increases IT adaptability; changes of the IT technology and evolution of standards, are easily integrated;
  - Improves service providers' response to sector evolution as changes in organisations and processes are easily accommodated.
- Modularity
  - Enables incremental development, deployment, and maintenance, reducing the need for large "one shot" software developments;
  - Decreases development effort and time by reducing complexity;
  - Reduces the total number of processes;
  - Fosters and facilitates the deployment of new services as the service orchestration supports the streamlining of processes and the assembly of existing services into new more complex ones.
- Neutrality
  - Allows legacy services and existing investment in IT assets to be leveraged.
- Resilience
  - Lowers risk and development effort; minimising the need to rewrite and test existing applications and services.
- Standards-based
  - Largely platform (hardware and software) independent allows service development and provision practically independent of IT infrastructure strategies;

- Allows economies of scale as the same technology can be applied to support and deliver a wide range of services;
- Reduces complexity and fragmentation resulting from use of proprietary technologies.

4.2 SSE: Service Support Environment for Earth Observation and Other Domains

4.2.1 Motivation and User Requirements

The ESA Service Support Environment (SSE) implements a neutrally managed, open, service-oriented and distributed environment that enables the orchestration and integration of EO, meteorological and geospatial data and services.

The three high level requirements that this infrastructure fulfils are:

- Simplify the access to multiple sources of EO data.
- Facilitate the extraction of information from EO data.
- Reduce the barrier for the definition and prototyping of EO Services.

Users of EO data require accessing multiple data sources from different providers. The analysis of the value-adding services has revealed that more than 60% of the efforts involved in the creation of such services are devoted to accessing the EO data.

The European Space Agency's Oxygen project [5] indicated that EO Services, whether commercial or public, are about the provision of the right information at the right moment to the proper user. The need to define the interoperability standards to ease the EO data access in Europe is as well a priority. In fact, during the initial phase of the joint ESA and EC GMES programme, the Agency shall provide harmonised access to ESA, national, EUMETSAT and other third party Earth Observation Missions for the so-called GMES Fast Track Services, and therefore provide the required capacity to satisfy the GMES space-based observation needs. In order to deliver the high-level operational services which are needed, it is necessary to integrate EO products, space data, with all kinds of other data and information.

The complexity of this next generation of integrated services may also require establishing:

- a distributed digital library of geospatial services, as well as,
- a network of centres able to support the partners who will contribute to the production and delivery of data access and information services.

It is therefore necessary to develop tools to support the orchestration of data acquisition and handling, transformation of formats, geomatic functions and the required data access, processing and value-adding services chains. To this end, the identification of a set of common EO related standards and the support of a neutral and open service-enabling environment becomes mandatory to respond to the need for EO services and "information products" closer to user expectations and processes (easily understandable and ready-to-use).

4.2.2 The Methodology for Architecture Design

The proposed Service Oriented Architecture is designed and described making use of the RM-ODP model (Reference Model of Open Distributed Processing - RM-ODP - SO/IEC 10746-1:1998) [6]. The RM-ODP model has been modified to take into account the objective of addressing a digital library of distributed services rather than a distributed processing system for which the RM-ODP was originally defined.

The RM-ODP model analyses open distributed systems through 5 different views of the system and its environment:

- The enterprise viewpoint: focuses on the purpose, scope and policies for the system.
- The information viewpoint: focuses on the semantics of the information and information processing performed.
- The computational viewpoint: enables distribution through functional decomposition of the system into objects which interact at interfaces.
- The engineering viewpoint: focuses on the mechanisms and functions required to support distributed interaction between objects in the system.
- The technology viewpoint: focuses on the choice of technology in that system.

In the design of the proposed architecture, the RM-ODP was tailored by replacing the computational viewpoint with a service viewpoint as detailed in the following paragraphs.

4.2.3 Enterprise Viewpoint

The enterprise viewpoint is concerned with the business activities of the Service Support Environment. These activities can be represented by two sets of use cases related to respectively the end-user of services and the service provider (or service owner).

- End-users benefit from this environment as it brings together distributed EO services and EO products offered by multiple service providers. Via this common access point (accessible from the Web Portal or from a service registry or catalogue), the end-user can more easily discover services and products matching his exact requirements. EO product, collection and service catalogues for multiple missions of different satellite operators are offered within a single environment and are linked with data access, programming (planning of sensors' acquisitions), ordering and processing services hereby offering a one-stop solution for users of EO data and services.
- The environment empowers service providers by offering them the possibility to advertise and integrate their new and existing services within this one-stop Portal for EO data and EO services. They are provided with cost-effective tools based on open standards allowing them to publish and manage their services as well as monitor their use whilst keeping control over the services backend on their local infrastructure. These services can be combined with

other services that may possibly be provided by third parties hereby facilitating the definition of advanced value-adding services on EO imagery by distributing the processing steps over different specialists. Service providers can offer these services via the Portal pages or via a machine-to-machine "data access integration layer" allowing discovery of their services via a virtual service and data registry or catalogue.

The enterprise viewpoint thus addresses following high level objectives:

- provide a neutrally managed overarching infrastructure enabling the interactions among service providers and with end-users;
- permit service interaction whilst avoiding the service de-localisation (i.e. services remain on the service provider infrastructure);
- allow easy publishing and orchestration (i.e.: chaining of services into more complex ones) of synchronous and asynchronous EO services for online and offline processes;
- support "subscription" type services and standing orders (e.g. fires active monitoring and alerting);
- support the evolution and maintenance of services;
- allow easy identification of, and access to requested services and products, with progress follow-up until completion;
- integrate services from multiple domains, e.g. geospatial, meteorological, in-situ, to exploit multi-domain synergies;
- minimise service provider investments by building on open standards.

As the objective is to define an environment capable of supporting multiple scenarios, we envisage implementing a scalable digital library environment. This will allow the services to be deployed at different scales as depicted in the figure 4.1: global (e.g. European or worldwide), national, regional or thematic and local.

The Service Support Environment aims at providing a consistent experience within which the user will be able to discover and access a variety of services offered by numerous disparate providers. At the same time the content and behaviour of these services should be predictable allowing the user to anticipate the results and use the services through a normal Internet connection. This idealised approach is represented in figure 4.2 as an "Internet bus" approach [7].

4.2.4 Information Viewpoint

The information viewpoint specifies the modelling of all categories of information that the proposed architecture deals with, including their thematic and spatio-temporal characteristics as well as their metadata. Within ESA's SSE infrastructure, these information models are based upon Open Standards where these are available.

Service Metadata Information Model

Within SSE, being a portal that provides access to distributed geospatial services, one of the main information categories relates to the descriptions or

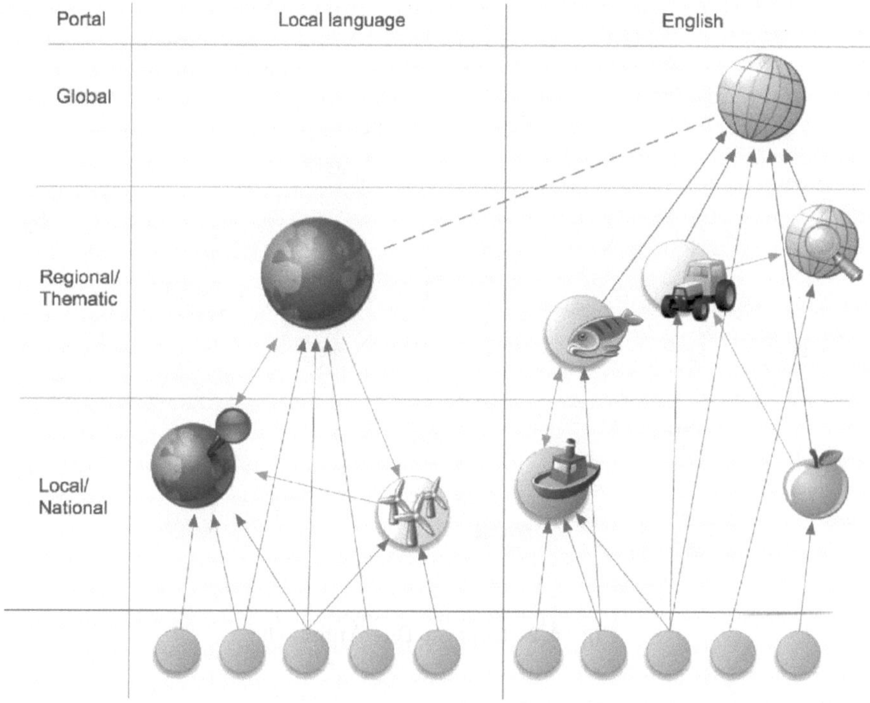

Fig. 4.1. Distributed Service Support

metadata of these services. This service metadata needs to provide the details to allow for machine-to-machine communications but also contain descriptive information targeted at human readers. This service metadata can be modelled according to ISO 19119 [8], as shown in figure 4.3. Each service may be defined by:

- Identifying properties of the service itself: the type of the service, its title and abstract, its usage restrictions, its region/time period of applicability;
- Identifying information of the service owner: the point of contact;
- The service operations (e.g. GetCapabilities, GetMap, GetFeasibility, SubmitOrder, ...) with their connection points and protocol bindings;
- The parameters associated with these operations and dependencies and chains of operations.

In addition this ISO 19119 information model offers the possibility to couple services to data metadata. This is required for services that are considered to be tightly coupled with datasets or datasets collections, as for instance EO product catalogue or ordering services that allow respectively the discovery and ordering of products pertaining to specific collections. This coupling is not required for

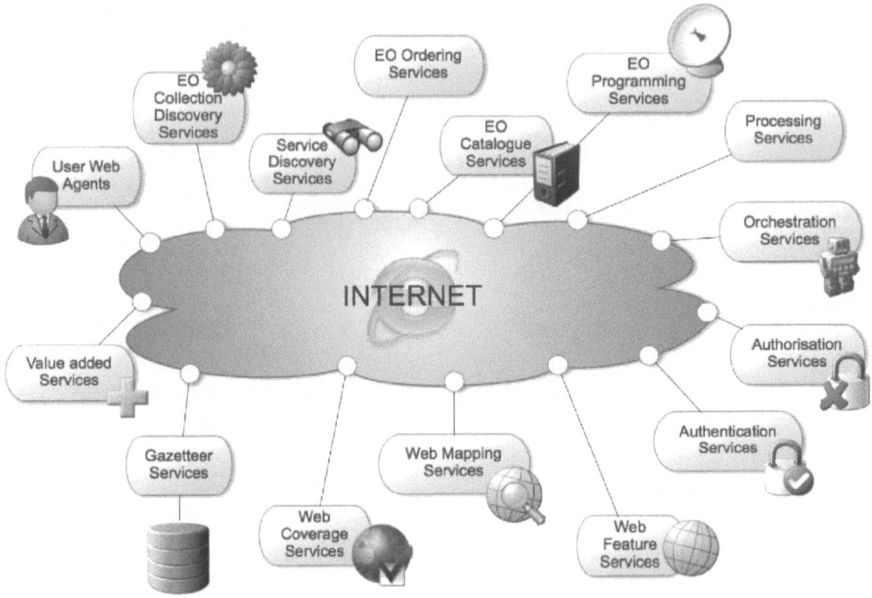

Fig. 4.2. Internet Bus Approach

so-called loosely coupled services that are not associated with specific datasets or dataset collections. Services that are mixed coupled are associated with a specific dataset or dataset collection but can also operate on other datasets as for instance a Web Coordinate Transformation Service that may work on EO Products from specific collections but may also work on other images. The coupling between the service and dataset(s) collections is done via an "OperatesOn" association. In addition individual service operations can be tied to the datasets by providing the "CoupledResource" Information.

For the discovery of EO related services, a minimal recommended subset of this ISO 19119 information model required for the discovery of services has been defined. This subset includes all mandatory ISO 19119 items and makes a number of optional elements mandatory. The optional operation parameters, dependencies and operation chain related elements are not included in this minimal set that is visualised within the figure 4.3.

This ISO 19119 standard defines the model for geographic services' metadata; it however does not specify the exact XML grammar to be employed. This is dealt with in the draft ISO 19139 standard [9] that describes the transformation of the abstract UML models into XML schema.

EO Collection Metadata Model

EO collections are collections of datasets sharing the same product specification. An EO collection can be mapped to "dataset series" in ISO terminology. In the

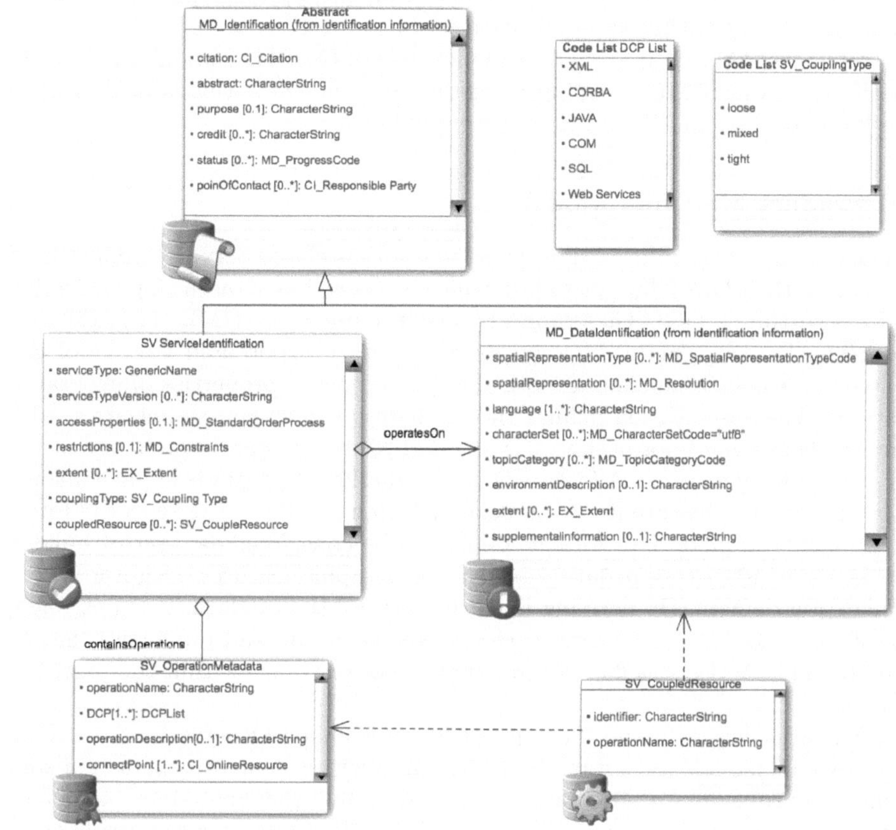

Fig. 4.3. ISO 19119 Service metadata

Earth Observation context, a collection typically corresponds to the series of datasets (i.e. products) derived from data acquired by a single sensor onboard a satellite or series of satellites and having the same mode of operation. Examples of EO collections are for instance "TerraSAR-X spotlight mode" or "ESA ENVISAT MERIS Full Resolution L1+2".

The metadata of EO collections can be described by employing the ISO 19115 standard for Geographic Metadata [10]. This 1SO19115 standard defines the schema required for describing geographic information. It provides information about the identification, the extent, the quality, the spatial and temporal schema, spatial reference, and distribution of digital geographic data. It defines a set of building blocks consisting of more than 300 elements (classes, attributes and relations) that a spatial data community can use in order to establish a community metadata profile that contains all elements required by the community. Out of the entire set of available metadata elements, there are a number of mandatory core elements that are required within each metadata. Apart from these core elements, there are a number of elements that are optional but that may be made mandatory within an application profile.

As for the services information model, this ISO 191115 standard does not specify the exact XML encoding of metadata documents, this is dealt with in the ISO 19139 standard that was referenced above.

EO Product Metadata Model

The most important information model in the EO-related Service Support Environment is the EO product metadata which has been based on the Open GeoSpatial Consortium - OGC Geography Mark-Up Language - GML [11]. GML is a modelling language and XML encoding for the transport and storage of geographic information, including both the geometry and properties of geographic features. The specification defines the mechanisms and syntax that are used to encode geographic information in XML and constitutes an interchange format for geographical features. A feature is an abstraction of a real world phenomenon; it is a geographic feature if it is associated with a location relative to the Earth.

More than a mere data format, GML can be considered as a set of building blocks for constructing a data model for geographic features within a specific application domain. By deriving from the base GML standard, it is possible to specify, in a so-called application schema, the structure and properties that are used to characterise specific features that are of relevance within the particular domain.

GML also defines a specific type of features called Observations. "A GML observation" models the act of observing, often with a camera, a person or some form of instrument. An observation feature describes the "metadata" associated with an information capture event, together with a value for the result of the observation. This covers a broad range of cases, from a tourist photo (not the photo but the act of taking the photo), to images acquired by space borne sensors. Next to the properties that are inherited from the base gml feature element, an observation is characterised by a mandatory time and result and optional properties that reference the sensor or instrument and the target of the observation. Obviously, these base properties are not sufficient to model the EO product metadata.

Hence a set of specific GML application schema have been defined that derive from the generic observation class and that add specific EO product-related properties [12]. These properties are described using definitions from the ISO 19115/19139 set of standards, for properties where such an ISO element is available. This is indicated in the figure 4.4 by the gmd schema layer. Derived from the base gml schemas and using the ISO element is the generic EO metadata schema that resides in the hma (Heterogeneous Missions Accessibility) namespace. At a higher level there are a number of other schemas that are extending the generic EO product metadata for the different types of missions or sensors, recognising the fact that not all sensors require the same elements in their metadata description.

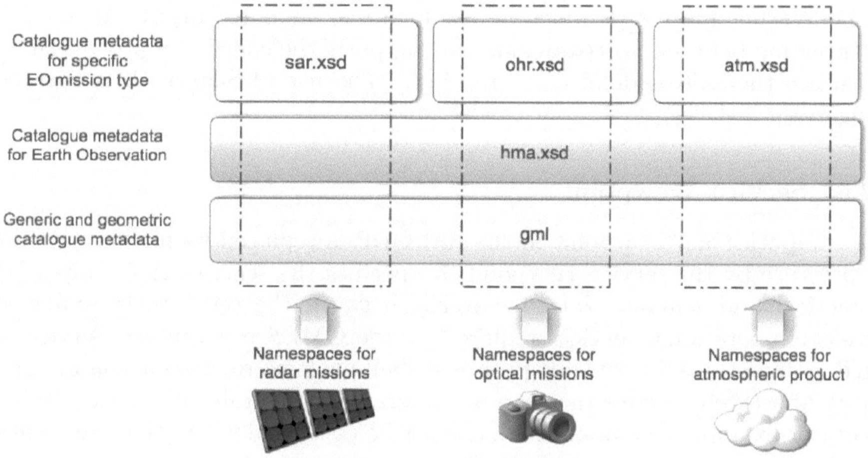

Fig. 4.4. Gmd schema layer

These sensor specific schemas: OHR for optical (high resolution), SAR for (synthetic aperture) radar and ATM for atmospheric products are shown near the top of the figure 4.4.

Other Information Models

Semantic Web Technology: Thematic instances of the SSE, e.g. an SSE dedicated to marine-related services, may also use emerging Semantic Web technology to facilitate the modelling and interlinking of information. It is envisaged to use W3C Resource Description Framework - RDF and Web Ontology Language - OWL to enable end-users to more easily identify and locate services related to their domain, e.g. algae bloom, oil spill etc. by interlinking the services with agreed ontologies and controlled domain vocabularies or thesauri. The ESA KEO [13] project and the European Commission projects InterRisk [14] and WIN [15] are working on such marine-related ontologies.

SensorML: The SensorML standard is being developed within the Sensor Web Enablement Initiative - SWE of the OGC. Within this activity a number of standards are being developed that are related to the interoperable discovery and collection of data from in-situ and remote sensors. Together these standards are intended to allow interoperable addressing of web-centric, open, interconnected, intelligent and dynamic network of sensors. They enable spatio-temporal understanding of an environment through co-ordinated efforts between multiple numbers and types of sensing platforms, both orbital and terrestrial, fixed and mobile [16]. The SensorML draft standard [17], still in the voting stage of the OGC approval process at the time of writing, establishes a standard schema for metadata describing sensors and sensor platforms. Within the context of SensorML, sensors and sensor systems are modelled as processes. Processes are entities that take one or more inputs and through the application of well-defined methods using specific parameters, result in one or more outputs. It also supports linking between processes and thus supports the concept of process chains, which are themselves defined as processes. The use of SensorML is currently being evaluated.

4.2.5 Service Viewpoint

The computational viewpoint in the RM-ODP is replaced within the proposed architecture by the service viewpoint: It specifies the services that support the syntactical and semantic interoperability between the services, including the high-level operational services required by the GMES programme. Service oriented architectures like the one proposed shall place no restrictions on the granularity of a (Web) service that can be integrated. The grain size can range from small (for example a component that must be combined with others to create a complete business process) to large (for example an application). It is envisaged to support two main categories of services:

- Basic services are limited services running on the service providers' local infrastructure. Basic services may be requested (ordered) via the Portal's user interface, or from within a composite service (or workflow).
- Composite services are services consisting of a combination of basic services or other composite services. A service provider using the graphical workflow definition tools provided by SSE can model composite services. Composite services can comprise services provided by different service providers.

Another way of dividing services into categories relates to the specific functions performed by the service. The following set of specific EO data access services has been defined to specifically support the GMES Programme:

- Collection and service discovery;
- Catalogue Service;
- Product Programming and Order;
- Online Data Access;
- Satellite Multicast Service;
- Identity (user) management;
- Service Orchestration;
- Processing Services.

This Service Viewpoint defines these different services from a functional point of view. The interfaces with which all of these services are implemented form the subject of the Technology Viewpoint that is described below.

Collection and Service Discovery

An end-user typically uses collection discovery to locate dataset collections meeting the needs of his application domain e.g. urban planning, precision farming etc. The service discovery service then provides access to the services that operate on these dataset collections, e.g. catalogue, ordering, data access or programming services.

Catalogue Service

The catalogue service allows a user to find datasets or products within a discovered dataset collection that meet specific search criteria such as time, geographic extent, cloud cover, snow cover, polarisation etc. and gives access to all dataset metadata available in a catalogue. As explained within the information viewpoint, these product metadata vary depending on the type of mission: optical, radar or atmospheric.

Product Programming and Order

A user accesses the ordering service to order datasets referenced from within the (distributed) catalogue service. He can also order future products, not yet in the catalogue by using the programming service.

Online Data Access

Various on-line data access services provide access to ordered datasets via the Internet. Such services typically use the File Transfer Protocol (FTP) for allowing access to EO data, but also more advanced methods such as OGC Web Services for data delivery and visualisation are supported by the SSE architecture:

- Web Coverage Services (WCS) for access to EO datasets,
- Web Feature Services (WFS) for access to features information derived from EO imagery (e.g. land cover classification),
- Web Map Services (WMS) for visualisation and evaluation purposes.

In the future, WCS access to satellite imagery may be combined with JPEG2000 compression technology and emerging Geo Digital Rights Management (GeoDRM) approaches.

Satellite Multicast Service

The ESA Data Dissemination System (DDS) complements the in-orbit intersatellite link between ENVISAT and Artemis as shown in figure 4.5. This satellite multicast service is used for data circulation and transfer within the ESA Ground Segment. Its spare capacity may be used by the SSE, as shown in figure 4.6, to multicast SSE service results to end-users having limited terrestrial Internet capacity using Eutelsat's Ku band in Europe and C-band in Africa.

Identity (user) management

The current SSE supports access by the following types of users:

- Anonymous users can activate services for which the service provider does not require user information.

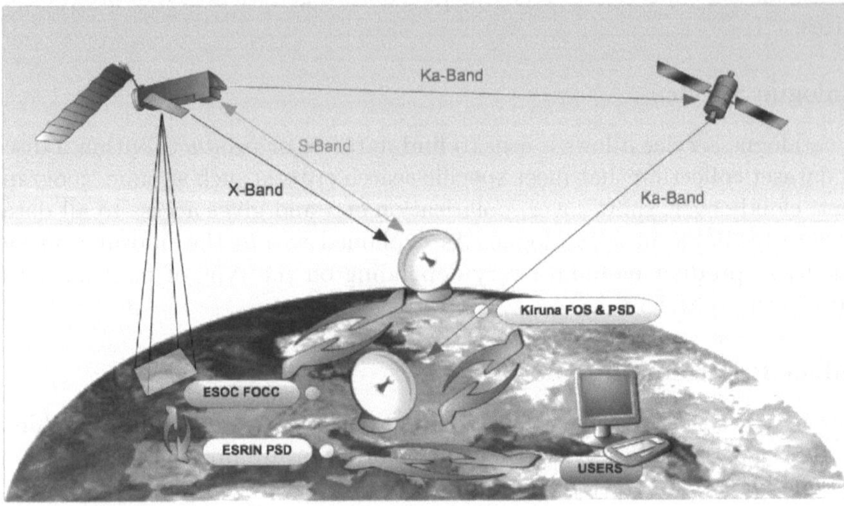

Fig. 4.5. Artemis satellite relaying ENVISAT data to ground

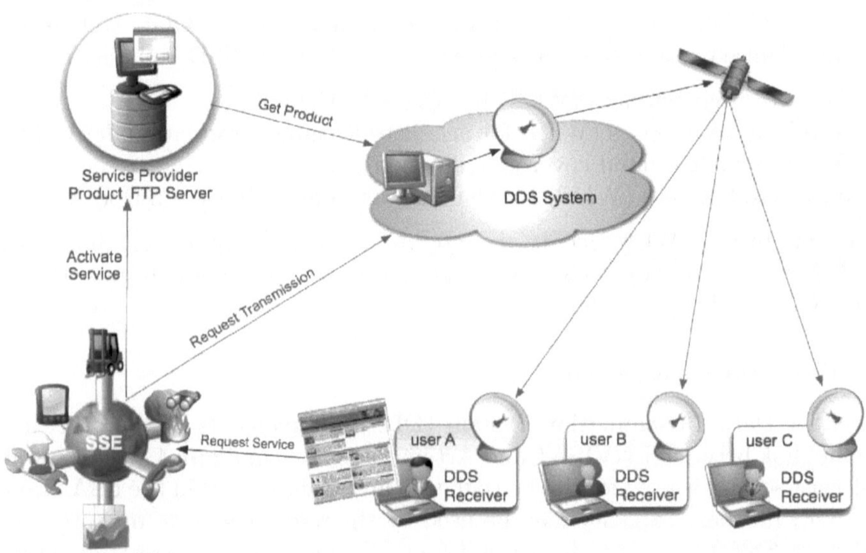

Fig. 4.6. DDS use via SSE

- Registered users can activate services for which the service provider has not restricted the access.
- Service providers can in addition publish services on the SSE Portal and deploy composite services implemented as workflows as well as monitor the

execution of these services via a console. A service provider can also specify his service to be access-restricted and thereby only allow specific registered users to access his service.
- The Administrator performs the administration of the system and can assign service provider rights to a user.

The next version of the SSE will allow for a single sign-on to the SSE by users registered in the various EO ground segments by providing a federated identity across participating ground segments. It will also support explicit definition of service access policies to the published Web services. A prototype of this is being prepared, which will rely on the OASIS Security Assertion Markup Language - SAML [18], OASIS ws-security [19] and possibly also the W3C ws-policy [20] specification. Internally, the user profiles will be stored in an LDAP registry.

Service Orchestration

Service orchestration, based on the OASIS Business Process Execution Language - BPEL [21], allows for composition of Web services, typically from multiple service providers or ground segments which themselves become then available as Web services. The orchestration engine is the service used to execute the resulting composite services. The resulting workflows can also contain interactive steps, e.g. to model a service provider's internal authorisation processes.

Processing Services

Extensive tests have been performed within the ESA Grid infrastructure leading to a preliminary definition of a Grid-based Processing Service. This with the objective of reducing the burden caused by the transfer of large EO coverages by transferring instead the processing algorithms on the Grid that hosts within its storage element the coverages to be processed.

The OGC Web Coordinate Transformation Service (WCTS) is another processing service that was experimented with, in combination with a Spot Image Web Coverage Service as described in the OGC OWS-3 Imagery Workflow Experiments [22] and OWS-4 EO Sensor Planning Service experiments [23].

A set of simple data manipulation services that perform generalisation, reprojection, format conversion and clipping on both EO imagery and vector datasets that may be derived from these are integrated as horizontal services into the SSE. These form basic services that are available to other service providers to allow product delivery that exactly meets the end users' format requirements.

4.2.6 Engineering Viewpoint

The engineering viewpoint specifies the mapping of the service specifications and information models to the chosen service and information infrastructure. The Service Support Environment (SSE) provides a reusable architecture for the integration of services in the Earth Observation (EO) and geospatial domains.

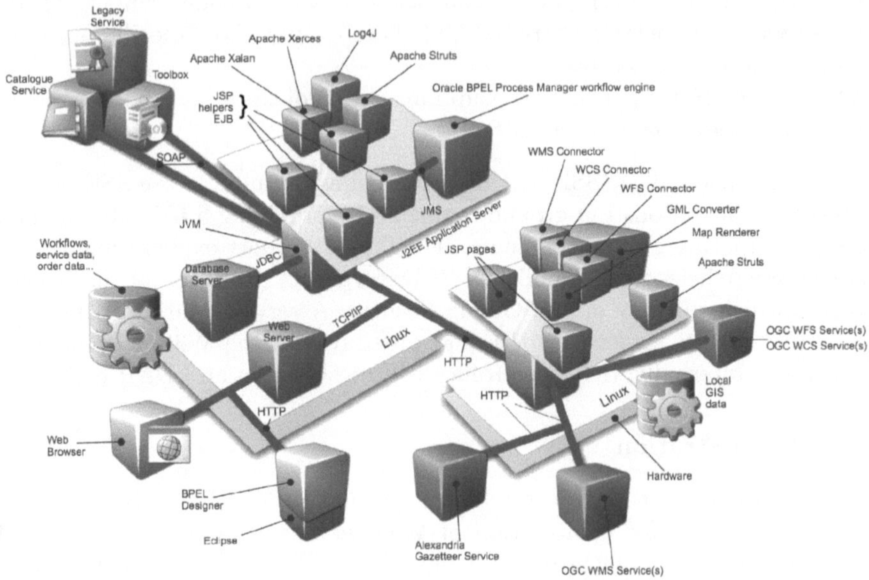

Fig. 4.7. Top-level view of the SSE Portal and its external interfaces

The figure 4.7 shows the top-level view of the SSE Portal and its external interfaces. The main entry point of this architecture is the SSE Internet Portal where users, EO and GIS service providers and EO data providers are integrated in a coherent supply chain.

The server design is based on a Java 2 Enterprise Edition - J2EE Application server. A J2EE-based architecture is a component-based architecture for enterprise applications, which is highly scalable and platform independent.

Java Server Pages - JSP technology is used to implement the presentation layer of the application. The Apache Struts framework that implements the Model-View-Controller design pattern was used for decoupling data access and business logic from data presentation and user interaction. This framework serves as a major input for the emerging JavaServer Faces - JSF specification which is currently being defined by Sun Microsystems.

The Java Message Service - JMS API is an asynchronous communication mechanism that the SSE business logic employs for communicating with the workflow engine. It would in principle allow for the workflow engine to be run on a dedicated server.

A Java Database Connectivity - JDBC interface between the SSE software and the RDBMS which implements the object persistency allows for easy replacement of the RDBMS component.

The WebMapViewer cartographic interface, used by end-users of services for delineating areas of interest as input into services and by service providers for displaying geo-referenced outputs of services, is implemented as a thin client based entirely on JSP/Struts and Dynamic HTML. This architecture does not depend on a Java Applet or ActiveX control and hence does not require the use of a specific Java Virtual Machine within the browser which makes this solution more performing and portable than the applet-based solution used during the early stages of the project. Seamlessly integrated within the SSE portal, this WebMapViewer allows the display of GML, read from files or delivered by Web Feature Services, layers from Web Map Services and raster data delivered by Web Coverage Services.

4.2.7 Technology Viewpoint and Standards

The technology viewpoint specifies the technological choices and the standards adopted for the Service Support Environment and the operational issues of the infrastructure. For example Basic Services are connected over the Internet as Web services, using the Simple Object Access Protocol - SOAP, the Web Service Description Language - WSDL and File Transfer Protocol - FTP for large data flows.

Open Standards

Definition and importance

Open Standards form the design basis for all internal and external interfaces of the Service Support Environment. The concept of Open Standards is defined by the European Interoperability Framework (EIF) developed within the IDABC community programme of the EC's Directorate-General for Informatics. IDABC stands for Interoperable Delivery of European eGovernment Services to public Administrations, Businesses and Citizens [24].

According to the EIF [24], an Open Standard must minimally adhere to the following criteria:

- The standard is adopted and will be maintained by a not-for-profit organisation, and its ongoing development occurs on the basis of an open decision-making procedure available to all interested parties (consensus or majority decision etc.).
- The standard has been published and the standard specification document is available either freely or at a nominal charge. It must be permissible to all to copy, distribute and use it for no fee or at a nominal fee.
- The intellectual property - i.e. patents possibly present - of (parts of) the standard is made irrevocably available on a royalty-free basis.
- There are no constraints on the re-use of the standard.

The prime objective of the use of Open Standards is interoperability. The EIF [24] formulates interoperability as "the ability of information and communication technology (ICT) systems and of the business processes they support to exchange data and to enable the sharing of information and knowledge". In other words, interoperability is the fact of letting different systems developed by different suppliers co-operate together by employing common definitions, information models, encodings and interfaces.

The use of Open Standards tends to lead to flexible modular solutions instead of vertical stovepipe systems as the functions and interfaces between different components are typically well described in Open Standards based designs.

As a consequence of the use of Open Standards, one also reduces vendor lock-in. As one is not dependent any longer on the complete solution of one particular vendor, one becomes more flexible in the choice of a particular component. Because of these advantages, it is expected that the use of Open Standards will reduce the cost of systems developments.

Major governmental initiatives foster the growth of Open Standards and legislation is coming in to place that force the use of Open Standards. The INSPIRE initiative of the European Commission for establishing an Infrastructure for Spatial Information in the European Community is only one of these. Others include the already referenced EIF that states in its recommendations that Open Standards is one of main principles to be considered for any e-Government services to be set up at a pan-European level.

Standardisation organisations

The Service Support Environment is built upon the standards produced by the following organisations:

- W3C: The mission of the World Wide Web Consortium - W3C is "to lead the World Wide Web to its full potential by developing protocols and guidelines that ensure long-term growth for the Web." W3C primarily pursues its mission through the creation of Web standards and guidelines. Since 1994, W3C has published more than ninety such standards, called W3C Recommendations. The W3C standards form the core of the World Wide Web: HTTP, XML, SOAP, WSDL, XSL, ... are only a few of the most well-known standards.
- OASIS: The Organisation for the Advancement of Structured Information Standards is a not-for-profit, international consortium that drives the development, convergence, and adoption of e-business standards. The consortium produces more Web services standards than any other organisation along with standards for security, e-business, and standardisation efforts in the public sector and for application-specific markets. Founded in 1993, OASIS has more than 5.000 participants representing over 600 organisations and individual members in 100 countries. The OASIS Standards are mainly Web Service related standards as ebXML, UDDI, SAML, BPEL, WS-Security, WS-Notification.
- ISO: The International Standards Organisation of which the Technical Committee 211 on Geographic Information/Geomatics deals with standardisation in the field of digital geographic information. Members of ISO are the national standardisation organisms.
- OGC: The Open GeoSpatial Consortium is an international industry consortium of more than 300 companies, government agencies and universities participating in a consensus process to develop publicly available interface

specifications to support interoperable solutions that "geo-enable" the Web, wireless and location-based services, and mainstream IT.
- IETF: The Internet Engineering Task Force is an open international community of network designers, operators, vendors, and researchers concerned with the evolution of the Internet architecture and the smooth operation of the Internet. IETF has standardised numerous Internet protocols such as HTTP, LDAP, etc.
- CEN: the European Committee for Standardisation that "contributes to the objectives of the European Union and European Economic Area with voluntary technical standards which promote free trade, the safety of workers and consumers, interoperability of networks, environmental protection, exploitation of research and development programmes, and public procurement".

Service Discovery

The OGC CS-W 2.0.1 ISO Metadata Application Profile [25] will be used for service discovery. This application profile of the Catalog Services for the Web (CS-W) specification provides ISO 19139 compatible XML Schema for Catalog Search Result sets for ISO 19115 (data) and ISO 19119 (service) metadata. In addition, it defines the available operations and the supported query criteria (queryables) that catalog clients can use for interacting with the catalog service. It enables users to discover and retrieve metadata about available Earth Observation related services hereby providing sufficient information to assess the usefulness of a service for the task at hand. Furthermore, it offers the binding information for accessing this service (e.g. HTTP binding together with a hyperlink).

In addition, the SSE architecture supports service discovery via Universal Description, Discovery and Integration (UDDI) [26] from OASIS with the advantage of allowing mainstream IT tools from large vendors (Oracle, IBM, Microsoft) to discover the exposed services.

We envisage in the near future to also support a service catalogue based on the CS-W ebRIM Application Profile [27] and use the Cataloguing of ISO Metadata - CIM extension package [28].

Collection Discovery Service

As for the discovery of services, the CS-W ISO Application Profile [25] is used for the discovery of EO product collections of which the metadata is encoded within an ISO 19115 and ISO 19139 based model and encoding. This offers the following advantages:

- The possibility to discover EO Collections and Services using the same discovery protocol facilitates the implementation of the common use case of discovering services that operate on particular EO Product collections.
- The use of a pre-existing standard, allows the utilisation of existing Commercial Of The Shelf and Open Source Software tools as the "GeoNetwork

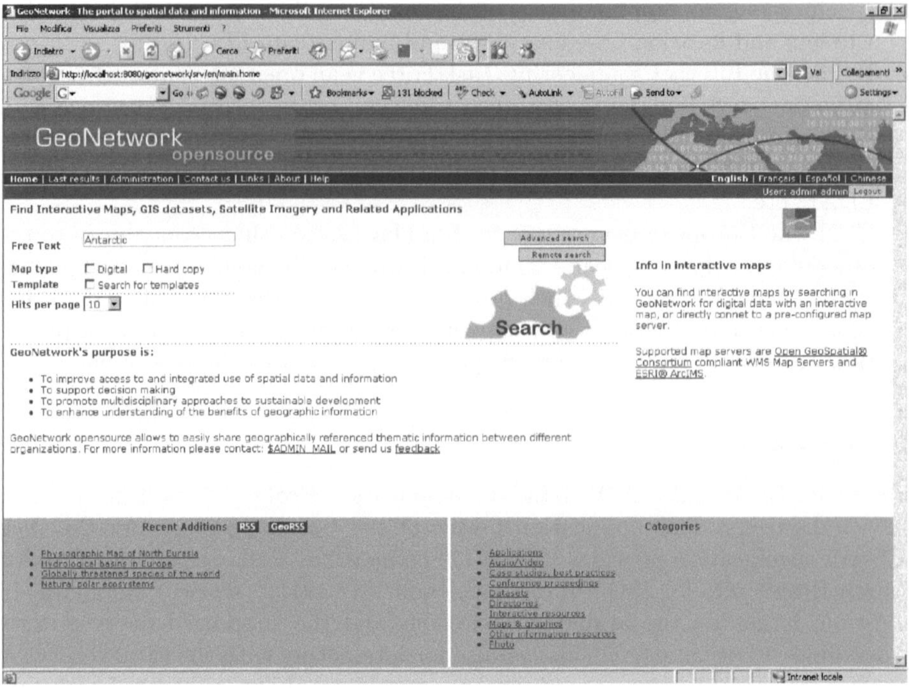

Fig. 4.8. GeoNetwork site

Opensource" catalogue solution of the Food and Agriculture Organisation of the United Nations - FAO for storing and discovering dataset collection metadata as well as the associated services (see figure 4.8).

The fact that an existing standard of the Geospatial Information - GI domain is used for EO collection discovery permits interoperability with Catalogue Services and Catalogue Clients from within the GI domain. As an example, a service is currently being integrated within the SSE that allows the discovery of dataset collections from the National Geographic Institute of Belgium, the dataset collections of a commercial GIS data distributor and the ESA Product collections. We envisage in the near future to also support a collection catalogue based on the CS-W ebRIM Application Profile [27] and use the CIM extension package [28] once open-source software supporting it becomes available.

Catalogue Service

For the "dataset" (product) catalogue service the CS-W Application Profile for Earth Observation [29] was initially proposed to take into account the EO specific product metadata.

Since then, we have migrated to the ebRIM Application Profile of CS-W [27] as this allows using an application profile that is also in use outside the EO domain. This is in-line with the recent OGC decision to recommend the use of the ebRIM model as information model for future catalogue specifications. The ebRIM information model for EO data is defined in the EO extension package for ebRIM Profile of CSW2.0 [30].

In both cases, the metadata are specified as a GML application schema defined within OGC 06-080 and described within the Information Viewpoint [12].

Programming and Ordering Services

The SSE provides interfaces to access programming and ordering services for EO products. The programming interface constitutes an EO profile of the OGC Sensor Planning Service (SPS) [31]. The ordering service interfaces are defined in the Ordering Services for Earth Observation Products paper [32].

Online Data Access

The following set of geospatial services defined by the OGC will be used for user interaction and/or data access and information visualisation.

Web Map Service - WMS

This OGC implementation specification (and ISO standard) [33] defines the interactions between client and server applications for the delivery of maps, rendered geographic information, over HTTP. The usefulness of using the Web Map Service specification is that it allows the seamless overlay of map layers that are derived from a number of different servers operating different software components using the same protocol. Within the SSE architecture, this standard is employed for both supplying background maps for the cartographic display and for visualising outputs of services. Of specific Interest for Remote Sensing applications is the recently published EO Profile of WMS [34] that tightly defines how Web Map Services are to be used for EO Products. It specifies amongst other elements how the high level Earth Observation information model consisting of collections, products and image bands/derived products should be mapped on WMS concepts as Layers and Time and Sample Dimensions. In summary, each product collection is offered as a WMS layer where the Time dimension parameter is used to distinguish individual products and the sample dimensions are used for specifying the different wavelengths or the different derived parameters. This EO Profile of WMS is specifically useful to allow the evaluation of EO Products with respect to, for instance, cloud cover after the discovery of the products metadata within a Catalog but before the actual ordering of the product.

Web Feature Service - WFS

WFS is another OGC implementation specification (and draft ISO standard) that deals with the remote querying and updating of geographical data stores over the Web [35]. It is employed within the SSE as the protocol that allows end-users to query geographical referenced service results. For example, the results of an EO imagery processing service that determines geographic features (e.g. bush fires, oil spills, land cover objects, ...) from satellite images can be stored inside a WFS data store and then interactively interrogated by end users of these services. Features are delivered to the client in GML and optionally alternative formats. The geometry information contained within this GML can be used to present the features on a map display and the attribute information can be easily converted to HTML by the use of XSLT style sheets. Transactional Web Feature Services are being evaluated as a mean for capturing geographic feature edits. Example use cases are the storage of specific EO derived features into remote databases or the delineation of areas of interest within SSE. Also the use of the gazetteer service profile of this specification is being investigated to allow the remote interrogation of gazetteers, geographical dictionaries of place names and their corresponding coordinates.

Web Coverage Service - WCS

WCS is the OGC implementation specification [36] for the delivery of coverages, in casu raster data files, across the Internet. It is used as the interface by which clients can download and visualise EO products. This WCS can be considered as an alternative to the current way of working with data transfer of EO Products via FTP or HTTP. The advantage of employing WCS is the fact that by using Open Standard interfaces for online data access the transfer of (references to) data inside services chains is facilitated. In addition, the WCS standard offers the facility to specify formatting information as supported image formats, spatial reference systems, image bands, spatial and temporal extent and pixel dimensions as part of the data access request. Out of the box software solutions exist in both commercial off the shelf and Open Source Software worlds that do offer such formatting operations hereby combining on-line data access with basic processing functionality. The use of Web Coverage services in conjunction with the GML in JPEG2000 for Geographic imagery encoding standard - GMLJP2 [37] is being investigated as the format for EO product delivery via WCS where the JPEG2000 compression technology is augmented with GML based encoding of the imagery geocoding information, product metadata, image annotations and features extracted from the images.

Sensor Observation Services - SOS

Within OGC's Sensor Web enablement initiative, the Sensor Observation Service standard is being developed that allows interoperable access to sensor measurements over HTTP [38]. Sensors can be described in SensorML and sensor measurements can be encoded within the Observation and Measurements application profile of GML. The SSE is currently being extended to allow access to time series of in-situ measurements at discrete point locations via SOS.

Service Orchestration

The workflow engine is the component executing the workflows within a Service-Oriented Architecture. It executes business processes based on the Business Process Execution Language for Web Services - BPEL standard. When the Portal goes down, and comes back up, active workflow instances are resumed automatically. The workflow engine utilises the RDBMS (Oracle 9i) for workflow persistency. The workflow engine is implemented with the Oracle BPEL Process Manager.

The service orchestration is implemented using BPEL for most of the services which are part of SSE. It supports both synchronous and asynchronous services. Possibly the most interesting example of a synchronous service is the product cross-catalogue search service. This service (initially implementing [39], subsequently implemented using the EO Profile [29] and finally to be ported to ebRIM [30]) allows simultaneous access to the ESA multi-mission product catalogue and the catalogue of Spot Image. The BPEL process that implements this service accesses both catalogues in parallel and returns the combined search results to the Portal. Other applications of service orchestration in the HMA project implement access via composite services to ordering and programming services for multiple satellite ground segments.

SSE provides a workflow management console that can be used by each service provider to monitor execution of his workflows. This tool is Web-based and integrated in the SSE Portal. It gives the service provider a complete overview of currently active and completed workflows and it allows access to all input and output messages. In addition, it supports the service provider in deploying new workflows that he has prepared off-line within the workflow editor. Service providers prepare the BPEL service workflows with a workflow editor which is available free of charge. A BPEL Designer based on Eclipse can be downloaded from the SSE Portal, but other editors such as Oracle JDeveloper can be used as well.

Internal interfaces

The SSE Interface Control Document - ICD provides an in depth description of the application of the following protocols and standards and their use within the SSE architecture.

- The user connects to the SSE Portal via a Web browser. Both HTTP [40] and HTTPS [41] (using certificates) are supported. The user registration for instance is done via HTTPS to keep the user information confidential. FTP [42] is used by the user in case he has to download or upload large amounts of data as input or output of a service. These 3 specifications are defined by the Internet Engineering Task Force (IETF).
- The Business Process Execution Language (BPEL) [21] has been standardised by OASIS. It provides a vendor independent format to orchestrate Web

services. The workflow tools, i.e. workflow editor, workflow engine and workflow monitoring console are all built around this open standard. Details on how BPEL is being used in the SSE are provided by [43].

- Message-based SOAP (Simple Object Access Protocol [44]) over HTTP or HTTPS for secure communication is used as protocol between the SSE Portal and the service providers. In some cases, this may be simplified and FTP or (structured) e-mail communication may be used instead. SOAP is firewall-friendly, and platform independent. It is thus well-suited to integrate services in a heterogeneous environment.
- WSDL (Web Services Description Language 1.1 [45]) is used to define the SOAP interface of a service in a formal way that can be processed by the workflow editor. It is equivalent to CORBA IDL.
- XML Schema [46, 47, 48] is used by the service provider to define the structure of the RFQ and Order input and output XML messages. The SSE Portal provides a predefined XML schema. However, some services may have service specific input/output parameters that the service provider can then model by providing his custom XML schema. When registering his service on the Portal, the service provider provides the URL of the XML schema and of the WSDL file, which are then uploaded by the SSE Portal. The SSE ICD provides a more detailed description of the use of XML Schema.
- XSLT [49] is used by the SSE Portal to render the request for quotation (RFQ) and service outputs (expressed in XML and GML) into formatted HTML or XHTML [50] rendered with Cascading Style Sheets - CSS [51] that is displayed on the SSE Portal. XSLT is also used to render the data entry forms.
- XPath [52] defines an "expression" notation allowing to extract an element from an XML Document Object Model - DOM object. It is used to extract information from XML messages inside the workflows.
- OGC Geography Markup Language (GML) [11] is used where needed to encode geolocated feature information. GML defines the base data model and encoding for working with geographic feature data. For use within the SSE Portal specific GML application schema deriving from the base GML schema files were defined for both area of interest and service result features. GML is also the default output format for Web Feature Services. A subset of the GML3.1.1 standard provided by the Simple Features Profile of GML [53] in its "Level 0 compliancy level" is supported by SSE's WFS client. Furthermore the metadata of EO products is modelled as an application profile of this GML3.1.1 standard as described above.
- OGC Styled Layer Descriptor Language (SLD) [54] is used to determine the symbols with which geographic features encoded in GML are portrayed on the SSE WebMapViewer.
- OGC Web Map Context (WMC) [55] is the XML format and encoding for storing the state of the WebMapViewer. This Web Map Context document holds the service provider's map configuration being the Spatial Reference

System, the spatial extent, the list of WMS layers, FeatureTypes and Coverages together with their associated connection information. Service providers can interactively configure the map that is offered to their users and this map information is encoded within the WMC format. Likewise, end users can export this map composition in this format and can later on re-import this.
- OGC Filter Encoding Specification [56] is the query language that is used for interrogating OGC Catalogs and Web Features Services.
- Ws-inspection [57] is used to allow for service discovery. The workflow editor can discover the services and associated XML schemas which are deployed on the SSE via a WSDL file published on the SSE Portal.
- The SSE Portal contains an optional UDDI v2 [26] registry which allows for discovery of registered services using any UDDI client.
- WS-Addressing [58] is used to correlate asynchronous message exchanges in a standard way instead of an ad-hoc content-based mechanism.
- All SSE geographic service results can also be viewed with the Google Earth client. SSE automatically transforms its GML Application schema to Google's Keyhole Markup Language (KML) to publish for instance satellite imagery footprints resulting from a catalogue search together with its browse images provided via a WMS interface into a Google Earth compliant format.
- The various application profiles of OGC CSW are used to connect to product and collection catalogues as explained before.
- WS-Security [19] and SAML [18], both from OASIS, will be used in the new version of the SSE which will allow for a federated identity provision across various EO Ground Segments.
- The on-line help system is provided as an Eclipse Infocenter plugin which uses an XML notation to define the structure of the HTML documentation.
- News are published in RDF Site Summary (RSS) format.

4.2.8 Examples of SSE Services

SSE provides access to numerous distributed services based on existing or new applications. These services are related to environment, science, decision making (e.g.: before, during and after a crisis), public or commercial activities, like:

- Image processing: geocoding, ortho-correction, histogram stretch, classification;
- Data conversion: image reprojection, generalisation, raster and vector format conversions, clipping, feature extraction;
- Data archives: e.g. Spot Vegetation S10 archive, Landsat, IKONOS VHRSI coverage over Flanders;
- EO product and service catalogues: ESA Multi-Mission, Spot Image Dali/ Sirius, RadarSat, Quickbird, KompSat, NASA ESG, NEODC.

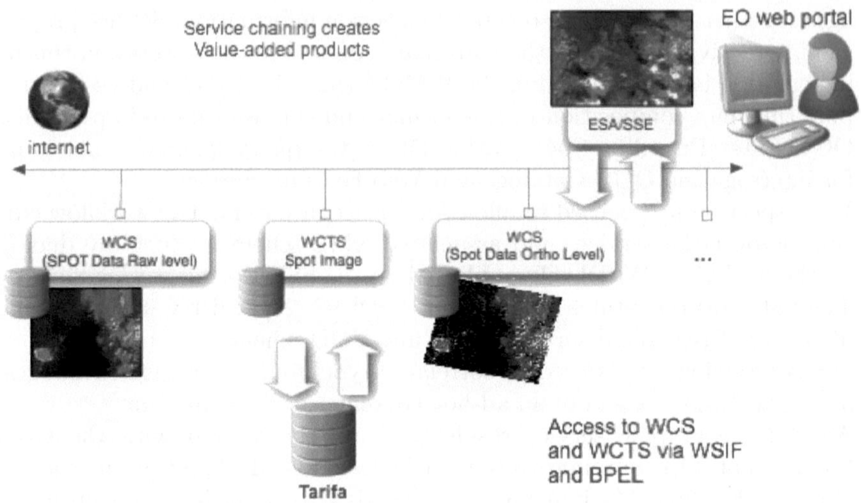

Fig. 4.9. Chaining services

- Coastal and sea monitoring: oil spill detection, ocean colour maps, tidal currents etc.;
- Air pollution: ozone service;
- Desertification: apparent green cover;
- Flood monitoring: flood mapping, snow cover mapping, shoreline extraction;
- Forest monitoring: burnt areas, dry matter productivity, bush fire detection, forest fire risk maps and carbon absorption maps;
- Oil, gas and mineral exploitation: offshore basin screening;
- Urban monitoring: sealed area maps, green area detection;
- GIS data services: geodata catalogues, topographical maps subscription service, geocoding of addresses and reverse geocoding of coordinates etc;
- Simulation models;
- In-situ data access services.

In addition, these "services" can be combined by creating "chains" of services (see figure 4.9), or by overlaying service results with layers coming from other sources (see figure 4.10).

A European SSE Portal (see http://services.eoportal.org) is installed at ESA-ESRIN (Frascati, Italy). It is operated by the Earth Observation Ground Segment Department. Regional SSE Portals are similar to the Global Portal, but integrate services covering a specific region or country, and may support the local language. The regional Portals offer a localised user interface. The SSE Portal at the Centre de Suivi Ecologique (CSE) in Dakar (Senegal) is an example of a regional Portal (see http://cse.services.sn).

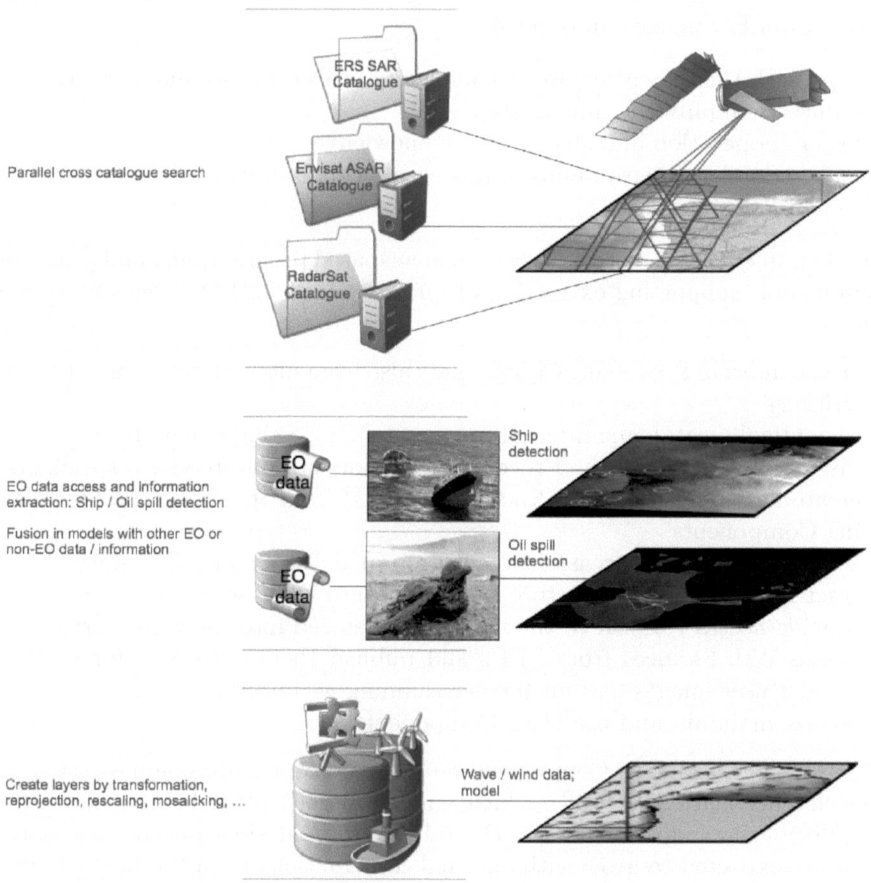

Fig. 4.10. Overlaying results from various services

4.3 KEO: A Component-Based Programming Environmet for Earth Observation

4.3.1 The Project

Image Information Mining (IIM) is one of the elements in the direction of empowering the user (e.g.: researcher, service provider, decision maker) with the capability to identify relevant information from EO data by interactive or batch processing of images. Examples of goals to be achieved in the long term via intuitive interfaces can be the identification of desertification trends, the forecast on iceberg routes, the global changes in chlorophyll concentrations. As a step in the above direction, ESA started the project "Knowledge-centred Earth

Observation" (KEO), aimed at supporting more automated extraction of information from EO images, in order to:

- cope with the increasing size / complexity of EO images and archives
- reduce / simplify the mining steps
- foster cooperation and share / reuse knowledge
- make complex programming / processing capabilities easily accessible also to non EO experts

KEO implements a distributed component-based programming and processing environment, supporting extraction of information from EO images by permitting to:

- create internal Processing Components also from interactive training (no code writing)
- semantically catalogue internal or external Processing Components
- discover semantically the Processing Components suitable for a specific task
- create Feature Extraction Processors (FEPs) from one or a chain of Processing Components
- start systematic or occasional execution of FEPs against EO images to extract information and to store the results into OGC servers
- provide access via SSE to the information stored into the OGC servers
- create Web Services from FEPs and publish them onto SSE for their use against user images and for further chaining within SSE
- create, maintain and use Data Components

Within KEO, a Data Component can be part of a more complex structure, the Reference Data Set (RDS), which can contain also other heterogeneous data and information (e.g.: documents, Digital Elevation Models, photos, processors). RDSs are expected to grow with external contributions from the partners using them (e.g.: new algorithms, validation data, reports). KEO makes use of many of the approaches and standards presented before. Therefore they are not described again, but only mentioned when relevant.

4.3.2 The Vision

The source requirement for KEO was that *The information extraction process must be replaced by more automated, direct and human centred methods.*

- "Automated" is a consequence of the data size / complexity and of the efficiency / cost of the non-automatic process.
- "Direct" responds to the need to reduce the steps between the user and the information, since currently the user expresses his needs to a chain of one or more experts, who possibly get in touch with experts in other domains to collaborate in the extraction of the information.
- "Human centred" brings the focus on systems that could be managed also by non-EO experts via semantic interactions, like those between human beings.

Extraction of information from images is one of the elements in the direction of empowering the user (researcher, service provider, end user, etc.) with the capability to easily identify relevant data (even from large, distributed and multi-domain sources) and to request, via advanced interfaces, their further processing to extract information. Examples can be the identification of desertification trends, the forecast on iceberg routes, the global changes in chlorophyll concentrations, etc.

In summary, the information extraction process should, in the long term, rely on a very powerful (knowledge-based and learning) and extremely easy (supporting semantic interactions based on user domain knowledge) component-based programming environment.

Within this vision, KEO should permit to an expert to identify available components suitable for his application, to create new components mainly aimed at the extraction of information from images, to chain all these components and to create services from chained processing execution or extracted data.

Automated / Direct Information Extraction

Extraction of information from satellite images in KEO relies on two types of methods:

- Feature Extraction Algorithms (FEAs) [59];
- Probabilistic Information Mining (PIM) [60, 61].

The first type includes methods that are specific for each feature (usually involving the analysis of pixels, looking for a predefined pattern) and therefore more robust and precise. These methods can provide fully or partially automatic information extraction (partially when the user must intervene to define parameters or take decisions).

The second type is more generic. It works by extracting and storing the basic characteristics of image pixels and areas during the ingestion of a set of images. The user then trains the system by selecting the representative basic characteristics from sample images. When satisfied with the training, the user requests to apply it to the entire image set in order to identify all the images containing the searched characteristics. This method has the following advantages:

- no need to re-process the entire image set when searching for new features (if compatible with ingestion parameters);
- the selected feature can be closer to the user expectations and perceptions (the same feature can have different meanings to different users: e.g. a forest for an environmentalist, a forest guard, a geologist, a urban planner, ...);
- the system can learn from experts and make this knowledge available to other users.

The second type permits a "direct" interaction, closer to user perception, and does not require, during use, specific knowledge of the EO domain.

In summary, information extraction, automated or direct, should permit or encompass:

- interactive exploration of single images: Scene Understanding;
- interactive exploration of entire image collections: Information Discovery;
- identification of relevant images also from their information content (and not only for their position in space and time): Content Based Image Retrieval;
- systematic processing of images to automatically extract features and make them available as maps or GIS objects: Feature Extraction.

Probabilistic Information Mining is provided by the Knowledge-based Information Mining (KIM) tool [62]. Related test systems are available on Internet, at the time of writing, at: http://kes.acsys.it/kes or http://kes.esrin.esa.int/kes.

Human Centred

A human centric system must support semantic interactions. This requires the coding of specific data, information and knowledge. The coding must also support computer processes like automatic checking or reasoning.

Users

In KEO, two types of external users are addressed:

- **Remote Sensing Expert**, with EO, signal and image processing know-how, using KEO to create FEPs for new information mining capabilities.
- **Domain Expert**, not necessarily an EO, using KEO to interactively explore a collection of images or to run an available FEP against own images.

Ontology

Ontology implies knowledge encoding, through the definition of concepts and their relations. In order to make easier the use of EO data, the KEO ontology must include EO domain specific concepts, but also concepts linking the EO domain with other domains, in order to permit non-EO experts to identify useful KEO Data Components, Processing Components or FEPs [63, 64]. Its implementation is based in the use of Semantic-Web XML technologies for knowledge representation, transformation and visualisation.

Semantic / Terminology

The implementation of the ontology (concepts and relations) using semantic terms, simplifies user interaction. In order to easy the access to KEO components by non-EO domain experts, it is necessary to add terms which link EO and non-EO terminologies. For example, the system should permit to identify "Processing Components for oil spills" as well as related source products. This requires a higher level of abstraction, where the "oil spill" term must be linked to a Processing Component capable to operate on a "SAR product". The terminology must be consistently implemented across all Processing and Data Components using a shared vocabulary.

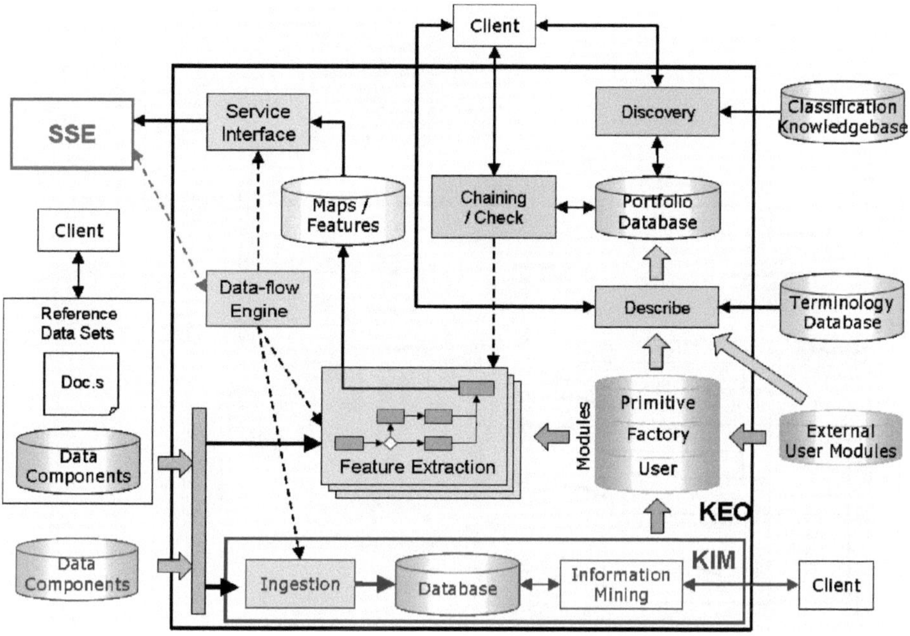

Fig. 4.11. KEO concept

Easiness of Use

Dataflow Visual Programming is the paradigm behind the KEO easiness of use. A Domain or Remote Sensing Expert must not have complex programming skills. Therefore the KAOS Client allows to graphically design FEPs in an easy way. It permits to place on a sketchpad the components (available from semantically searchable / browsable palette) and to interconnect them by simply drawing lines between them. The system also automatically checks the consistency of the dataflow using the semantic information associated to each component.

4.3.3 KEO Elements

The KEO system is a complex SOAP, J2EE and OGC based system with complexity hidden to the users thanks to its KAOS Client. This is a stand-alone, Java Integrated Development Environment permitting the Administrator to manage the system and the users to act on the key elements of KEO, which are(see figure 4.11):

- Knowledge-based Information Mining (KIM);
- Feature Extraction with associated engine and semantics;
- Reference Data Sets.

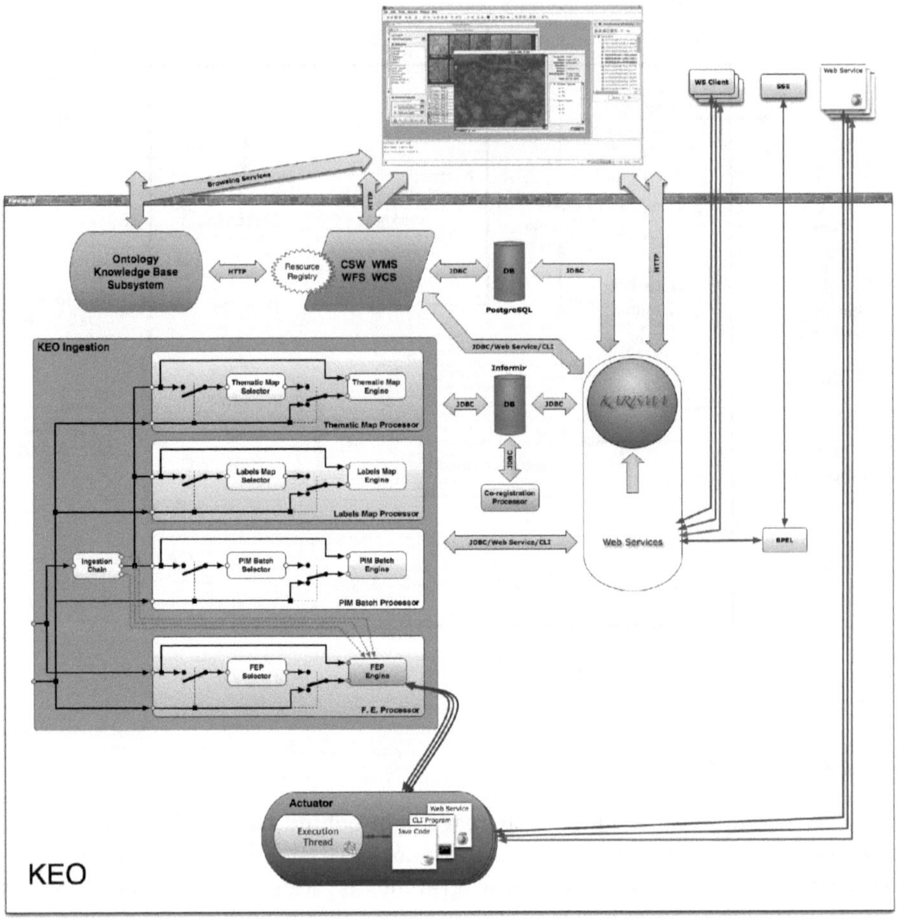

Fig. 4.12. KEO system overview

The KAOS client interacts with the various parts of the system in order to permit to browse Reference Data Sets and Data Components, to define Feature Labels through KIM, to create Processing Components and to chain them into FEPs for feature extraction. All these activities are supported by a semantic knowledge / data base subsystem simplifying user interactions and permitting discovery of available modules as well as automatic chaining verification by the system.

The Feature Extraction Processors (FEPs), set-up through the KAOS client to extract features, are executed when new images are available in the KEO input. In this phase the Processing Components of the FEPs are distributed over the available processing nodes (computers) depending on the ingestion chain policies and on the availability of FEP actuators. The extracted features are then stored into OGC servers (WMS [33], WFS [35], WCS [37] and CS-W [25, 27, 29, 39]).

The figure 4.12 provides an overview of the system and shows the main standards and protocols used.

Knowledge-Based Information Mining

The Knowledge-based Information Mining (KIM) tool is based on J2EE technology and is open for integration into Service Oriented Architectures. Kim has two main parts:

- Image Ingestion
- Interactive Information Mining

Image Ingestion

Image Ingestion permits the Administrator to:

- define image collections (and associated characteristics, models, procedures / parameters, databases, etc.);
- manually or automatically start ingestion of images from off-line media, archives or SSE (i.e.: user provided images) into the collections' databases (preparing them for interactive information mining);
- monitor and control the ingestion process and results.

The collection definition process affects the type of information (mainly linked to the feature's size) that can be detected in the images, depending on image and ingestion resolution (lower resolutions reduce storage requirements but increase number of images that can be handled as well the minimum size of identifiable objects). Therefore the method can be applied against increasing image resolutions (and therefore to a decreasing number of images in order to keep internal databases within reasonable sizes) for Content Based Image Retrieval, Information Discovery or Scene Understanding.

Interactive Information Mining

Interactive Information Mining (see figure 4.13) permits authorised users to:

- train the system to explore image collections for specific features;
- obtain relevant image identifiers or feature maps / objects;
- store the trained Feature Label for reuse;
- search available Feature Labels defined by other users.

After having selected a collection and the associated models, the user trains the system on the searched feature by pointing on pixels of system-provided sample images (small but of high quality), looping until satisfied. At each iteration the training is applied to the entire collection and potentially relevant images are presented to the user. For satisfactory images, the user can obtain directly from the client the image identifiers as well as feature maps or feature objects in GIS formats. The search and information extraction are supported as appropriate also by segmentation and geometry modules, to be activated at ingestion time.

The training result can be associated to a semantic term and saved as Feature Label for future private, group or public use, as specified by the owner (public

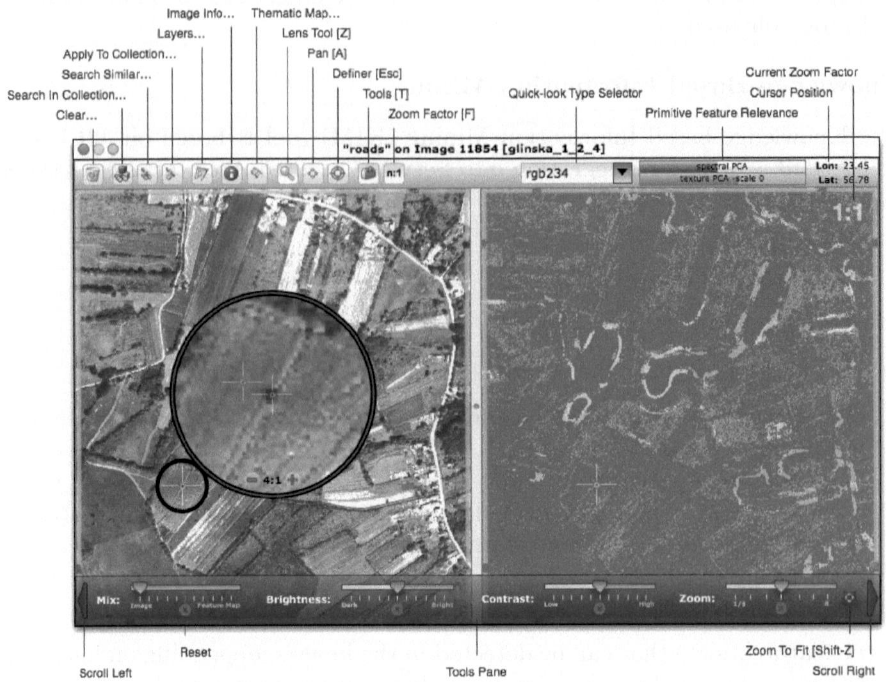

Fig. 4.13. KIM user interface example

use must be authorised by the Administrator). The Feature Labels for public use must be semantically described in the local Portfolio Database and linked at least to a Service Category for discovery and navigation from relevant concepts (e.g.: to identify Feature Labels relevant for pollution).

KEO Extensions

It is possible to wrap some Feature Labels as Web Services (creating batch processors from KIM training) and publish them for restricted or public use:

- locally, for use as user modules (Processing Components);
- into SSE, for use as SSE services.

When trying to identify available (public) Feature Labels, experts from different domains are supported by discovery and navigation tools, which operate on the centralised Classification Knowledgebase and on the local Portfolio Database:

- during discovery (search), the expert enters semantic terms of his domain to identify relevant products or services: the system applies the provided terms first to the Classification Knowledgebase and then to the local Portfolio Database and presents the results in relevancy order;

- during navigation, the expert can explore, via an intuitive and graphical interface, the combined path of the Classification Knowledgebase and the Portfolio Database.

KIM Values

KIM represents a novel and unique theoretical concept and frame of collaborative methods for:

- the extraction and exploration of the content of image collections or of other multidimensional signals;
- supporting image interpretation;
- establishing the link between user needs, knowledge and information content of images;
- better exploration and understanding of Earth structures and processes;
- communicating at high semantic abstraction between heterogeneous sources of information and users with a very broad range of interest;
- increasing the accessibility and utility of EO data.

Feature Extraction

The KEO system architecture allows the execution of Feature Extraction Processors (FEPs) that extract features from images and store them as maps or objects into OGC servers (WMS, WFS, WCS and CS-W). These FEPs are described by dataflow graphs (see figure 4.14) as chains of Processing Components, each one triggered by the availability of data at its input gates (called inlets), coming from other components' outlets or from the KEO input portlets.

The Processing Components used to build a FEP can be:

- *primitive modules*, performing simple operations on simple data type values (i.e. adding two integer numbers, regular expression processing of strings);
- *factory modules*, acting on complex data type values (i.e. image format conversion, image warping into a new coordinate system);
- *user modules*, performing custom feature extraction algorithms on any type of data;
- other FEPs (available combinations of previous types).

FEP execution, including debugging and monitoring support, is the responsibility of the Feature Extraction Processor Engine subsystem (see figure 4.15). The FEP Engine is a data-flow engine that executes Feature Extraction Processors and coordinates the activities of their Processing Components distributed over a Wide Area Network (WAN).

FEP Actuators (see figure 4.16) make platform-dependent modules (Java coded, CLI programs, Web Services) available to the FEP Engine. Three threads compose an actuator:

- the *Data Listener*, whose purpose is to continuously listen for inlets data coming from the source data provider;

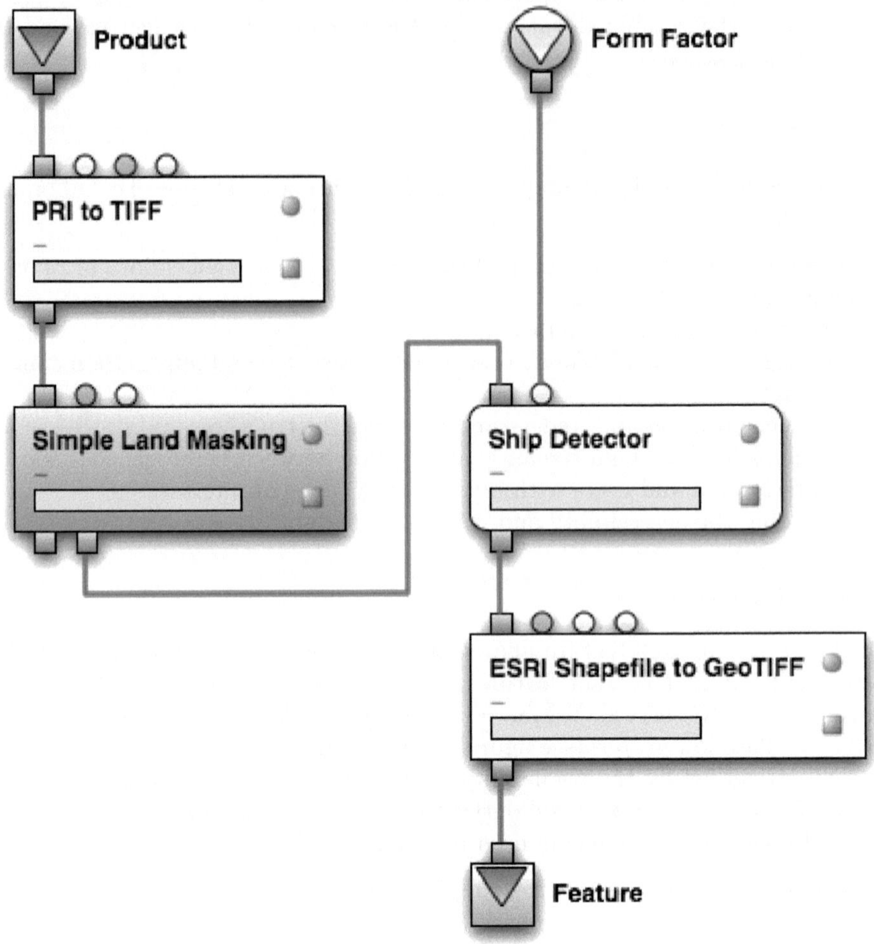

Fig. 4.14. KEO data flow graph

- the *Execution thread*, that executes the module's processing algorithm (invoking corresponding Java code or Web Service function, or starting a CLI program): it is normally sleeping and is waken up when all inlets have compatible data;
- the *Data Producer*, that notifies registered listener when outlet's data is produced: it is normally sleeping and is waken up as soon as new data is produced in any module's outlet.

Actuators have a local database containing the list of modules that can be executed. The modules of a FEP are ingested into the FEP Engine and loaded in all the actuators compatible with module type (Java coded, CLI programs, Web Services) and platform.

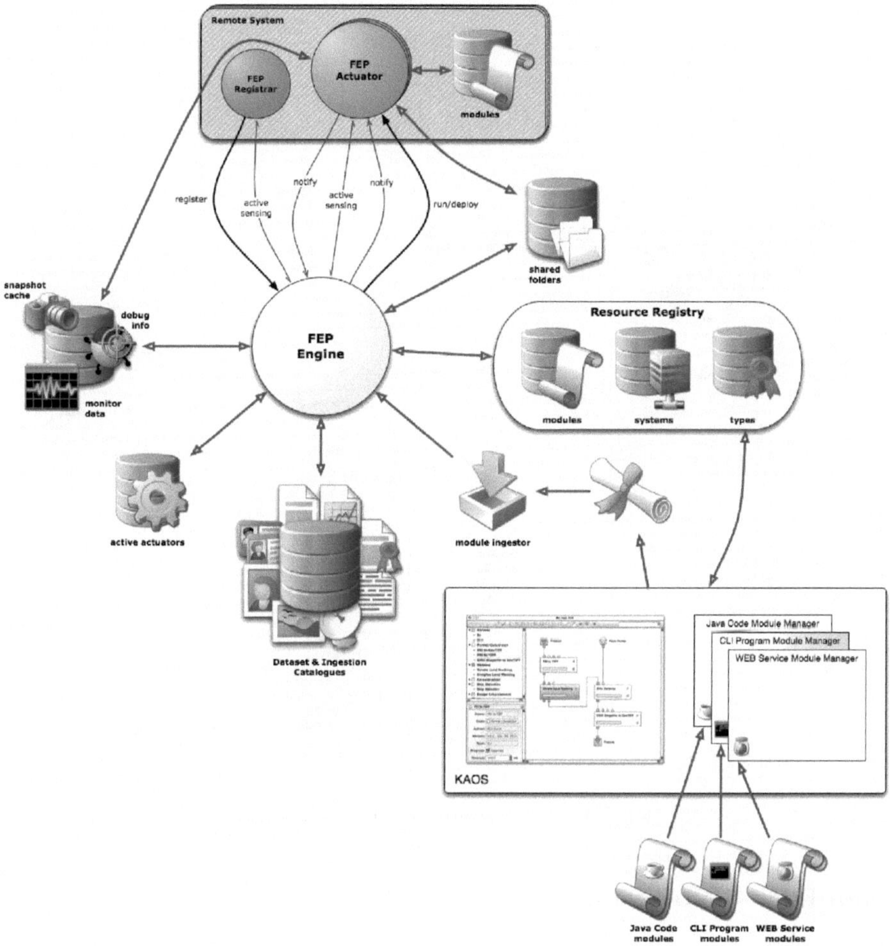

Fig. 4.15. FEP engine and related elements

Data Flow - Input

The input to Processing Components could be images or auxiliary data obtained from:

- output of a Ground Segment;
- Reference Data Sets (see below), which can be used to:
  - select needed data;
  - download data into the "KEO pick-up point";
  - test a FEP.

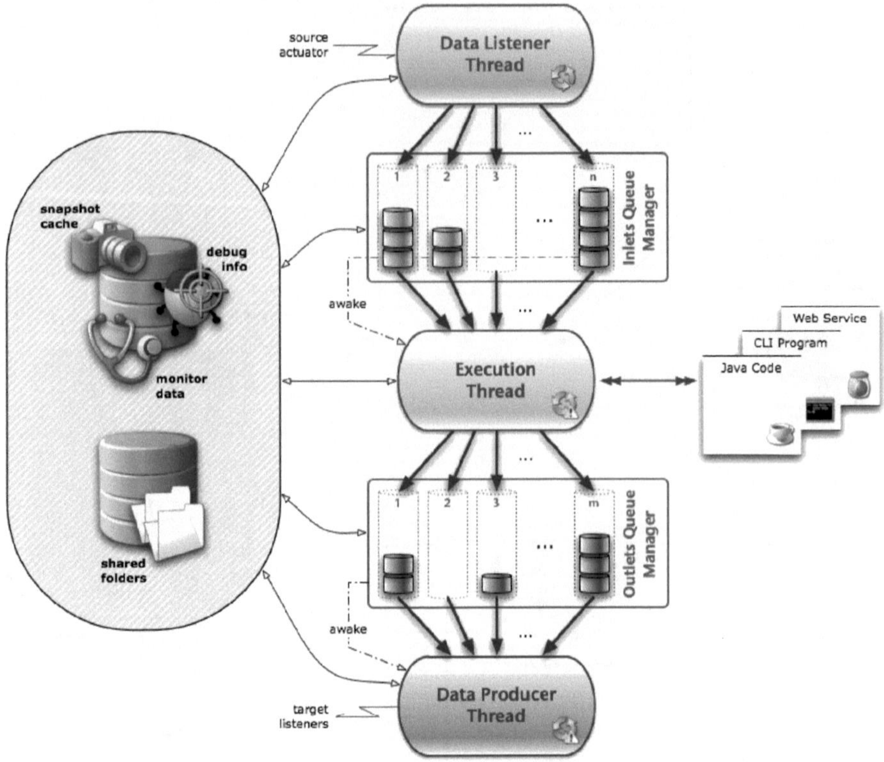

Fig. 4.16. Elements of FEP Actuators

Data Flow - Processing

Processing is performed a Feature Extraction Processor, which can be made of a number of Processing Components, provided in the form of Web Services, command line executable or Java modules, each one performing a specific task, like format con-version, image cutting, vegetation index computation, urban / not-urban classification, etc. Processing Components can also be created from wrapping a KIM training, which provides a probabilistic classifier, into a Web Service.

Input and output data of a Processing Component are univocally assigned to a specific type, in order to permit automatic chaining check at FEP design and avoid mismatches that might block correct processing at run-time.

Data Flow - Output

The output of the last Processing Component in the FEP should be in the form of geo-referenced raster or vector data, for publishing onto OGC servers:

- Web Map Server (WMS) and Web Coverage Server (WCS) for raster data;
- Web Feature Server (WFS) for vector data.

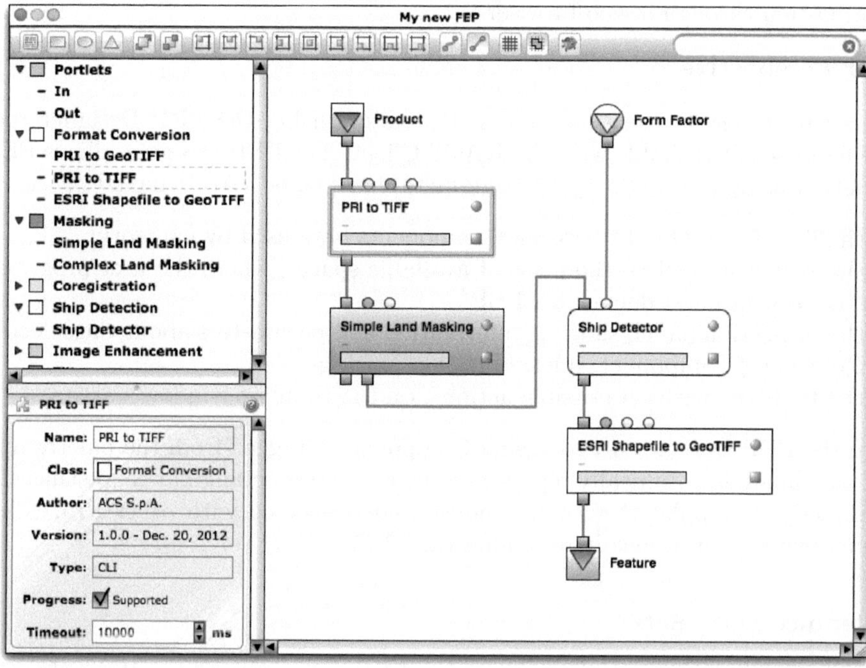

Fig. 4.17. FEP Designer interface

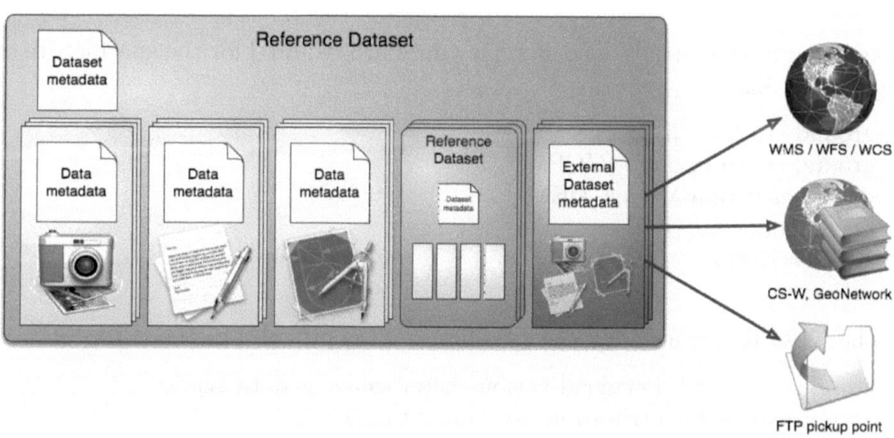

Fig. 4.18. Reference dataset

Related OGC services are exposed through SSE, for accessing the extracted informa-tion through a web browser.

User Perspective

Users can design Feature Extraction Processors using the FEP Designer tool (see figure 4.17) available with the KAOS Client. The FEP Designer allows FEP definition using a *visual programming* metaphor [65, 66, 67]. It includes:

- the list of available Processing Components organised by category;
- the *sketchpad* (filling the most of available space), where the user programs (by drawing) and debugs his FEP;
- the inspection panel, showing information and parameters about the selected Processing Component, inlet, portlet or outlet;
- the tool-bar, with all possible actions, on the topmost window's area.

In the FEP Designer a Processing Component, dragged from the palette into the sketchpad, is graphically represented by a pseudo-rectangular shape that can be directly manipulated with the mouse and connected with other Processing Components for graphically designing the FEP.

Reference Data Sets

Reference images are necessary to test a FEP. They must be suitable for the application and have precise and controlled characteristics, such as:

- good radiometric calibration,
- precise ortorectification,
- etc.

In addition to the images, also auxiliary data are required for the specific site or application, like:

- ground truth data,
- ground control points (GCPs),
- digital elevation Model (DEM),
- documentation,
- site description,
- etc.

The KEO Reference Data Set (see figure 4.18) provides methods to:

- store, search and download various heterogeneous data types;
- accept external contributions for content creation.

KEO Reference Data Sets are currently based on the GeoNetwork standards, which is a free and open source catalogue application designed to manage spatially referenced resources through the web. On the top of the GeoNetwork CS-W, the knowledgebase will be used to improve the semantic search and navigation of data.

SSE

The SSE portal is used as the KEO web front-end, permitting to:

- publish and execute KEO services
- access, via the OGC services, the information systematically extracted by the KEO FEPs and stored onto the OGC servers.

4.4 Conclusions

This chapter covers several apparently un-related research topics: service oriented architecture, service publishing, service orchestration, knowledge based information mining, information and feature extraction, and content based information retrieval. Each of the above mentioned topics in itself could be developed in more specialised subjects. The objective here has been to provide an overview of the systems which can be put in place to support various kinds of user needs and to show how they relate each other, as well as how they relate with higher level user requirements.

SEE and KEO, including the KIM tool, have been designed following the needs expressed by the concerned communities (service providers and researchers), after many years of research and technological development, started since the year 2000 for some aspects. SSE is operationally available since 2006, it is continuously enhanced with new features and it is being upgraded to the results of the Heterogeneous Mission Accessibility project and test-bed. KIM is providing initial results and test systems are available at http://kes.acsys.it/kes or http://kes.esrin.esa.int/kes. KEO is in the middle of its development: an initial prototype is available and the first complete release is expected for the beginning of 2008. In addition, a separate technology project aims at upgrading KEO and at creating Processing Components for the handling of image time series. These systems complement each other. KEO is more focused on supporting the cooperative creation, mainly by researchers, of new algorithms for the extraction of basic information for Content Based Image Retrieval or from Scene Understanding, Information Discovery or Feature Extraction. This basic information should simplify the following task of service providers. They can cooperate through SSE to combine this basic information (from KEO or other systems) also with data from other domains, in order to extract the final information required by the end users (e.g.: decision makers), through a cooperative process, where each one focuses on own core expertise.

The joint use of KEO and SSE permits the creation and provision of systematic or on-request. Information Based Services, like the MERIS Cloud Free Service provided from rolling or permanent archives (identification of MERIS images with cloud cover below a user defined threshold within the area of interest). Since they are provided via SSE, these services can be chained into more complex ones. Both environments might expand the use of EO data, but might have impacts on related archives. For interactive or batch extraction of basic information, or even just to identify relevant images, the basic products must be available. Increasing processing power and storage capability are progressively permitting systematic processing of all EO data for the generation of the basic products. Where this capability is not reached, systematic generation should be put in place at least for specific regions or time ranges. In addition, as automatic IIM improves in reliability (see the progresses in basic classification), the service providers should be offered more and more basic information products rather than image products. In order to simplify their use by service providers and researchers, such information products might require standardisation in format and used servers, which might be brought to their technological limits when loaded with the results of an entire mission. The basic products should be radiometrically and geometrically (e.g.: orthorectification) corrected in order to ensure reliable and consistent extraction of the basic information, as well as its comparability over time and compatibility with the GIS world. Such corrections and basic information extraction might be provided by additional facilities close to the archives (at least where very fast links are not available). These facilities should also foresee the possibility to host user processing, in order to permit personalised extraction of information, which might require reprocessing of large data sets with specialised algorithms.

References

1. Reeves, H., Lenoir, F.: Mal de Terre, Editions du Seuil, 2003-03
2. Achache, J.: Les sentinelles de la Terre, Hachette, 2004-04
3. EU FP6 STREP STREAM, `http://stream.etro.vub.ac.be/home.html`
4. Joint Interoperability Workshop EC/ESA, Towards a single information space for Environment in Europe, Frascati (April 3, 2007)
5. Achache, J.: A New Perspective for Earth Observation - The Oxygen (O2) Project. ESA Bulletin (116), 22–33 (2003)
6. Reference Model of Open Distributed Processing, ISO/IEC 10746-1:1998 (1998)
7. United Nations Spatial Data Infrastructure: Vision, Implementation Strategy and Reference Architecture, DRAFT DISCUSSION PAPER (October 2006) (referenced in RFI Response from Mick Wilson, UNEP, 28 December 2006)
8. RD 29 ISO19119:2005, Geographic Information - Services
9. ISO 19115:2003, Geographic Information -Metadata
10. ISO 19139 (2006, V1.0), Geographic Information -Metadata -Implementation specification
11. Cox, S.: OpenGIS Geography Markup Language Encoding Specification, Version 3.1.1, OGC 03-105r1, 2004-02-07
12. Gasperi, J.: GML 3.1.1 Application schema for Earth Observation products, Version 0.1.4, OGC 06-080r1, 2007-02-07
13. KEO Project, `http://earth.esa.int/rtd/Projects/KEO/index.html`
14. EU FP6 STREP Project Interoperable GMES Services for Environmental Risk. Management in Marine and Coastal Areas of Europe, `http://interrisk.nersc.no/`
15. EU FP6 Integrated Project Wide Information Network, `http://win-eu.org/`
16. Botts, M.: OGC SensorWeb Enablement Architecture Document, OpenGIS discussion paper, 2006-03-04, Version 1.0.0, OGC 06-021r1

17. Botts, M.: OpenGIS Sensor Model Language (SensorML) Implementation Specification, 2007-04-16, Version 1.0.0 OGC 07-000
18. Security Assertion Markup Language (SAML) v1.1, OASIS standard set, 2003-08
19. Web Services Security v1.0 (WS-Security 2004), OASIS standard set, 2004-01
20. Bajaj, S., et al.: Web Services Policy 1.2- Framework (WS-Policy), W3C Member Submission, 2006-04-25
21. Web Services Business Process Execution Language Version 2.0, OASIS Committee Draft, 2005-12-21
22. Coene, Y.: OWS-3 Imagery Workflow Discussion Paper, OGC 05-140r1, Version 1.0, 2006-04-20
23. Hoa Nguyen, T., et al.: OWS-4 Sensor Planning Service for EO, OGC 07-008r1, 2007-01-25
24. European Interoperability framework for Pan-European eGovernment services, Version 1.0, `http://ec.europa.eu/idabc/servlets/Doc?id=19528`
25. Vogues, U., Senkler, K.: OpenGIS Catalogue Services Specification 2.0.1 (with Corrigendum) - ISO Metadata Application Profile, Version 1.0, OGC 04-038r4, 2006-03-23
26. Universal Description, Discovery and Integration (UDDI) v2.0, OASIS standard set, 2002-08
27. Martell, R.: OpenGIS Catalogue Services - ebRIM (ISO/TS 15000-3) profile of CSW, Version 1.0.0, OGC 05-025r3, 2005-10-17
28. Lesage, N.: OGC Cataloguing of ISO Metadata (CIM) using the ebRIM profile of CS-W, Version 0.1.0, OGC 07-038, 2007-03-22
29. Coene, Y., Gilles, M.: OpenGIS CSW Application Profile for Earth Observation, Version 0.2.0, OGC 06-079r2, 2007-01-26
30. Primavera, R.: EO Products Extension Package for ebRIM (ISO/TS 15000-3) Profile of CSW 2.0, Version 0.1.1, OGC 06-131r1, 2007-02-26
31. Mrigot, P.: SPS Application Profile for Earth Observation Sensors, Version 0.9.3, OGC 07-018, 2007-03-22
32. Marchioni, D.: Ordering services for Earth Observation Products, Version 1.2.0 Draft, OGC 06-141, 2007-02-19
33. de La Beaujardire, J.: OpenGIS Web Map Service Implementation Specification, Version 1.1.1, OGC 01-068r3, 2002-01-16
34. Lankester, T.: Web Map Services Application Profile for EO Products, Version 0.1.0, OGC 06-093, 2006-07-05
35. Vretanos, P.: OpenGIS Web Feature Service Implementation Specification, Version 1.1.0, OGC 04-094, 2005-05-03
36. Sonnet, J.: OpenGIS Web Map Context Documents Implementation Specification, Version 1.0, OGC 05-005, 2005-01-19
37. Kyle, M., et al.: OpenGIS GML in JPEG 2000 for Geographic Imagery (GMLJP2) Encoding Specification, Version 1.0.0, OGC 05-047r, 2006-01-20
38. Priest, M.: OGC Sensor Observation Service, Version 1.0.0; OGC 06-005r3, 2007-04-16
39. Martin, J., et al.: OpenGIS Catalogue Services - Best Practices for Earth Observation Products, Version 0.3.0, OGC 05-057r4, 2006-03-20
40. Fielding, R., et al.: HTTP Hypertext Transfer Protocol - HTTP/1.1, W3C RFC 2616 (1999)
41. Rescorla, E.: HTTP over TLS, W3C RFC 2818 (2000)
42. Postel, J., Reynolds, J.: File Transfer Protocol, W3C RFC 959 (1985)

43. Gaur, H., et al.: BPEL Cookbook, Best Practices for SOA-based integration and composite applications development, July 2006. PACKT Publishing (2006), ISBN 1904811337
44. Gudgin, M., et al.: Simple Object Access Protocol (SOAP) Version 1.2, W3C Recommendation, 2003-06-24
45. Chinnici, R., et al.: Web Services Description Language (WSDL) Version 2.0, W3C Candidate Recommendation, 2006-03-27
46. Paoli, J., et al.: Extensible Markup Language (XML) 1.0 (Fourth Edition), W3C Recommendations, 2004-02-04
47. Thompson, H., et al.: XML Schema Part 1: Structures Second Edition, W3C Recommendation, 2004-10-28
48. Biron, P., et al.: XML Schema Part 2: Data Types Second Edition, W3C Recommendation, 2004-10-28
49. Berglund, A.: Extensible Stylesheet Language (XSL), W3C Recommendation, 2006-12-05
50. Baker, M.: XHTML Basic, W3C Recommendation, 2000-12-19
51. Bos, B., et al.: Cascading Style Sheets, level 2, W3C Recommendation, 1998-05-12
52. Clark, J.: XML Path Language (XPath) Version 1.0, W3C Recommendation, 1999-11-16
53. Vretanos, P.: OpenGIS Simple Features Profile of GML 3.1.1, Version 1.0, OGC 06-049, 2006-04-25
54. Lalonde, W.: OpenGIS Styled Layer Descriptor Implementation Specification, Version 1.0.0, OGC 02-070, 2002-09-19
55. Evans, J.: OpenGIS Web Coverage Service Implementation Specification, Version 1.0.0, OGC 03-065, 2003-08-27
56. Vretanos, P.: OpenGIS Filter Encoding Implementation Specification, Version 1.1.0, OGC 04-095, 2005-05-03
57. Ballinger, K., et al.: Web Services Inspection Language (WS-Inspection) 1.0, 2001-11
58. Box, D., et al.: Web Services Addressing (WS-Addressing), W3C Member Submission, 2004-08-10
59. Thygesen, J.H., Ølykke (ROVSING A/S), K.R., Sørensen, M.K., Rasmussen (GRAS), M.S., Skriver (ØrstedáDTU), H., Johansen, P., Pece (DIKU), A.: Feature Manipulation Methods: Current Status and Research Trends, `http://earth.esa.int/rtd/Documents/SURF_TN.pdf`
60. Datcu, M., Daschiel, H., Pelizzari, A., Quartulli, M., Galoppo, A., Colapicchioni, A., Pastori, M., Seidel, K., Marchetti, P.G., D'Elia, S.: Information Mining in Remote Sensing Image Archives - Part A: System Concepts, Geoscience and Remote Sensing. IEEE Transactions 41(12) (December 2003), `http://earth.esa.int/rtd/Articles/IIM_IEEE_2003.pdf`
61. Colapicchioni, A., D'Elia, S., Datcu, M.: Information Mining in Remote Sensing Image - The KIM. In: KES and KIMV projects PV 2004 Conference Proceedings (2004), `http://earth.esa.int/rtd/Articles/IIM_PV-2004.doc`
62. Datcu, M., Seidel, K.: KIM - Knowledge Driven Information Mining in Remote Sensing Image Archives - Executive Summary, `http://earth.esa.int/rtd/Documents/KIM_Executive_Summary.doc`
63. Varas, J., Busto, J., Torguet, J.R.: EO Domain Specific Knowledge Enabled Services (KES-B). In: Lacoste, H., Ouwehand, L. (eds.) Proceedings of ESA-EUSC 2004 - Theory and Applications of Knowledge-Driven Image Information Mining with Focus on Earth Observation (ESA SP-553), Madrid, Spain, March 17-18, 2004, p. 17.1 (2004) (Published on CDROM as ESA Special Publication no. 553)

64. Pereira, A., Ribeiro, R.A., Sousa, P.A., Pantoquilho, M., Bravo, P., Falcão, A., D'Elia, S.: An ontology to support knowledge enabled services on earth observation. In: ESA-EUSC 2004 proceedings (2004) (Published on CDROM as ESA Special Publication no. 553), http://earth.esa.int/rtd/Events/ESA-EUSC_2004/Ar18_Pereira.pdf
65. Andrenacci, P., Favreau, E., Larosa, N., Prestigiacomo, A., Rosati, C., Sapir, S.: MARS: RT20/EDIT20 Ñ Development Tools and Graphical User Interface for a Sound Generation Board. In: Jose, S. (ed.) International Computer Music Conference (ICMC), October 14-18, 1992, pp. 340–343 (1992)
66. Telea, A.C., van Wijk, J.J.: Simulation and Visualization in the VISSION Object Oriented Dataflow System. In: Proc. ASCI 1999 Conference, The Netherlands, June 15-17 (1999)
67. Zicarelli, D., Taylor, G.: Max 4.6 Fundamentals. In: Cycling '74 (2006)

5

On Content-Based Image Retrieval Systems for Hyperspectral Remote Sensing Images

Miguel A. Veganzones, José Orlando Maldonado, and Manuel Graña

Computational Intelligence Group, UPV/EHU
www.ehu.es/ccwintco

Summary. This chapter includes a review of some key elements of Content Based Image Retrieval Systems (CBIR). We place into this context the the works found in the literature regarding remote sensing images. Our own focus is on hyperspectral images. The approach we are pursuing is that of characterizing the spectral content of the image through the set of endmembers induced from it. We describe some ideas and numerical experiments for a system that would perform CBIR on hyperspectral remote sensing images . We propose a spectral image similarity to guide the search to answer the queries.

5.1 Introduction

The interest in fast and intelligent information retrieval systems over collections of remote sensing images is increasing as the volume of the available data grows exponentially. For instance a single in-orbit Synthetic Aperture Radar (SAR) sensor collects about 10 – 100 Gbytes of data per day, giving about 10 Tbytes of imagery data per year. New sensing instruments can produce an information amount of one order of magnitude higher than this. Current retrieval systems offer to their users raw images, thematic maps and ancilliary data in response to very precise queries requiring a detailed knowledge of the structure of the stored information. However there is a growing need for more flexible ways to access the information based on the image computed intrinsic properties, i.e. texture and induced semantic contents. In CBIR systems, the images stored in the database are labeled by feature vectors, which are extracted from the images by means of computer vision and digital image processing techniques. In CBIR systems, the query to a database is specified by an image. The query's feature vector is computed and the closest items in the database, according to a similarity metric or distance defined in feature space, are returned as the answers to the query. This is the low level, semantic free, definition of the CBIR systems, that does not take into account the semantic gap between the user expectations and the system response. We are specially interested in hyperspectral images and the ways to characterize them for CBIR tasks.

The fields of applicability are extensive, geoscience tools are demanded for disasters prevision, enviromental analysis, security. The investigation of new tools capable of predict, monitor and assist in preparation of strategies to minimize

M. Graña and R.J. Duro (Eds.): Comput. Intel. for Remote Sensing, SCI 133, pp. 125–144, 2008.
springerlink.com

the damage produced by natural and human-made disasters, would contribute to a better quality of life and to the reduction of economical losses. The ability to perform automatic and intelligent searches is a key tool for these goals. The field of CBIR is already a mature area of research [33]. Some facts, like the Semantic Gap between low level representations and high level semantic world models, have been acknowledged leading to new design and research paradigms. A fruitful track of research and source of innovative approaches introduces user interaction to obtain some retrieval feedback that allows the building and refinement of user world models [3, 7, 8, 13, 14, 15, 16, 18, 19, 20, 24, 25, 31, 34, 36].

The attempts to define CBIR strategies for remote sensing hyperspectral images are scarce and partial. In [1] the authors use the spectral mean and variance, as well as a texture feature vector, to characterize hyperspectral image tiles. The approach searches for specific phenomena in the images (hurricanes, fires, etc), using an interactive relevance strategy that allows the user to refine the search. In [7] some of the texture features from the first spectral PCA image are used as feature vectors.

We propose the characterization of the hyperspectral images by their so-called endmembers. Endmembers are a set of spectra that are assumed as vertices of a convex hull covering the image pixel points in the high dimensional spectral space. Endmembers may be defined by the domain experts (geologists, biologists, etc.) selecting them from available spectral libraries, or induced from the hyperspectral image data using machine learning techniques. In [10, 11] we have proposed an automated procedure that induces the set of endmembers from the image, using AMM to detect the morphological independence property, which is a necessary condition for a collection of endmembers. The goal in [10, 11] was to obtain a method for unsupervised hyperspectral image segmentation. There the abundance images were of interest. In this paper, the goal is to obtain a characterization of the image that can be useful for CBIR. The collection of endmembers serves as a summary of the spectral information in the image.

In section5.2 we introduce some ideas about the general structure of CBIR systems, referring them to the systems found in the literature of remote sensing images. In section 5.3 we present some ideas about the evaluation of CBIR systems, discussing how systems found in the literature have been validated. In section 5.4 we discuss our own ideas and proposal for a CBIR system based on the spectral characterization of the images.

5.2 CBIR Systems Anatomy

A CBIR system is a query resolving system over image collections that use the information inherently contained in the image. The CBIR system has to be able to extract quantitative features from the images that allow the system to index the image collection and to compute a distance between images. The simplest typical use of CBIR systems is described in figure 5.1. The user interacts with the system by a querying interface, usually a web page, where the query is defined and sent to the CBIR engine. In the process described by figure 5.1 the query

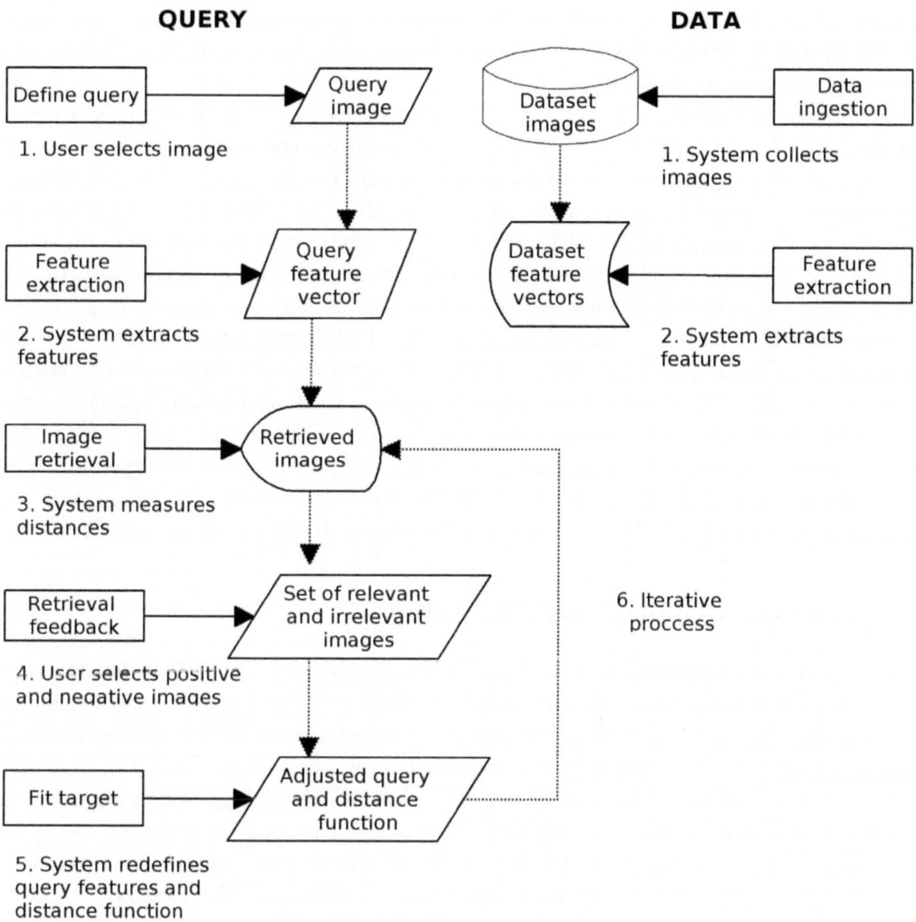

Fig. 5.1. Typical use of a CBIR system

is represented by an image provided by the users, asking the CBIR system for a list of the most similar images in the database. To resolve the query, the CBIR engine computes the image features which correspond to a point in the metric space defined by the system. Each image in database has a representative in this metric space so a distance to the query image could be computed for each image, using a similarity (or disimilarity) function. This produces a list ordered by similarity (or disimilarity) to the query image, that is presented to the user as response. In real world applications, computing the distance to each and every image in the database has an unbereable cost. Therefore, the database is indexed grouping the images into clusters. Then heuristic algorithms are used to retrieve the group of most similar images. The retrieved images are inspected by the user, selecting the ones that are relevant to his target search and those that are irrelevant. The sets of relevant and irrelevant images are used to redefine the query so the features quantifying the query information fits better the target in the mind of the user. The search is iterated until the set of retrieved images satisfies the user's expectatives.

From the previous description, it is natural to say that a CBIR system is composed of three subsystems: the query definition, the feature extraction and the image retrieval. Query definition is the proccess by which the user defines the query. It could be a graphical design interface or a web page. It can be as simple as selecting an image or it can be more sofisticated, i.e. the user can select regions of interest on the query image and interact with the retrieval proccess in a more expert mode. The feature extraction is the proccess of quantifying image information as a collection of low level characteristics, denoted as the feature vector. Image Retrieval encompasses the methods that use the feature vectors to select from the database the images that satisfy the query, in the sense of being the most similar ones. To prevent the Semantic Gap, the image retrieval may involve the interaction with the user to improve the query in any way. We review some ideas about CBIR systems, while omitting others, like the visualization function [33], which are not relevant to the work in this chapter.

5.2.1 Query Definition and Interaction

In image information retrieval defining the query may be a critical design issue. The most common query specification method in CBIR search engines is the Query By Example(s), where the query is represented by an example image (or a set of images) related to the search target. The *Page Zero Problem* [38] arises when the user has not any image to start querying. When the user is only interested in an specific region of the images, he may specify the query by selecting a region of interest (ROI), which is known as the Query By Region(s). In the Query By Sketch, the user can draw an sketch of the desired target to specify the query. Conventional Query By Text could be usefull in situations where relevant information about images is collected as textual metadata, as for example, geografical coordinates, time of adquisition or sensor type. Text may be used as a complement to the visual query, or may solve the Page Zero Problem.

Formally, the Query Space [33]is the goal dependent 4-Tuple $\{D_Q, F_Q, S_Q, Z_Q\}$ where:

- D_Q is the subset of active images from dataset collection D within which the retrieval system is going to perform the search. For example, the user could select a subset of images belonging to an specific sensor in a concrete date of adquisition instead of querying the full image collection.
- F_Q is a selection of the quantitative image features $F_Q \subset F$. Usually the user doesn't have the knowledge to select the most relevant features for resolving the query, and the system automatically decides what features are going to be used.
- S_Q is the similarity function. This similarity function should be parameterized to allow adaptation to diverse goals and data sets. Very often the user

is not capable of decide which similarity function is the most appropiate for the formulated query and S_Q is selected automatically by the system.

- Z_Q is a collection labels to capture the goal dependent semantic concepts. For example, if the user wants to retrieve images containing rivers, Z_Q could be defined as $Z_Q = \{'water','river'\}$. Conventional CBIR systems have not the capacity of incorporate semantic concepts to the query resolving system, so they use exclusively the Feature Vector information as the criteria for searching the query's answer.

Queries can be specified either in an *exact* or *approximate* way. The exact specification requires that the answer $A(q) \subseteq I_Q$ to the query contains the images satisfying certain criteria, which can be given by some relational equations or propositions. The approximate specification requires that the answer $A(q)$ to the query provides the images ordered relative to the query satisfaction. Exact specification may consist of spatial, image or group predicates. The approximate specification may consist of spatial relations (skeetches) or image examples.

The interaction of the user with the system can be formalized as a sequence of Query Spaces $\{Q^1, \ldots, Q^n\}$. The transition between those query spaces is the effect of the user interaction, which can take several forms, from indicating the relevance of the answer images up to modifying the distance between images.

Regarding the kind of search, the following types can be distinguished [33]:

1. Search by category: the user wants to retrieve images belonging the same class.
2. Search by target: the user wants to retrieve images containing an object.
3. Search by association: user has not an specific target in mind and its only aim is to browse through the database.

In the context of remote sensing image databases, the proposed systems cover some of these possibilities. In [1, 20] a basic relevance approach is applied to multispectral images. In [18] the topology of the feature space is used for relevance feedback. In [31, 6, 7] a semantic model of the cover type desired by the user is constructed as a bayesian network from the interaction with the user.

5.2.2 Feature Extraction

The feature extraction process produces the low dimensional (relative to image size) quantitative description of the image which allows the computation of similarity measures, the definition of an ordering of the images and their indexing for the search processess. For remote sensing images, some approaches use the spectral information, while other emphasize the spatial features. Most of the works reported in the literature have been performed on Landsat images, of similar sensor images, whose number of bands is small compared to modern hyperspectral sensors.

In [12] the image features are computed as multispectral distribution invariants from a physically-based image formation model. The result of dimensional reduction procedures, such as the Orthogonal Subspace Projection [30] can be also used as fatures for classification and search. Eigenviews are proposed by [29] for object recognition. A collection of image features is computed offline in the system structure proposed in [6, 5]. These features are the parameters of Gauss Markov Random Fields modelling texture and spatial fatures in the case of optical and SAR images. In [31] a similar process is proposed to build up a vocabulary of image features, previous to the interactive construction of a Bayesian semantic model of the user target. This vocabulary is the result of a clustering performed over the parameter data from the texture model estimation.

Unmixing algorithms

The hyperspectral imagery could be seen as mixing model where an hyperspectral image is the result of the linear combination of the pure spectral signature of ground components, named endmembers, with abundance matrices. Let $E = [e_1, \ldots, e_n]$ be the pure endmember signatures where each $e_i \in R^d$ is a d-dimensional vector. Then, the hyperspectral signature $h(x,y)$ at the pixel with coordinates (x,y) is defined by the expression:

$$h(x,y) = \sum_{i=1}^{n} e_i \Phi_i(x,y) + \eta_i(x,y) \tag{5.1}$$

where $\Phi(x,y)$ is the n-dimensional vector of fractional abundance at pixel (x,y) and $\eta(x,y)$ is the independent additive noise component. There are two constraints in the equation 5.1, the abundance non-negatice constraint 5.2 and the abundance sum-to-one constraint 5.3, respectivily defined as

$$\Phi_i(x,y) \geq 0, \text{ for all } 1 \leq i \leq n \tag{5.2}$$

$$\sum_{i=1}^{n} \Phi_i = 1 \tag{5.3}$$

These restrictions require careful numerical methods for the computation of the fractional abundances [32] when the endmembers are known. The image spectral signatures could be a good characterization of the image but, additionaly, the abundance matrices could be used as spatial information about the image. The idea of using the spectral unmixing fractional abundances as a kind of feature extraction for classification purposes was introduced in [9]. The feature vector for each pixel (x,y) in the image is the vector $[\Phi_1(x,y), ..., \Phi_n(x,y)]$ which is the set convex coordinates of the pixel relative to the enmembers. We extend this idea to the design of CBIR systems for hyperspectral images. In fact, we will assume the set of endmembers E as the image features. The key problem then is the definition of the set of endmember spectral signatures. A library of known pure ground signatures or laboratory samples could be used. However this poses several problems, such as the illumination invariance, the difference in sensor intrinsic parameters and the *a priori* knowledge about the material composition of the scene. There are diverse approaches in the literature for the estimation of endmembers from the image data. Early approaches [37] searched for the minimum simplex covering the image data. A method based on morphological operators is proposed in [26, 27]. In [2]an statistical approach is proposed. In [10, 11] we proposed a method based on morphological independence and the use of Associative Morphological Memories (AMM). Recently, [4] proposed a fast Pixel Purity Index procedure. An approach based on ICA is proposed in [23].

5.2.3 The Similarity Function

The retrieval process provides a subset of images answering the query on the basis of the similarity between the images, computed over the feature vector extracted from each image. There are two basic strategies to compute the answer A_Q, either an ε-similarity or an KNN-similarity retrieval. In ε-similarity retrieval, the system selects only those images in D_Q whose similarity to the query criteria falls below a given threshold ε. In KNN-similarity retrieval, the system ranks D_Q according to the similarity to the query given a similarity function, S_Q, and retrieves the first K images. These retrieval strategies are similar to the exact and approximate searches described above. There are two essentially different retrieval paradigms: the metric and the probabilistic.

Metric paradigm

Most of the retrieval systems found in the literature are based in the metric paradigm. The image information is quantified as low level feature vectors in a high dimensional metric space. $F: I \rightarrow \Re^r$. Therefore, an image is described as a point in a r-dimensional space. The distance in feature space corresponds to the image similarity function $S: IxI \rightarrow \Re^+$.

The equation 5.4 expresses the similarity function between two images $S(I_1, I_2)$ as a function of the distance $d(F_1, F_2)$ between their corresponding feature vectors. The function Ψ is a positive, monotonically nonincresing function.

$$S(I_1, I_2) = \Psi(d(F_1, F_2)) \tag{5.4}$$

The Euclidean distance (eq. 5.5) is the simplest and most used distance function. Other useful metrics are the Manhattan (eq. 5.6) and the Maximum (eq. 5.7) distances.

$$d_E(F_1, F_2) = \sqrt{\sum_{i=0}^{r-1} (F_{1i} - F_{2i})^2} \tag{5.5}$$

$$d_M(F_1, F_2) = \sum_{i=0}^{r-1} |F_{1i} - F_{2i}| \tag{5.6}$$

$$d_{Max}(F_1, F_2) = \max_i \{|F_{1i} - F_{2i}|\} \tag{5.7}$$

where F_{1i}, F_{2i} are the i-th feature of the respective feature vectors. When scale invariance is needed, the Correlation-Based distance (eq. 5.8) can be used.

$$d_{Corr}(F_1, F_2) = \frac{\sum_{i=0}^{r-1} F_{1i} F_{2i}}{\sqrt{\sum_{i=0}^{r-1} F_{1i}^2} \sqrt{\sum_{i=0}^{r-1} F_{2i}^2}} \tag{5.8}$$

In a relevance-guided iterative retrieval proccess [1, 8, 14, 15, 19, 20] the user feedback is specified through the identification of a set of relevant and irrelevant images in the answer set $A(q)$, aiming to better approach the target that the user has in mind. In the metric paradigm this goal can be reached from two different points of view. The first modifies the query definition while second one modifies the similarity function. Redefining the query is equivalent to move the query point across the feature vector space. One way of doing that is the following one: in each iteration the user identifies the relevant images and their relevance degree; then the query is redefined as the weighted centroid of the set of relevant images weighted by their relevance degree. Let η_j be the relevance degree of the j-th relevant image. Each component of the new feature vector corresponding to the redefined query, F_i', is given by the following equation:

$$F_i' = \frac{\sum_j \eta_j F_{ji}}{\sum_j \eta_j}$$

Another kind of query redefinition is based in both the relevant and irrelevant set of images selected by the user. In [31] the relevance feedback guides the estimation of a Bayesian Network parameters. Let F_{ji} be the i-th feature of the j-th image on relevant set. Let $\bar{F}_{ki}$ be the i-th feature of the k-th image on irrelevant set. Let N_F and $N_{\bar{F}}$ be the number of images in the relevant and irrelevant set respectively, and α, β, γ are suitable constants; the following equation defines the query movement along the feature space at each retrieval feedback iteration.

$$F_i' = \alpha F_i + \beta \left(\frac{1}{N_F} \sum_{j=0}^{N_F - 1} F_{ji} \right) + \gamma \left(\frac{1}{N_{\bar{F}}} \sum_{k=0}^{N_{\bar{F}} - 1} \bar{F}_{ki} \right)$$

The second relevance feedback class of methods works with the similarity function, updating it at each relevance feedback iteration having into account the user preferences. Weighting the Euclidean distance so the user's information could be incorporated into the retrieval proccess can be achieved in two different ways, depending on whether there is any interdependency among the feature vector components or not. If the feature vector components are mutually independent, the relevance of each feature element is modelled by a diagonal matrix, $\Lambda = diag \cdot (\lambda_1, \lambda_2, \ldots, \lambda_r)$ where λ_i represents the individual relevance of the i-th feature component. The Weighted Euclidean Distance (WED) is defined by equation 5.9. If the feature components are not independent, we have a Generalized Euclidean Distance (GED), defined by equation 5.10, where W is a symetric matrix representing the mutual dependent of the feature components.

$$d_{WED}(F_1, F_2) = (F_1 - F_2)^T \Lambda (F_1 - F_2) \tag{5.9}$$

$$d_{GED}(F_1, F_2) = (F_1 - F_2)^T W (F_1 - F_2) \tag{5.10}$$

Both WED and GED are instances of the Mahalanobis distance. The values of Λ's diagonal can be estimated as $\lambda_i = \frac{1}{\sigma_i}$, where σ_i is the feature vector component standard deviation computed over the relevant set of images selected by the user in the answer to the query. If the standard deviation for the i-th component is high, this feature component has not information about the user target. Other approach to set Λ is trying to maximize the likelihood of the answer to the query, minimizing the distance between the previous query feature vector and the feature vectors of the selected relevant images. The minimized objective function is defined by $J(\Lambda) = \sum_{j=0}^{K-1} d_{WED}(F_Q, F_j)$, where F_Q is the feature vector of the previous query definition and F_j the feature vector of each one of the images in the relevant set. The $\sum_{i=0}^{r-1} \lambda_i^2 = 1$ constraint is imposed to avoid the trivial all zeros solution, and the estimation problem is stated as the minimization problem, $\hat{\Lambda} = argmin J(\Lambda)$. The estimation of the parameters of the matrix W for the GED is similar to the optimal estimation of the WED parameters. In this case, the minimization problem is defined by $\hat{W} = argmin J(W) = argmin \left\{ \sum_{j=0}^{K-1} d_{GED}(F_Q, F_j) \right\}$ under the constraint $det(W) = 1$. All these ideas can be extended to Correlation Based distances, which we will not elaborate here.

Probabilistic paradigm

In the probabilistic paradigm [31] the user target or semantic interpretation of the search is modelled by a probability distribution defined on the feature space. According to that, the process of image retrieval is an inference proccess that asigns a probability value to each image in database. The retrieval user feedback aims to fit the best probalibity distribution for the query, using the positive and negative relevance sets of images. Then, this probability distribution is used to asign *a posteriori* probability for each image in the database, which will be used as the distance measure for the query answer retrieval. Although in many cases the same strategy is used for positive and negative examples, is known that the negative images are not related to the target distribution, but they are isolated and independent, so different methodologies appear to treat the negative set of images product of the retrieval feedback iterations.

The Gaussian distribution for the semantic target features, $F \rightarrow N(\mu_F, \Sigma_F)$, the probability density function is given by:

$$g(F) = \frac{1}{(2\pi)^{d/2} |\Sigma_F|^{1/2}} e^{-\frac{1}{2}(F - \mu_F)^T \Sigma_F^{-1} (F - \mu_F)}$$

where d is the feature vector dimension, μ_F is the average feature vector for the Gaussian distribution and Σ_F is the covariance matrix whose determinant$|\Sigma_F|$ is used to normalize. Using the bayesian aproximation, the log posterior probability of image I denoted by its feature vector F belonging to target model M is defined by:

$$p_M(F) = \ln P(M|F) \propto \ln p(F|M) + \ln P(M)$$

$$= -\frac{1}{2}\frac{(F-\mu_F)^T}{\Sigma_F} + \ln P(M) + constant$$

and the distance function can be defined by:

$$d(I, M) = -p_M(F_I)$$

The estimation of the log posterior probability needs the estimation of μ_F, Σ_F and the *a priori* probability of the target model, $P(M)$. Using only the set of positive images, let F_j be the feature vector of the j-th image on relevant set and N_F the number of relevant images in the set. To estimate the parameters needed to value $p_M(F)$ the following conventional equations are used:

$$\mu_F = \frac{1}{N_F}\sum_j F_j$$

$$\Sigma_F = \frac{1}{N_F - 1}\sum_j (F_j - \mu_F)(F_j - \mu_F)^T$$

If the features are supposed to be mutually independent the covariance matrix can be simplicited as a diagonal matrix of standard deviations, $\sigma_F^2 = diag \cdot (\sigma_1^2, \sigma_2^2, \ldots, \sigma_r^2)$. The *a priori* probability, $P(M)$, is assumed to be equal for all the possible classes in database son it remainds constant all along the retrieval proccess.

Instead of computing the parameters in each iteration using all the positive images, these can be estimated using the parameters of the previous iteration and the features of the new positive images in relevant set. Supposing features are mutually independent, let F_j^U be the feature vector of the j-th new positive image on the present iteration, U_F the number of new relevant images in the current iteration, N'_F the number of positive images in previous iteration and μ'_F, ${\sigma'}_F^2$ the parameters of the previous gaussian distribution. The new parameters μ_F and σ_F^2 are estimated using:

$$N_F = N'_F + U_F$$

$$\mu_F = \frac{N'_F \mu'_F + \sum_U F_j^U}{N_F}$$

$$\sigma_F^2 = N'_F {\sigma'}_F^2 + \frac{N'_F U_F {\mu'}_F^2 - 2N'_F \mu'_F \sum_U F_j^U - (\sum_U F_j^U)^2}{N_F} + \sum_U (F_j^U)^2$$

If the irrelevat set of images is used to estimate the distribution parameters, this can be seen as the difference of two distribution, the one estimated with the positive examples and the one with the negatives, been the resultant distribution, in the case of the Gaussian distribution, as an hyperbolic in the future space instead of the hiperelypsoid resultant if only the relevant set is used. Instead of using the negative examples to estimate the distribution parameters, the set can be used as a penalization for the distance image. For this, the final distance measure is:

$$d(I, M) = -p_M(F_I) + \sum_V p_V(d(F_I, F_V))$$

where the first term is the distance explained above and the second one is a penalization function where each image, V, in the irrelevant set is modelled as a gaussian distribution, and $p_V(d(F_I, F_V))$ is the log posterior probability of feature vectors distance, euclidean for example.

In [31] a fully Bayesian framework is proposed, that starts from the low level parameter inference (the Markov Random Field texture model) up to the labeling of the cover types in the image, and the labeling of the whole image, where the interaction process helps building a Bayesian model of the image semantics.

5.3 Evaluating CBIR Systems

In CBIR systems, the approapriate evaluation methodology is is an open question [33, 22, 5, 8, 35]. Most of the methods used in CBIR evaluation come from the Information Retrieval (IR) field where the researchers have great experience evaluating information retrieval systems. From this area of knowledge appears two standar methods, the Precision and Recall measures (below defined), that are long used over CBIR systems. But working with image collections instead of textual databases has some inherited difficulties that made IR methods insufficient. First of them is that there is not a standar image collection researchers could work with, so there is not a way to compare the results of different investigations.

Suppose an image database D and a query Q are given and that the database can be divided in two groups, the relevant set of images for the query Q, R_Q, and its complement, the set of irrelevant images $\overline{R}_Q$. Suppose the CBIR System retrieves a set of images in D for the query Q, A_Q. The Precision measure, denoted p, is defined as the fraction of the set of images retrieved by the system that are relevant for the query. Another point of view is the Error measure, denoted e, assumed to be the fraction of the set of images retrieved by the system that are irrelevant for the query. The Recall measure, denoted r, is the fraction of relevant images returned by the query. Usually this information is presented as a Precision (or Error) versus Recall graph, PR (or ER) graphs.

$$p = \frac{|A_Q \cap R_Q)|}{A_Q} \qquad e = \frac{|A_Q \cap \overline{R}_Q|}{A_Q} \qquad r = \frac{|R_Q \cap A_Q|}{R_Q}$$

But PR graphs have limitations and other analyses based on precision and recall measures are used too as, for example, the recall measure where the precision falls below 0.5, the non-interpolated mean average precision or the precision after the first N_K relevant images are retrieved where N_K is the number of relevant images in a K-NN based query. Moreover, the precision and recall measures may be misleading due to the great difficulty of defining the relevant set of images. Because of these limitations other measures have been defined for evaluating CBIR systems. One of them is the First Rank, $rank_1$, and the Average Rank, $\widetilde{rank}$, being respectively, the rank at which the first relevant image is retrieved and the normalized average rank of relevant images defined by equation 5.11 where N_D is the number of images in dataset, N_R is the number of images in the relevant set R_Q and $rank_i$ is the rank at which the ith relevant image is retrieved. A $\widetilde{rank} = 0$ indicates a perfect performance, while the performance goes worst as $\widetilde{rank}$ approximates 1. A random retrieval would give a evaluation of $\widetilde{rank} = 0.5$.

$$\widetilde{rank} = \frac{1}{N_D N_R}\left(\sum_{i=1}^{N_R} rank_i - \frac{N_R(N_R-1)}{2}\right) \tag{5.11}$$

Another point of view is not to use any query answer relevant image set. Given a query Q for the database D, an ideal ordering of the database $Z = [z_1, \ldots, z_k]$ is assumed. An ideal retrieval systems will provide a relevance measure $S(I_j) \in [0, 1]$ for each image. The system under evaluation will retrieve for query Q an ordering $Z_Q = [z_{\pi_1}, \ldots, z_{\pi_k}]$, where $[\pi_1, \ldots, \pi_k]$ is a permutation of $[1, \ldots, k]$. The weighted displacement, denoted ω, can be computed by equation 5.12 and can be used as a measure of system performance without the need of defining a relevant set, althought the problem of obtaining an ideal ordering Z remains open.

$$\omega = \sum_j S(I_j)\,|j - \pi_j| \tag{5.12}$$

5.3.1 Evaluation in the Literature of Remote Sensing

Early works, such as [12] contain as evaluation processes the gathering of a collection of images and the counting of relevant positive answers by the observer. The works of Datcu and colleagues have been quite influential in the last years. In [31] the evaluation consists on the exploration of a database of some aerial images and a Landsat TM image through a web site. The authors suggest some statistical measures of the labellling process, such as the variance of the labelling probabilities for a given class, or the information divergence. In [5] a complete evaluation protocol is proposed that includes the quality of the queried results by precision and recall values, as well as the probability to overretrieve and forget. They also analyze man-machine interaction. The Human Centered Concepts for remote sensing image database exploration [7], are tested over a collection of heterogeneous data from different sensor types, performing complicated tasks, such as risk analysis in mined areas and detection of objects and structures.

5.4 A Case of Hyperespectral Image Retrieval

In this section we will present some ideas about hyperspectral image retrieval which have been already introduced in [21]. The idea is that the endmembers induced from the image by any suitable method are to be used as the image features for image indexing and retrieval. For our own computational experiments we use the method proposed in [10, 11]. This is an appropriate characterization of the spectral image content. We propose an image similarity function defined on the basis of the distance between endmembers. In the next subsection we present this image similarity function, before commenting some computational results over a database of synthetic images.

5.4.1 Endmember Based Image Similarity

Let $E^k = [e_1^k, \ldots, e_{n_k}^k]$ be the image endmembers obtained applying any of the methods discussed in section 5.2.2 over the k-th image $f_k(x,y)$ in the database. Given two hyperspectral images, $f_k(x,y)$ and $f_l(x,y)$, their endmember sets E^k and E^l may have different number of endmembers, i.e. $n_k \neq n_l$. That implies that the feature spaces of the images are different, whithout any functional relation between them. The computation of the similarity between the images starts with the computation of the matrix of Euclidean distances between all the possible pairs of endmembers from each image:

$$D_{k,l} = [d_{i,j}; i = 1, ..., n_k; j = 1, .., n_l]$$

where

$$d_{ij} = \left\| e_i^k - e_j^l \right\|.$$

We compute the vector of row minimum values

$$\mathbf{m}_k = \left[m_i^k = min_j \left\{ d_{ij} \right\} \right],$$

and column minimum values

$$\mathbf{m}_l = \left[m_j^l = min_i \left\{ d_{ij} \right\} \right].$$

We compute the similarity between the hyperspectral images as follows:

$$d(f_k, f_l) = (\|m_k\| + \|m_l\|)(|n_k - n_l| + 1). \tag{5.13}$$

The endmember induction procedure may give different number of endmembers and endmember features for two hyperspectral images. The similarity measure of eq. 5.13 is a composition of two asymmetrical views: each vector of minimal distances measures how close are the endmembers of one image to some endmember of the other image.

The worst case is when all the endmembers E^k of image f_k are very near to a subset E^l_* of the endmembers E^l of the image f_l. Then the magnitude of the vector of row minimum distances will be very small, because all the rows will have a very low minimum value. However, the magnitude of the vector of column minimum distances will be much greater, because the columns corresponding to the subset of endmembers $E^l - E^l_*$ will have a relatively large minimum value. Thus the similarity measure of eq. 5.13 can cope with the asymmetry of the situation. Besides, the formulation of eq. 5.13 avoids the combinatorial problem of deciding the optimal matching of endmembers. When the number of endmembers is different from one image to the other, their difference is introduced as an amplifying factor. The measure is independent of image size and, as the endmember induction algorithm described in [10, 11] is very fast, it can be computed in acceptable time. The endmember set poses no storage problem. Our approach does not use spatial features, but the endmembers give a rich characterization of the spectral content of the image. A further work on our approach may be the study of spatial features computed on the abundance images produced by the spectral unmixing, solving the question of band selection or dimension reduction prior to spatial feature computation.

5.4.2 Experimental Results

The hyperspectral images used for the experimental results are generated as linear mixtures of a set of spectra (the ground truth endmembers) applying synthesized abundance images. The ground truth endmembers were randomly selected from a subset of the USGS spectral libraries corresponding to the AVIRIS flights. Figure 5.2 show some spectra in used in 5 endmember images. The synthetic ground truth abundance images were generated in a two step procedure, first we simulate each as an gaussian random field with Mattern correlation function of parameters varying between 2 and 20. We applied the procedures proposed by [17] for the efficient generation of big domain gaussian random fields. Second to ensure that there are regions of almost pure endmembers we selected for each pixel the abundance coefficient with the greater value and we normalize the remaining to ensure that the abundance coefficients in this pixel sum up to one. It can be appreciated on the abundance images that each endmember has several region of almost pure pixels, that appear as almost white regions when visualizing the abundance images. Image size is 256x256 pixels of 224 spectral bands . We have generated collections of images with 2 to 5 ground truth endmember/abundances, for total number of 400 images, 100 images for each number of predefined endmembers. This is the biggest synthetic hyperspectral image database we know of.

The experiment performed on these images consists on the following steps:

1. Compute the distances between the images in the database, on the basis of eq. 5.13, using the ground truth endmembers. The distances are computed between images with the same number of ground truth endmembers, and with all the remaining images.

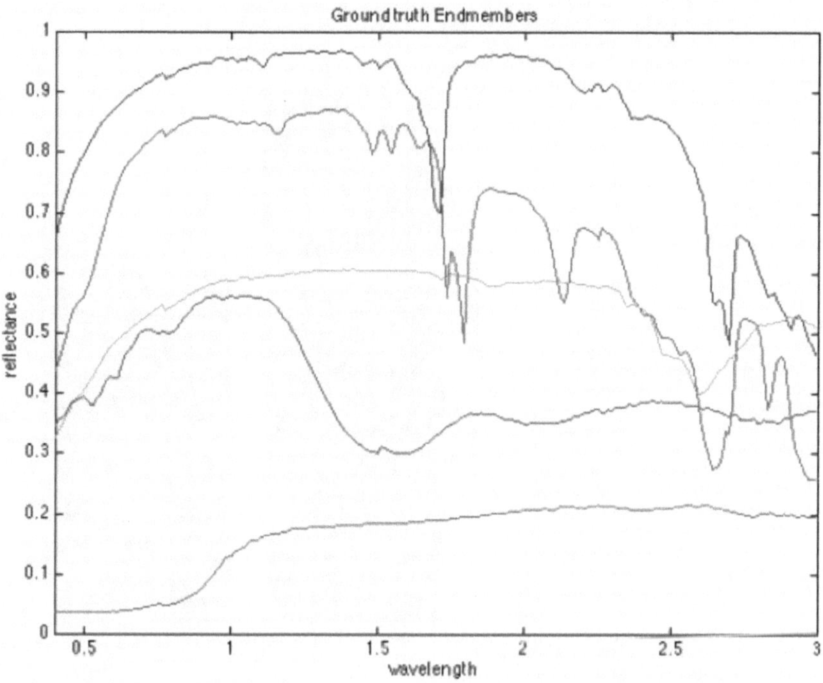

Fig. 5.2. Endmembers from the USGS repository used to synthesize an image

2. Extract the endmembers from the images using the approach described in [10, 11]. The value of the noise gain was set to a=0.5.
3. Compute the distances between the images in the database, on the basis of eq. 5.13, using the morphologically independent induced endmembers. The distances are computed between images with the same number of ground truth endmembers, and with all the remaining images.
4. We consider the R closer images to each image in each case (ground truth and morphologically independent induced endmembers) as the responses to a potential query represented by the image.
5. The images that appear in both responses (based on the ground truth and the morphologically independent induced endmembers) are considered as relevant response images, or correct responses.

In table 5.1 we present the results from the experiment with the 400 images, in terms of the average number of correct responses. First row presents the results when we pool together all the images, regardless of the number of ground truth endmembers. The next rows present the results when we only restrict the search to the subcollection of images with the same number of endmembers as the query image. Each entry in the table corresponds to the average number of relevant (as defined before) images obtained in the response to the query.

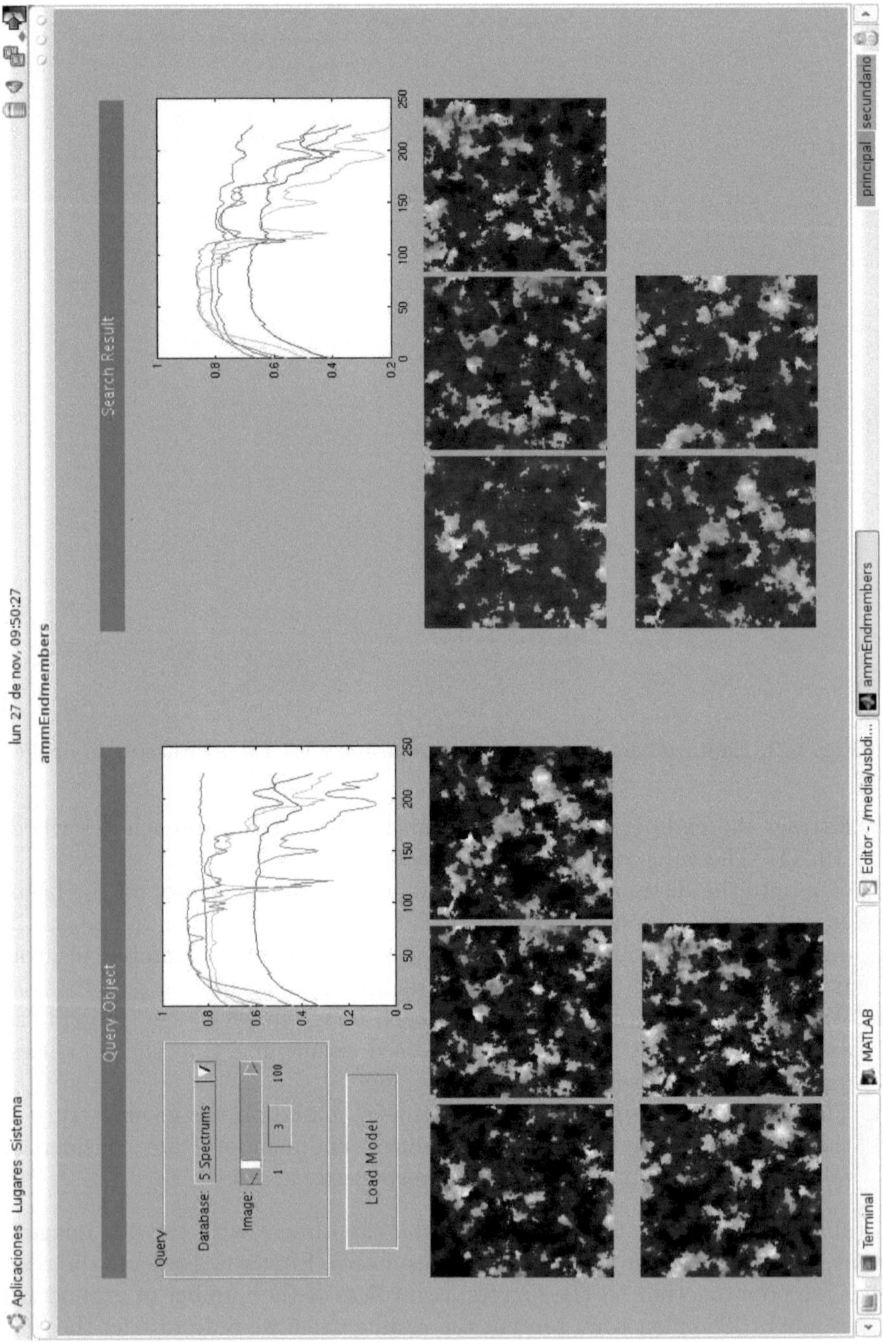

Fig. 5.3. An instance of the hyperspectral CBIR interface: a query and its closest image according to the set of induced endmembers

Table 5.1. Relevance results of the experimentation retrieving images from the synthetics image database

	R=1	R=3	R=5	R=10
All images	0.94	1.21	1.61	2.96
2 endmembers	0.81	1.55	2.27	4.67
3 endmembers	0.98	1.44	2.21	4.96
4 endmembers	0.99	1.53	2.36	4.81
5 endmembers	1.00	1.57	2.37	4.74

In table 5.1 it can be appreciated that the consideration of all the images as responses to the query introduces some confusion and reduce the average number of relevant images obtained in the query. This effect can be due to the fact that the morphological independence algorithm may find a number of endmembers different from the ground truth number of endmembers, making it possible for the image to match with images outside its natural collection of images. Then images with different ground truth numbers of endmembers may become similar enough to enter in their respective response sets.

When we restrict the search to the collections with the same number of ground truth endmembers as the query, all the results improve, except when R=1. We have that near 50% of the responses are significative when R¿1. The case R=1 can be interpreted as the probability of obtaining the closest image in the database according to the distance defined in eq. 5.13 or the probability of classification success. It can be seen that it is very high, close to 1 for all search instances, except for the case of 2 ground truth endmembers.

5.5 Conclusions and Further Work

We have proposed an approach to CBIR in homogeneous databases of hyperspectral images based on the collection of endmembers induced by an algorithms that searches for morphologically independent vectors. We have performed an experiment of exhaustive search on a collection of simulated hyperespectral images. The results are encouraging: almost 100% success in providing the closest image in terms of the ground truth endmembers.

However these results only confirm the ability of the AMM based algorithm to find a good approximation to the ground truth endmembers. We have still to test the effect of additive noise on the results, and maybe to perform comparison with other endmember extraction algorithms. Previous experiments comparing the AMM based algorithm with Independent Component Analysis (ICA) and other algorithms have been favourable to the AMM algorithm [10, 11].

It is possible to define other distances based on the endmembers extracted by the AMM (or any alternative algorithm). For example, the Euclidean distance between individual endmembers may be substituted by max/min distances. The whole set of endmbembers may be evaluated according to the Haussdorf distance.

There are also diverse ways to evaluate the diverse number of endmembers found in the images, introducing penalization terms.

We have not included yet any spatial element in the distance. One of the obvious ways to do it is to compute the correlation between the abundance images, matched according to the relative match of their corresponding endmembers. The distance between images would include a spatial correlation term, with an specific weight that must be tuned. If the images are homogeneous in their definition of the image domain, this correlation can be computed in the straightforward manner. However if they are of different sizes and/or the original capture can not be assumed to be exactly registered, there is a combinatorial problem to be solved, that of finding the optimal image domain transformation (translation, rotation, scaling) to register the two images being spatially matched. The cost of this operation may be prohibitive in the context of large database indexing. Additional information, such as image georeferencing coordinates, will serve to reduce this computational problem. The spatial information may also serve as the basis for user relevance feedback. Relevance may be easier to specify for the user on the abundance images, because she can focus on spatial structures.

Assuming that the image collection is homogenous we have avoided the problem of matching images from different sensors or images with different frequency bands missing due to image preprocessing and noise. This is a main difficulty for our algorithm, because we can not deal at present with missing data. The only available way is to interpolate assuming local continuity of the observed spectra. However, this solution may be hardly accepted by the remote sensing user (geologist, agriculture engineer, biologist, etc).

Acknowledgements

The Spanish Ministerio de Educacin y Ciencia supports this work through grant VIMS-2003-20088-c04-04.

References

1. Alber, I.E., Xiong, Z., Yeager, N., Farber, M., Pottenger, W.M.: Fast retrieval of multi- and hyperspectral images using relevance feedback. In: Proc. Geosci. Rem. Sens. Symp., IGARSS 2001, vol. 3, pp. 1149–1151 (2001)
2. Berman, M., Kiiveri, H., Lagerstrom, R., Ernst, A., Dunne, R., Huntington, J.F.: ICE: A statical Approach to Identifying Endmembers in Hyperspectral Images. IEEE Transactions on Geoscience and Remote Sensing 42(10), 2085–2095 (2004)
3. Brahmi, D., Ziou, D.: Improving CBIR Systems by Integrating Semantic Features. In: Proceedings of the First Canadian Conference on Computer and Robot Vision (CRV 2004). IEEE, Los Alamitos (2004)
4. Chang, C., Plaza, A.: A Fast Iterative Algorithm for Implementation of Pixel Purity Index. IEEE Geoscience and Remote Sensing Letters 3(1), 63–67 (2006)
5. Daschiel, H., Datcu, M.: Information Mining in Remote Sensing Image Archives: System Evaluation. IEEE Transactions on Geoscience and Remote Sensing 43(1), 188–199 (2005)

6. Datcu, M., Daschiel, H., Pelizzari, A., Quartulli, M., Galoppo, A., Colapicchioni, A., Pastori, M., Seidel, K., Marchetti, P.G., D'Elia, S.: Information Mining in Remote Sensing Image Archives: System Concepts. IEEE Transactions on Geoscience and Remote Sensing 41(12), 2923–2936 (2003)
7. Datcu, M., Seidel, K.: Human-Centered Concepts for Exploration and Understanding of Earth Observation Images. IEEE Transactions on Geoscience and Remote Sensing 43(3), 601–609 (2005)
8. Doulamis, N., Doulamis, A.: Evaluation of Relevance Feedback Schemes in Content-Based in Retrieval Systems. Signal Processing: Image Communication 21(4), 334–357 (2006)
9. Gillis, D., Bowles, J., Winter, M.E.: Using Endmembers as a Coordinate System in Hyperspectral Imagery. In: 16th Annual International Symposium on Aerospace/Defense Sensing, Simulation and Controls, Orlando FL (2002)
10. Graña, M., Gallego, J., Hernandez, C.: Further results on AMM for endmember induction. In: Proc. IEEE Workshop on Adv. Tech. Anal. Remotely Sensed Data, Washington D.C., October 2003, pp. 237–243 (2003)
11. Graña, M., Gallego, J.: Associative morphological memories for endmember induction. In: Proc. Geosci. Rem. Sens. Symp., IGARSS 2003, Tolouse, July 2003, vol. 6, pp. 3757–3759 (2003)
12. Healey, G., Jain, A.: Retrieving Multispectral Satellite Images Using Physics-Based Invariant Representations. IEEE Transactions on Pattern Analysis and Machine Intelligence 18(8), 842–848 (1996)
13. Hoi, C., Lyu, M.R.: Group-Based Relevance Feedback With Support Vector Machine Ensembles. In: Proceedings of the 17th International Conference on Pattern Recognition (ICPR 2004). IEEE, Los Alamitos (2004)
14. Hsu, C., Li, C.: Relevance Feedback Using Generalized Bayesian Framework With Region-Based Optimization Learning. IEEE TRansactions on Image Processing 14(10), 1617–1631 (2005)
15. Jiang, W., Er, G., Dai, Q., Gu, J.: Hidden Annotation for Image Retrieval With Long-Term Relevance Feedback Learning. Pattern Recognition 38(11), 2007–2021 (2005)
16. King, I., Jin, Z.: Integrated Probability Function and its Application to Content-Based Image Retrieval by Relevance Feedback. Pattern Recognition 36(9), 2177–2186 (2003)
17. Kozintsev, B.: Computations With Gaussian Random Fields, PhD Thesis, ISR99-3 University of Maryland (1999)
18. Kwak, J.W., Cho, N.I.: Relevance Feedback in Content-Based Image Retrieval System by Selective Region Growing in the Feature Space. Signal Processing: Image Communication 18(9), 787–799 (2003)
19. Li, B., Yuan, S.: A Novel Relevance Feedback Method in Content-Based Image Retrieval. In: Proceedings of the International Conference on Information Technology: Coding and Computing (ITCC 2004). IEEE, Los Alamitos (2004)
20. MacArthur, S.D., Brodley, C.E., Kak, A.C., Broderick, L.S.: Interactive Content-Based Image Retrieval Using Relevance Feedback. Computer Vision and Image Understanding 88(2), 55–75 (2002)
21. Maldonado, O., Vicente, D., Graña, M.: CBIR Indexing Hyperspectral Images. In: Geoscience and Remote Sensing Symposium, IGARSS 2005. Proceedings 2005 IEEE International, July 2005, vol. 2, pp. 1202–1205 (2005)
22. Müller, H., Müller, W., Squire, D.M., Marchand-Maillet, S., Pun, T.: Performance Evaluation in Content-Based Image Retrieval: Overview and Proposals. Pattern Recognition Letters 22(5), 593–601 (2001)

23. Nascimento, J.M.P., Bioucas Dias, J.M.: Does Independent Component Analysis Play a Role in Unmixing Hyperspectral Data. IEEE Transactions on Geoscience and Remote Sensing 43(1), 175–187 (2005)
24. Ng, C.U., Martin, G.R.: Automatic Selection of Attributes by Importance in Relevance Feedback Visualization. In: Proceedings of the 8th International Conference on Information Visualisation (IV 2004). IEEE, Los Alamitos (2004)
25. Park, G., Baek, Y., Lee, H.: Re-Ranking Algorithm Using Post-Retrieval Clustering for Content-Based Image Retrieval. Information Processing & Management 41(2), 177–194 (2005)
26. Plaza, A., Martínez, P., Pérez, R., Plaza, J.: Spatial/Spectral Endmember Extraction by Multidimensional Morphological Operations. IEEE Transactions on Geoscience and Remote Sensing 40(9), 2025–2041 (2002)
27. Plaza, A., Martínez, P., Pérez, R., Plaza, J.: A Quantitative and Comparative Analisys of Endmember Extraction Algorithms from Hyperspectral Data. IEEE Transactions on Geoscience and Remote Sensing 42(3), 650–663 (2004)
28. Plaza, A., Martínez, P., Plaza, J., Pérez, R.: Dimensionality Reduction and Classification of Hyperspectral Image Data Using Sequences of Extended Morphological Transformations. IEEE Transactions on Geoscience and Remote Sensing 43(3), 466–479 (2005)
29. Ramanath, R., Snyder, W.E., Qi, H.: Eigenviews for Object Recognition in Multispectral Imaging Systems. In: Proceedings of the 32nd Applied Imagery Pattern Recognition Workshop (AIPR 2003). IEEE, Los Alamitos (2003)
30. Ren, H., Chang, C.: A Generalized Orthogonal Subspace Projection Approach to Unsupervised Multispectral Image Classification. IEEE Transactions on Geoscience and Remote Sensing 38(6), 2515–2528 (2000)
31. Schröder, M., Rehrauer, H., Seidel, K., Datcu, M.: Interactive Learning and probabilistic Retrieval in Remote Sensing Image Archives. IEEE Transactions on Geoscience and Remote Sensing 38(5), 2288–2298 (2000)
32. Shimabukuro, Y.E., Smith, J.A.: The Least-Squares Mixing Models to Generate Fraction Images Derived From Remote Sensing Multispectral Data. IEEE Transactions on Geoscience and Remote Sensing 29(1), 16–20 (1991)
33. Smeulders, A.W.M., Worring, M., Santini, S., Gupta, A., Jain, R.: Content-Based Image Retrieval at the End of the Early Years. IEEE Transactions on Pattern Analysis and Machine Intelligence 22(12), 1349–1380 (2000)
34. Tao, D., Tang, X.: Nonparametric Discriminant Analysis in Relevance Feedback for Content Based Image Retrieval. In: Proceedings of the 17th International Conference on Pattern Recognition (ICPR 2004). IEEE, Los Alamitos (2004)
35. Vogel, J., Schiele, B.: Performance Evaluation and Optimization for Content-Based Image Retrieval. Pattern Recognition 39(5), 897–909 (2006)
36. Wang, L., Gao, Y., Chan, K.L., Xue, P.: Retrieval With Knowledge-Driven Kernel Design: An Approach to Improving SVM-Based CBIR With Relevance Feedback. In: Proceedings of the 10th International Conference on Computer Vision (ICCV 2005). IEEE, Los Alamitos (2005)
37. Winter, E.M.: N-FINDR: an Algorithm for Fast Autonomous Spectral Endmember Determination in Hyperspectral Data. In: Proceedings of SPIE, Imaging Spectrometry V, vol. 3753 (October 1999)
38. Ziou, D., Boutemedjet, S.: An Information Filtering Approach for the Page Zero Problem. In: Gunsel, B., Jain, A.K., Tekalp, A.M., Sankur, B. (eds.) MRCS 2006. LNCS, vol. 4105, pp. 619–626. Springer, Heidelberg (2006)

6

An Analytical Approach to the Optimal Deployment of Wireless Sensor Networks

J. Vales-Alonso[1], S. Costas-Rodríguez[2], M.V. Bueno-Delgado[1], E. Egea-López[1], F. Gil-Castiñeira[2], P.S. Rodríguez-Hernández[2], J. García-Haro[1], and F.J. González-Castaño[2]

[1] Department of Information Technologies and Communications
Polytechnic University of Cartagena, Spain
{javier.vales,mvictoria.bueno,esteban.egea,joang.haro}@upct.es

[2] Department of Telematics Engineering
University of Vigo, Spain
{scostas,xil,pedro,javier}@det.uvigo.es

Summary. In this work we propose and investigate a novel research topic in Wireless Sensor Networks (WSNs): sensor deployment in order to maximize the interest of the gathered information. The target areas are characterized by zones with different interest levels. We model the interest variable as an "importance function" that assigns a quantitative reference to each point. Due to the nature of WSNs, the sensor deployment must guarantee that the information of all nodes can reach certain control nodes or *sinks*. Otherwise, the events captured by the nodes are lost. This condition is equivalent to ensuring the existence of a communication path from every node to at least one of the sinks. We propose a global optimization model that fulfills all the conditions and we solve it with a simulated annealing algorithm that determines optimal node placement. Moreover, we also characterize the effects of an aerial deployment, that is, the effect of node placement inaccuracy in network behavior and performance. We have found that small placement variations may lead to strong fluctuations in the quality of the connectivity properties of the network topology, and thus to a significant performance degradation in terms of captured interest.

6.1 Introduction

The thriving recent research on Wireless Sensor Networks (WSN) is due to a great extent to the potential benefits for a broad range of applications. Surveillance, environment monitoring and biological risk detection are among the benefits [1]. So far, many aspects of WSN have received research attention, particularly those related to networking issues and self-organization.

Besides, the quality of the sensing procedure and data acquisition is crucial for the network to efficiently perform its intended task, and has to be considered at an early design stage. Sensing quality depends on the *coverage* of the network. In a wide sense, this issue comprises many aspects regarding different properties of the network. In fact, in sensor nodes, the term "coverage" may refer to different concepts, for instance,

M. Graña and R.J. Duro (Eds.): Comput. Intel. for Remote Sensing, SCI 133, pp. 145–161, 2008.
springerlink.com

Fig. 6.1. Aerial sensor deployment over Mars surface (artist conception, credit: NASA/JPL)

- *Wireless coverage* is the area around a node were it can effectively communicate.
- *Sensing coverage* is the area around a node were it can effectively sense some physical magnitude.

The former is related to the connectivity properties of the network and to a variety of associated problems, connectivity being a constraint in most of them, i.e. minimizing some parameter of interest such as energy consumption. The latter determines the quality of the sensed information in general, which is the main topic of another line of research [2].

In this paper, we consider a particular class of sensing coverage problems, defined in [5] as *constrained coverage* problems, that is, finding a deployment configuration that maximizes the collective sensing coverage of the nodes, while satisfying one or more constraints. In this work there are two main differences with previous works, yielding a more realistic scenario:

1. Some zones in the target area are more "interesting" than others, unlike in previous models, which assume a *uniformly distributed* interest across the target area. In other words, we consider that some distribution function of the target area describes the probability of sensing (capturing) an event of interest.
2. We also consider a network deployment stage that introduces placement inaccuracies, and how it affects the global sensing capacity of the network.

Martian exploration helps us illustrate our discussion. So far, Martian probes and rovers have only been able to explore small areas. Which are the "interest areas" in this context? The best chances of finding traces of life byproducts seem to be linked to the past presence of water. Thus, a probe should look in terrains whose composition and properties suggest such presence.

Current spacecrafts are bulky because they must carry engine fuel, power supply systems for the high gain satellite link antennas, and so on. Wireless Sensor Networks may become a feasible alternative to extend their monitoring range, by deploying large quantities of sensors over the interest areas (see Fig. 6.1). The interest area is defined by the probability of occurrence of some event of interest (in Fig. 6.2, the colors of the areas indicate the probabilities of the interest events). The question we face then is how to spread the sensors over those areas to maximize the probability of capturing events of interest. This is the *sensing coverage goal*, for a (possibly) non-uniform distribution of the probability of events. In addition, since sensor nodes are limited devices, they must report the captured events to a special node, the *sink*. In the Martian exploration example, only the probe is able to set a long distance satellite link. Thus, the information must hop from sensor to sensor towards the probe. Therefore, the location of sensors must allow a global communication with the sink. If a node or a group of nodes are isolated, their events will be *lost*. This leads us to a constrained coverage problem, since a connected network is compulsory.

Fig. 6.3 illustrates the problem. Every sensor has a sensing range (r_s) and a communications range (r_t). The distance between hops must be less than r_t and node location should allow to capture as many events as possible subject to r_s. At this point, two possibilities can be considered to deploy the sensing nodes:

- Deterministic: we assume that sink and sensor position can be individual and precisely selected. The solution to the constrained coverage problem provides the set of node locations. If we consider an aircraft deployment, it is unlikely that the nodes will precisely fall into their selected optimal points, and thus the influence of actual sensor positions on performance metric must be studied.
- Random: we assume that the position of the probe can be precisely selected, but the nodes are randomly spread around it (the probe itself launches the nodes, which is called a *ballistic surface deployment*). In this case, the problem is much harder. The results of the optimization problem should provide the parameters that aim the shots (like elevation and speed).

In this work we focus on the *deterministic* case. We introduce a mathematical programming formulation where the objective function seeks to maximize the collective interest, and the constraints reflect the need of a connected network. The rest of this chapter is organized as follows: background is discussed in section 6.2. Section 6.3 provides a formal description of the problem, as well as the parameters and variables involved. Section 6.4 describes the objective function of a global optimization approach to the problem. Using it, section 6.5 presents results for two synthetic scenarios, with large convex areas and small non-convex

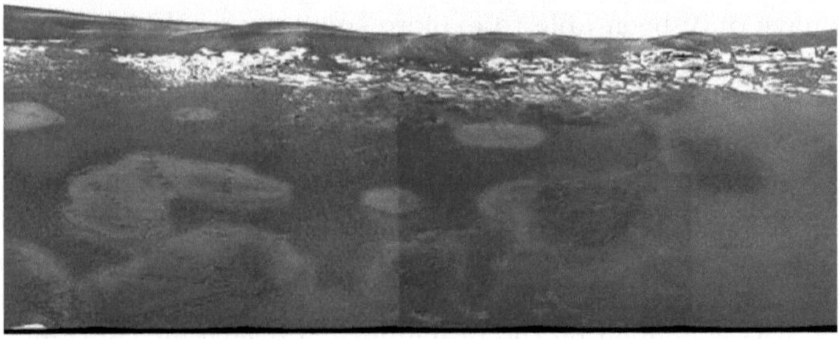

Fig. 6.2. NASA Opportunity Mars Rover image: MiniTES hematite measurements taken by the thermal emission spectrometer superimposed on an image from the panoramic camera. Credit: NASA/JPL.

areas, respectively. Then, section 6.6 studies the effect of random variations in the position of the nodes, to characterize aerial deployment. Finally, section 6.7 discusses the main conclusions of this paper.

6.2 Related Work

As mentioned in the introduction, WSN *coverage* refers to wireless and sensing range. Concerning the first, there is plenty of research on the influence of a variety of parameters in the network. In particular, the transmission power needed to guarantee connectivity is discussed in [8]. Other aspects like energy consumption or node failures have also been considered [9, 10]. In addition, diverse protocols that guarantee network connectivity when the nodes fail or sleep have been proposed [11, 12]. None of these works consider sensing coverage.

On the contrary, in [13] the relationship between sensing coverage and connectivity is discussed. The concept of K-*coverage* is also introduced: a region is K-covered if every location inside it is at least covered by K nodes. It is shown that, if $r_t \geq 2r_s$, 1-coverage implies connectivity. The authors further propose a Coverage Configuration Protocol (CCP), which configures the network to dynamically provide the required degree of coverage (these protocols are called *density control protocols*). The work in [6] is similar, including a density control protocol that considers energy consumption to maximize the network lifetime.

In [7], the problem of achieving K-coverage while conserving the maximum energy is addressed. The authors find the number of sensors that provide K-coverage for grid and random deployments. In those cases, it is assumed that all points have the same importance within the area of interest. Interestingly, in our work we model the importance of a point as a probability density function, and we find the locations that maximize such probability. Note that we do not employ the concept of K-coverage. In fact, our optimization model avoids overlapping of sensing areas. This approach is useful for applications requiring several sensor

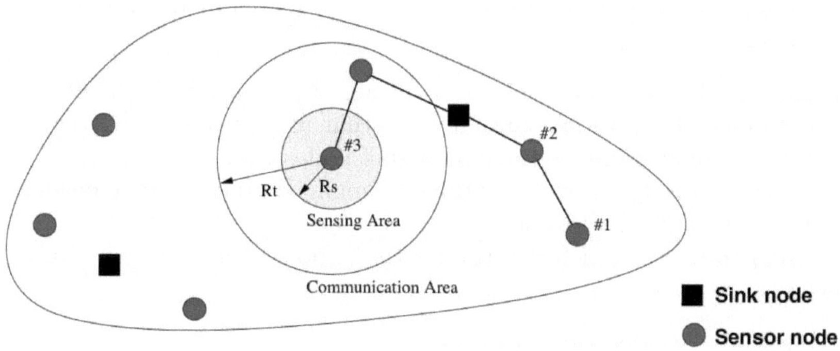

Fig. 6.3. Network deployment

readings of an area, as in vehicle tracking. Moreover, we introduce the concept of *K-survivability*, that is, we obtain the optimal locations that ensure connectivity in case of K sensor failures.

Regarding sensor deployment, reference [5] provides a self-deployment strategy for robots than move to optimal positions. The study in [14] is similar to ours. In it, the authors discuss the inherent uncertainty of sensor readings and assign a probability to sensor detections. These probabilities help to model terrain effects. The authors also describe an algorithm to place the sensors in a grid in order to provide adequate coverage. Instead, our algorithm selects positions in a continuous space.

A similar problem recently studied is optimal positioning of WSN base stations (sinks) [15]. Extending network lifetime or increasing responsiveness of data delivery have been used as performance metrics to obtain the optimal base station positioning. Static and dynamic techniques for positioning have been proposed. This work differs in both goal (maximizing the probability of capturing events) and domain of application (we look for the optimal position of a group of sensors, one of them being also a sink).

Finally, the issue of aerial sensor seeding has been covered in [16]. The algorithms in that work do not look for sensor positions, but control the movement of a flying robot to drop the sensors according to a planned topology. Thus, our proposal is complementary, in the sense that our results could be used as an input for the seeding robot.

6.3 Scenario and Problem Description

In this section we describe our Constrained Coverage Problem. As stated in the introduction, we aim at selecting the position of N sensors in a target area, by maximizing the probability of capturing events, or equivalently, by maximizing the interest (or importance) of the covered area. Henceforth, let us assume a bi-dimensional target area. Let us characterize our model by defining the following parameters:

- The interest of a position is defined by means of an *importance function* $\alpha(x)$, which maps each position x onto a value that indicates its importance.
- The total number of nodes (including sink nodes), N.
- A *transmission range* r_t, which is the maximum distance for two sensor transmitters to be mutually visible.
- A *sensing range* r_s, which determines the region where a sensor can collect measures.

The following considerations also hold:

- Sink nodes also have sensing capabilities, like conventional ones.
- For simplicity reasons, in each scenario there is a single sink node.
- The sensing range is much lower than the transmission range ($r_s << r_t$).
- The importance function around a point x has small variations. Therefore, $\alpha(x^*) \simeq \alpha(x)$, if x is close enough to $x*$.
- The topology of the scenario is known. More formally, all points x belong to a well defined *compact* set $Z \subset I\!R^2$.

Given any two points $x_1, x_2 \in Z$ let us denote $\delta(x_1, x_2) = ||x_1 - x_2||$ as the *Euclidean* distance between points x_1 and x_2. Let $S_i = \{x^* \in Z : \delta(x^*, x_i) < r_s\}$, $x_i \in X$ denote the open ball with center x_i and radius r_s. The joint importance captured by the N nodes in positions $(x_1, \ldots, x_N)$ is given by:

$$\begin{aligned} \alpha &= \int_{\forall x \in S} \alpha(x) \\ \text{being } S &= \bigcup_{i=1 \ldots N} S_i \end{aligned} \tag{6.1}$$

This can be simplified if we consider that sensing areas overlapping for neighbor sensors is unlikely. Thus, we obtain:

$$\alpha = \sum_{i=1}^{N} \int_{\forall x \in S_i} \alpha(x) \tag{6.2}$$

Let us simply denote $\alpha(x_i)$ as α_i. A further simplification is possible. If, in the sensing area of a sensor, the importance function has small variations[1], then equation (6.2) can be rewritten as:

$$\alpha = \sum_{i=1}^{N} \pi r_s^2 \alpha_i \tag{6.3}$$

[1] Let us remark that this does not imply that the importance function is uniform in the entire monitored area, but that the probability is *locally* uniform (around the node), which is reasonable if the physical magnitude to be sensed is continuous.

In fact, since πr_s^2 is constant, the problem is equivalent to maximizing:

$$\hat{\alpha} = \sum_{i=1}^{N} \alpha_i \tag{6.4}$$

That is, selecting the points in the interest area where the event probability is maximum, while satisfying the existence of at least one route from every sensor to the sink, even in case of node failure.

The resulting problem has local minima, and the objective function is not continuous due to the influence of the terrain (irregularly shaped regions of interest, line-of-sight obstacles) and the aforementioned connectivity constraint in case of errors. Therefore, it may be advisable to reformulate it as a global optimization problem (by taking the constraints to the objective function). Then it can be solved with proper heuristics. This approach is developed in the next section.

6.4 Optimization Model

Let us consider a synthetic area $Z \in \mathbb{R}^2$ with regions of interest, like the one sketched in figure 6.5. Clearly, in large/complex scenarios manual placement is not possible, specially when only a small number of sensors is available. Thus, we are interested in proposing a practical optimization model that approximates the solution of the problem we formulated in section 6.3.

As stated in the previous section, we want to deploy N sensors in a set of locations $X = (x_1, \ldots, x_N)$, $x_i \in Z$. Let α_i be the given importance of location x_i. Additionally, let β be a penalty on sensor coverage overlapping. We define the following quality functions:

- $f_{min}(X) : \mathbb{R}^{2N} \rightarrow \mathbb{R}$, $f_{min}((X \setminus \{x_i\}) \cup \{x_i + d\}) = \min(\delta(x_i + d, x_j))$, $x_i, x_j \in X$, $\forall i \neq j$. It measures the minimum separation between a given sensor whose position is shifted d units in $\mathbb{R}^2$ and all the others, and
- $f_s(X) : \mathbb{R}^{2N} \rightarrow \mathbb{R}$, $f_s(X) = \sum_{i=1}^{N} \alpha_i \cdot \text{area}(S_i) - \beta \cdot \text{area} \bigcap_{i=1}^{N} S_i$. It represents the joint importance of the measurements of the sensors, and penalizes coverage overlapping.

Our objective function is:

$$f^K(X) = \begin{cases} \gamma \cdot f_{min}(X) + f_s(X) & \text{if } X \text{ is } K\text{-survivable} \\ C & \text{otherwise,} \end{cases}$$

where C is a suitable constant. Set X is said to be K-survivable if, for any simultaneous failure of K sensors in X, every surviving sensor in x_i lies within the transmission coverage of another surviving sensor x_j ($\delta(x_i, x_j) \leq r_t$) at least, and every surviving sensor must communicate with a sink, either directly (single hop) or indirectly (multiple hops). It is assumed that at least one location in X with guaranteed survivability corresponds to a sink.

We define the *sensor placement optimization problem* as:

$$\begin{aligned} &\underset{X}{maximize}\ f^K(X) \\ &subject\ to: \{x_1,\ldots,x_N\} \in Z, \end{aligned} \tag{6.5}$$

which satisfies the following conditions:

A1. It is a box-constrained global optimization problem.
A2. $f^K(X) : I\!R^{2N} \rightarrow I\!R$ is upper bounded.
A3. $f(X)$ has directional derivatives $f'(X, d_i)$ everywhere defined along the canonical directions and their opposite (which positively span R^{2N}), and

$$\eta > 0 \Rightarrow \begin{cases} f'(X, \eta d_i) = \eta f'(X, d_i), \\ f(X + \eta d_i) = f(X) + \eta f'(X, d_i) + o(\eta) \end{cases} \tag{6.6}$$

Consequently, it is possible to apply the derivative-free method in [17], which converges to a stationary point of $f^K(X)$ regarding the directions given by the canonical base and its opposite. Alternatively, it is possible to approximate the solution with heuristics such as neighborhood search or simulated annealing. We have chosen the latter for the preliminary tests in this paper.

6.5 Numerical Tests

We generated two synthetic scenarios: "convex/large regions" and "non-convex/small" regions to test the performance of the optimization algorithm in the previous section. Figure 6.4 illustrates them. In order to assign importance levels, we cover each scenario with a 640×480 bitmap. Each bit has a weight from 0 (minimum interest) to 255 (maximum). Figure 6.4 represents it with a color scale (light→low interest, dark→high). To determine the importance of a sensor placement, we read the weight of the surrounding pixels. In the scenarios we spread $N = 30$ sensors with a coverage radius of $r_s = 10$ units and a transmission range of $r_t = 60$ units.

The initial location of all sensors is the center of the scenario x_c, in order to guarantee that we depart from a K-survivable set X. In order to speed up the calculations we do not compute $f_s(X_n)$ exactly, but approximate it as follows:

a) Set $f_s(X_n) = 0.0$
b) for $i = 1, \ldots, N$
   i) $f_s(X_n) = f_s(X_n) + \alpha_i S_i$
   ii) for $j = i + 1, \ldots, N$
      if $\delta(x_i, x_j) < 2r_s$ $f_s(X_n) = f_s(X_n) - (4r_s^2 - \delta(x_i, x_j)^2)$

The set X_n is K-survivable if every subset that results from extracting K locations from X_n is 0-survivable. To check K-survivability we proceed as follows:

We define three tables, *original*, *current* and *next*. It is assumed that sinks always survive. We assign all surviving nodes (simply *nodes*, hereafter) but the

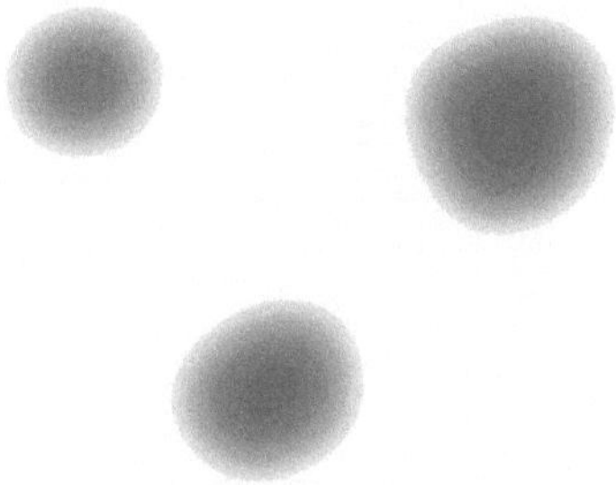

(a) Convex topology

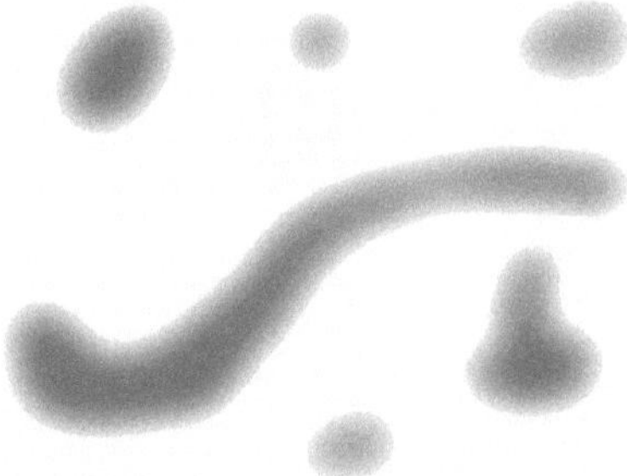

(b) Non-convex topology

Fig. 6.4. Scenarios in the numerical test

sinks to *original*. We assign all the sinks to *current*. Then we move all the nodes that fall within the coverage of a sink to *next*.

At that point, every node with direct communication with a sink belongs to *next*. The nodes in *original* may communicate with the nodes in *current* through some nodes in *next*.

As a last step, we delete all the nodes in *current*, move the nodes in *next* to *current* and repeat the entire process until *original* is empty (the set is survivable) or it is impossible to move any node from *original* to *next* (the set is not survivable).

Simulated Annealing We are now ready to define our simulated annealing algorithm:

Simulated annealing algorithm

Initialization: $T = T_0$, $i = 0$. Set $x_i = x_c$, $i = 1, \ldots, N$.
While $T > 0$:

a) $i = i + 1$. If $i > N$, $i = 1$ and $T = T - \triangle T$.
b) Let $v \in I\!R^2$ be a random direction so that $||v|| = 1$. Set $d = T \cdot r_t \cdot v$.
c) If $x_i + d \notin Z$ or $X^* = X - \{x_i\} + \{x_i + d\}$ is not K-survivable, goto a).
d) If $f^K(X^*) > f^K(X)$, $p = 1 - 0.5T$. Otherwise, $p = 0.5T$.
e) If $rand() < p$, $X = X^*$.

Constant C is implicit, in the sense that it is a lower bound for $f^K(X)$ that prevents from updating the set of locations. The other settings in our tests were $T_0 = 1$ and $\triangle T = 10^{-4}$. Note the importance of parameter γ in the definition of $f^K(X)$. A small value makes it difficult to colonize new regions of interest, since the nodes try to stick together as far as the value of $f_s(X)$ is large. A large value tends to disperse the nodes regardless of the importance of the regions they measure. Therefore, it is necessary to tune this parameter. We started from $\gamma = 0$ and kept increasing it in 0.1 steps until the nodes began to ignore the regions of interest.

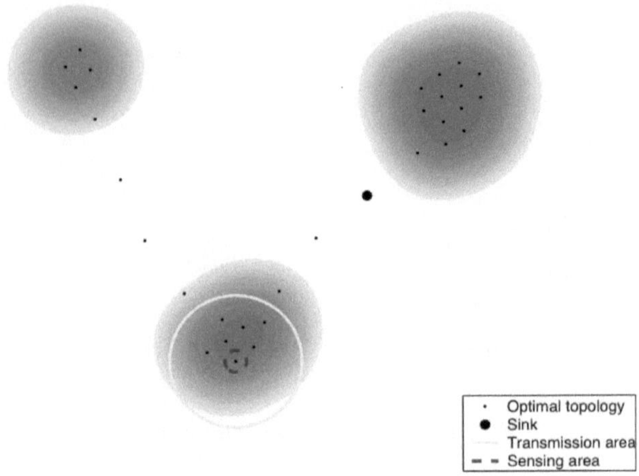

Fig. 6.5. Convex/large regions scenario, $K = 0$, $\gamma = 1.0$

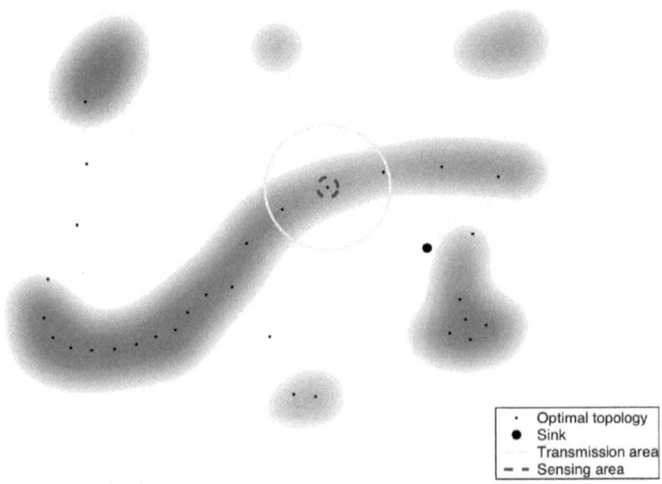

Fig. 6.6. Non-convex/small regions scenario, $K = 0$, $\gamma = 1.1$

Figures 6.5 to 6.8 show our results for $K = 0$ and $K = 1$. In each case we indicate the value of γ after tuning it (we started with $\gamma = 0$ and increased it in 0.1 steps. We chose the largest value that produced a non-degenerate case, which we identified visually). Elapsed times on a Pentium IV were under 10 seconds for $K = 0$ and less than 50 seconds for $K = 1$. It is interesting to observe that setting $K = 1$ produces alternative paths, but the algorithm tends to use as few nodes as possible to create "bridges" between regions of interest.

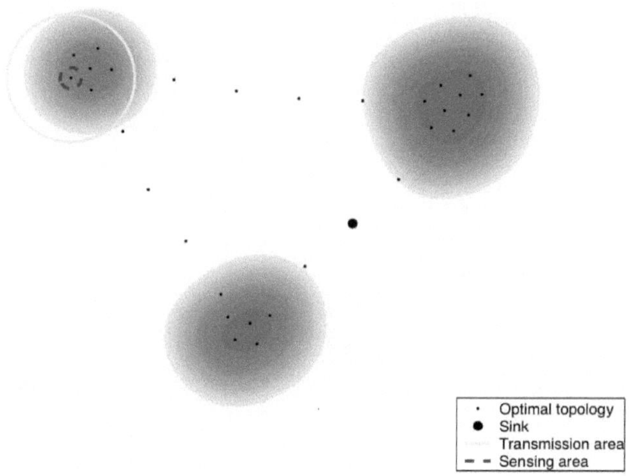

Fig. 6.7. Convex/large regions scenario, $K = 1$, $\gamma = 0.5$

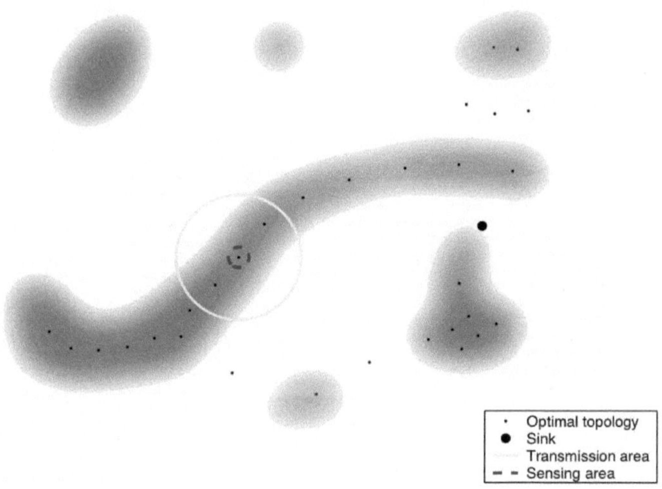

Fig. 6.8. Non-convex/small regions scenario, $K = 1$, $\gamma = 1.1$

6.6 Characterization of Aerial Deployment

To complement the results in the previous sections, we also model the impact of the inaccuracies of aerial deployment. We approximate them as random changes that determine the final position of nodes. Their effect is twofold. First, the captured importance may be lower in the real positions. Second, and more importantly, the network can be split due to communication disruptions. Therefore, the information from some node branches may not reach the sinks. Certainly,

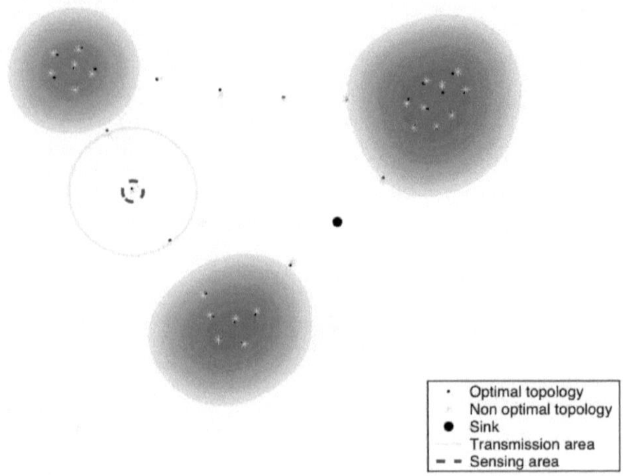

Fig. 6.9. Convex/large regions scenario, $K = 1$, aerial deployment

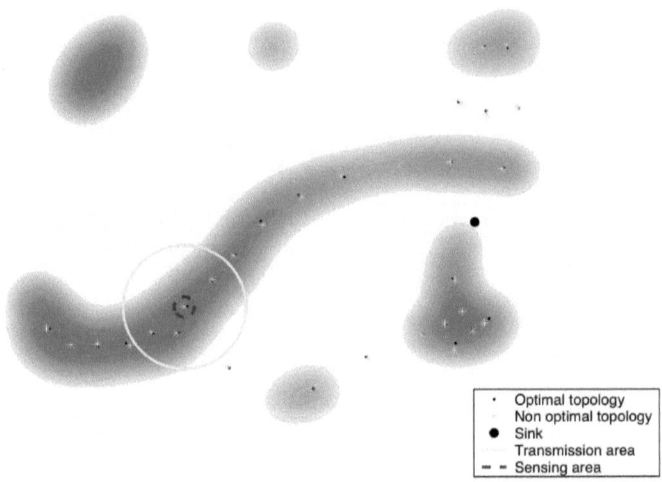

Fig. 6.10. Non-convex/small regions scenario, $K = 1$, aerial deployment

these effects grow with the variance of the positions. Thus, we are interested in characterizing information loss as a function of that variance.

Specifically, the initially selected position $x = (x^1, x^2)$ of a node is altered by a random variable $(\boldsymbol{X^1}, \boldsymbol{X^2})$. Thus, the final position of the node is $\hat{x} = (x^1 + \boldsymbol{X^1}, x^2 + \boldsymbol{X^2})$. Additionally, let us assume that position modifications of different nodes are independent.

Characterizing the random variable $(\boldsymbol{X^1}, \boldsymbol{X^2})$ is a complex task, since it must model many factors (for instance, density of atmospheric layers, wind, height of

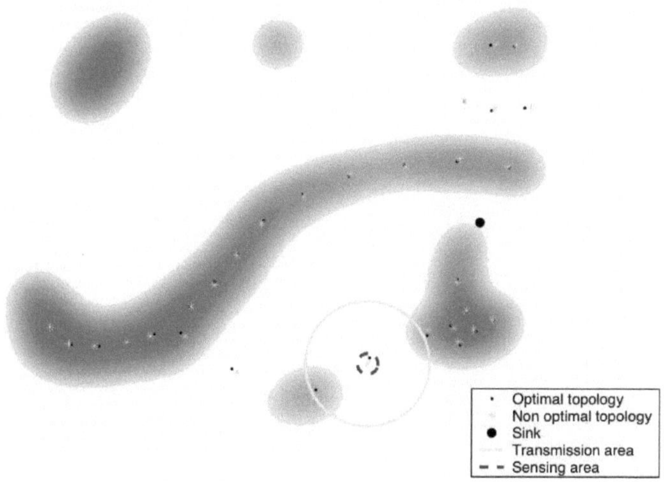

Fig. 6.11. Convex/large regions scenario, $K = 0$, aerial deployment, unconnected network

the launching position, etc.). For simplicity, in this section we will assume that $(\boldsymbol{X}^1, \boldsymbol{X}^2)$ is a normal bivariate distribution with null mean, null correlation coefficient (that is, $\boldsymbol{X}^1$ and $\boldsymbol{X}^2$ are independent) and known variance (σ^2). Therefore, σ controls the dispersion of the actual placement from the solution of the global optimization problem.

Figures 6.9 and 6.10 show the effect of aerial deployments on the scenarios of the previous section. Optimal placements are marked as dots, while actual ones are marked as asterisks. In this first example the network is connected, and the variation in information loss is minimal and exclusively due to the misalignment from the optimal placements. Indeed, the network in figure 6.11 is unconnected. A considerable amount of information is lost (from the isolated branches).

We have also evaluated the effect of σ in the captured importance (α). The following experiment was performed numerically: Starting with the optimal configuration, the position of each node is displaced along both the horizontal and vertical axis by two independent normal variables $N(0, \sigma)$. This computation is conducted to obtain statistics of the mean $\hat{\alpha}(X)$ value (see section 6.3), and the probability that the deployment yields an unconnected network. If the new network is connected (that is, there is a path from every node to the sink) the importance function $\hat{\alpha}(X)$ is computed, and that sample is added to the mean calculation. Otherwise, it is assumed that $\hat{\alpha}(X) = 0$, and the probability of obtaining an unconnected network is updated. This process was performed until the statistics had a quality of 0.95 or better with a relative tolerance of 0.05.

Figures 6.12 and 6.13 show the probability that the network is connected for the convex and non-convex scenarios, respectively. As it can be observed, this probability drops suddenly even for low values of σ. Therefore, aerial deployments may produce split networks, and it may be useless to take the optimal

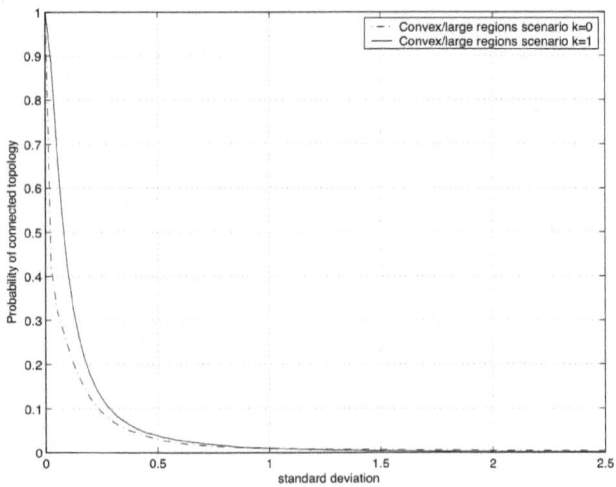

Fig. 6.12. Convex/large regions scenario, probability of connected topology versus σ

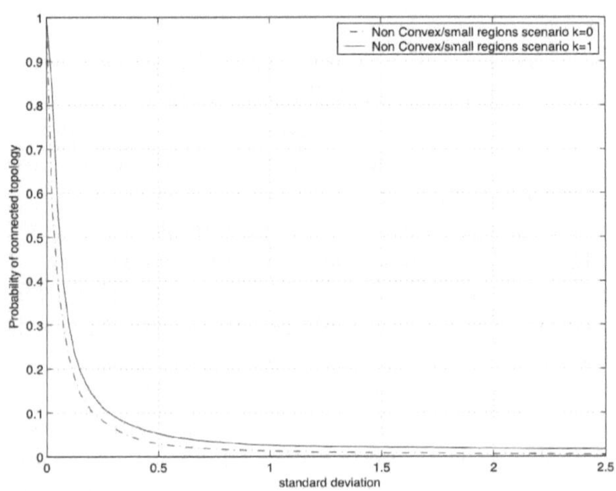

Fig. 6.13. Non-convex/small regions scenario, probability of connected topology versus σ

configuration as a reference. This effect is a consequence of the "bridges" between "information islands", as shown in figure 6.11. The objective function tends to separate nodes to the maximal distance in zones without importance. This is logical, since a small number of nodes are available, and they must be used to cover the area of interest. However, in the case of aerial deployments, there is a high probability that the network will become unconnected at those bridges.

Besides, figures 6.14 and 6.15 show $\hat{\alpha}(X)/\hat{\alpha}(X^*)$ as a function of σ, being X^* the optimal set of locations and X the suboptimal set. That is, the captured

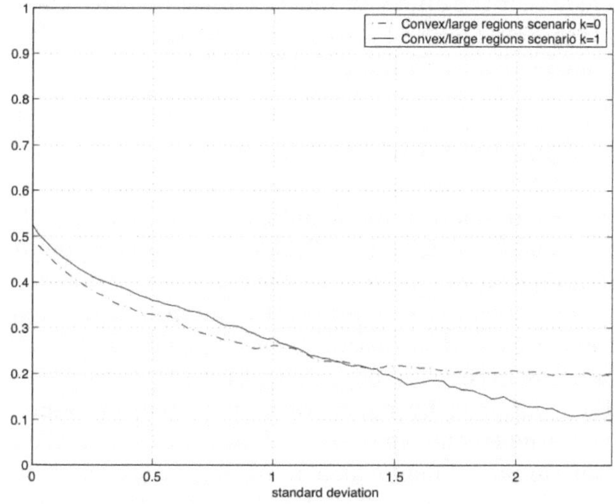

Fig. 6.14. Convex/large regions scenario, $\hat{\alpha}(X)/\hat{\alpha}(X^*)$ in connected topologies versus σ

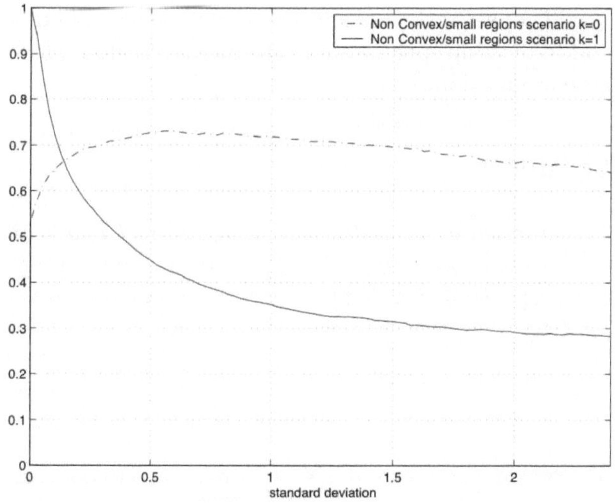

Fig. 6.15. Non-convex/small regions scenario, $\hat{\alpha}(X)/\hat{\alpha}(X^*)$ in connected topologies versus σ

relative importance versus the optimal one. In fact, note that selecting $\sigma = 0$ is equivalent to the optimal configuration in section 6.5, *i.e.* $X = X^*$. Thus, these curves show the effect of inaccuracies in the position of the nodes, which may result in creating disconnected topologies. As expected, the effect of disconnected topologies produces a fast decreasing of the captured information as σ increases.

We can conclude that the objective function must be redesigned if the scenario contains several separated interest areas to cope with the aerial deployment issue, while still maximizing captured importance.

6.7 Conclusions and Future Work

In this work we have developed mathematical programming tools that allow us to select the best placement for a set of nodes in an area of interest. In the formulation of the optimization problem we have modeled the main characteristics of WSNs and some realistic assumptions: the interest is non-uniformly spread over the monitored area, and the network must be connected, i.e. it must provide communication paths from every node to at least one of the sinks. The optimal placement is then obtained by means of a simulated annealing algorithm. In addition, we have studied the effect of random variations in the position of the nodes (illustrating an aerial deployment). We have discovered that, although it finds optimal placements, our formulation for the optimization problem tends to establish large hop bridges between highly interesting areas. Therefore, small variations in node positions may lead to network disconnection, preventing information from isolated nodes from reaching the sinks.

Our future work will be orientated towards formulating new optimization procedures that will consider the inaccuracy of real placements more realistically. Indeed, we intend to develop specific solutions for the random placement problem described in the introduction.

Acknowledgements

This work has been funded by *Ministerio de Educación y Ciencia* grant DEP2006-56158-C03, by *Ministerio de Industria, Turismo y Comercio* grant m:Ciudad (FIT-330503-2006-2, partially funded by FEDER), and by *Xunta de Galicia grant* PGIDIT05TIC00101CT, Spain.

References

1. Akyildiz, I.F., Sankarasubramaniam, Y., Cayirci, E.: Wireless Sensor Networks: a survey. Computer Networks 38(4), 393–422 (2002)
2. Meguerdichian, S., Koushanfar, F., Potkonjak, M., Srivastava, M.B.: Coverage Problems in Wireless Ad-Hoc Sensor Networks. In: Proc. IEEE 20th Annual Joint Conference of the IEEE Computer and Communications Societies, vol. 3, pp. 1380–1387 (2001)
3. Gage, D.W.: Command and control for many-robot systems. In: Proc. 19th AUVS Technical Symposium, pp. 22–24 (1992)
4. Marengoni, M., Draper, A., Hanson, A., Sitaraman, R.A.: System to Place Observers on a Polyhedral Terrain in Polynomial Time. Image and Vision Computing 18, 773–780 (1996)

5. Poduri, S., Sukhatme, S.G.: Constrained Coverage for Mobile Sensor Networks. In: Proc. IEEE Intl. Conf. Robotics and Automation, pp. 165–172 (2004)
6. Zhang, H., Hou, J.C.: Maintaining Sensing Coverage and Connectivity in Large Sensor Networks. Ad Hoc & Sensor Wireless Networks 1, 89–124 (2005)
7. Kumar, S., Lai, T.H., Balogh, J.: On K-coverage in a Mostly Sleeping Sensor Network. In: Proc. IEEE Conf. on Mobile Computing and Networking, pp. 144–158 (2004)
8. Gupta, P., Kumar, P.R.: Critical Power Asymptotic Connectivity in Wireless Networks. In: Stochastic Analysis, Control, Optimization and Applications, pp. 547–566. Birkhauser, Basel (1998)
9. Santi, P.M., Blough, D.: An Evaluation of Connectivity in Mobile Wireless Ad Hoc Networks. In: Proc. Intl. Conf. on Dependable Systems and Networks, pp. 89–98 (2002)
10. Shakkottai, S., Srikant, R., Shorff, N.B.: Unreliable sensor grids: coverage, connectivity and diameter. In: Proc. IEEE 22nd Annual Joint Conference of the IEEE Computer and Communications Societies, vol. 2(5), pp. 1073–1083 (2003)
11. Xu, Y., Heidemann, J., Estrin, D.: Geographic Informed Energy Conservation for Ad Hoc Routing. In: Proc. 7th ACM/IEEE Intl. Conf. on Mobile Computing and Networking, pp. 70–84 (2001)
12. Chen, B., Jamieson, K., Balakrishnan, H., Morris, R.: Span: And Energy-Efficient Coordination Algorithm for Topology Maintenance in Ad Hoc Wireless Networks. In: Proc. 7th ACM/IEEE Intl. Conf. on Mobile Computing and Networking, pp. 85–96 (2001)
13. Wang, X., Xing, G., Zhang, Y., Lu, C., Pless, R., Gill, C.: Integrated Coverage and Connectivity Configuration in Wireless Sensor Networks. In: Proc. First Intl. Conf. on Embedded Networked Sensor Systems, pp. 28–39 (2003)
14. Dhillon, S.S., Chakrabarty, K., Iyengar, S.S.: Sensor placement for grid coverage under imprecise detections. In: Proc. Fifth Intl. Conf. on Information Fusion, vol. 2, pp. 1581–1587 (2002)
15. Akkay, K., Younis, M.F., Youssed, W.: Positioning of Base Stations in Wireless Sensor Networks. IEEE Communications Magazine 45(4), 96–102 (2007)
16. Corke, P., Hrabar, S., Peterson, R., Rus, D., Saripalli, S., Sukhatme, G.: Autonomous deployment and repair of a sensor network using an unmanned aerial vehicle. In: Proc. IEEE Intl. Conf. on Robotics and Automation, vol. 4, pp. 3602–3608 (2004)
17. González-Castaño, F.J., Costa-Montenegro, E., Burguillo-Rial, J.C., García-Palomares, U.M.: Outdoor WLAN Planning via Non-Monotone Derivative-Free Optimization: Algorithm Adaptation and Case Study. In: Computational Optimization and Applications (in press)

7

Parallel Spatial-Spectral Processing of Hyperspectral Images

Antonio J. Plaza

Department of Technology of Computers and Communications
University of Extremadura, Avda. de la Universidad s/n
E-10071 Cáceres, Spain
aplaza@unex.es

Summary. Hyperspectral image processing has been a very active area in remote sensing and other application domains in recent years. Despite the availability of a wide range of advanced processing techniques for hyperspectral data analysis, a great majority of available techniques for this purpose are based on the consideration of spectral information separately from spatial information information, and thus the two types of information are not treated simultaneously. In this chapter, we describe several spatial-spectral techniques for dimensionality reduction, feature extraction, unsupervised and supervised classification, spectral unmixing and compression of hyperspectral image data. Most of the techniques addressed in this chapter are based on concepts inspired by mathematical morphology, a theory that provides a remarkable framework to achieve the desired integration. Parallel implementations of some of the proposed techniques are also developed to satisfy time-critical constraints in remote sensing applications, using NASA's Thunderhead Beowulf cluster for demonstration purposes throughout the chapter. Combined, the different topics covered by this chapter offer a thoughtful perspective on future potentials and emerging challenges in the design of robust spatial-spectral techniques for hyperspectral image analysis.

7.1 Introduction

Hyperspectral imaging is concerned with the measurement, analysis, and interpretation of spectra acquired from a given scene (or specific object) at a short, medium or long distance by an airborne or satellite sensor [1]. The concept of hyperspectral imaging originated at NASA's Jet Propulsion Laboratory in California, which developed instruments such as the Airborne Imaging Spectrometer (AIS), then called AVIRIS, for Airborne Visible Infra-Red Imaging Spectrometer [2]. This system is now able to cover the wavelength region from 0.4 to 2.5 μm using more than two hundred spectral channels, at nominal spectral resolution of 10 nm. As a result, each pixel vector collected by a hyperspectral instrument can be seen as a *spectral signature* or *fingerprint* of the underlying materials within the pixel (see Fig. 7.1).

The special characteristics of hyperspectral datasets pose different processing problems, which must be necessarily tackled under specific mathematical formalisms, such as feature extraction, classification and segmentation, spectral

M. Graña and R.J. Duro (Eds.): Comput. Intel. for Remote Sensing, SCI 133, pp. 163–192, 2008.
springerlink.com

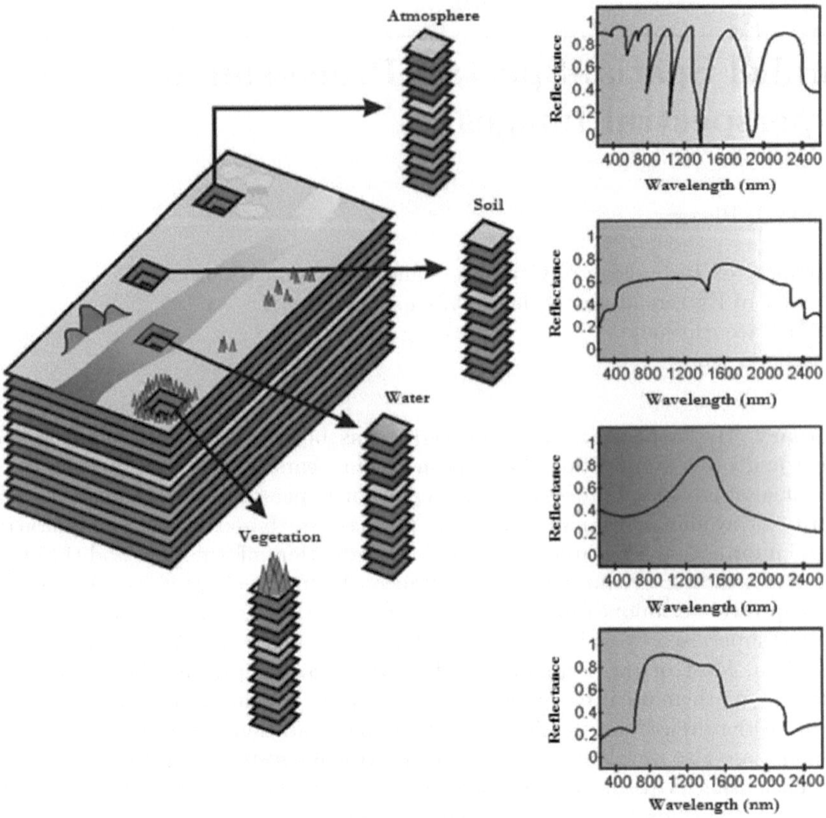

Fig. 7.1. The concept of hyperspectral imaging

mixture analysis, or data compression [3]. For instance, several computational intelligence-related techniques have been applied to extract relevant information from hyperspectral data during the last decade [4]. Taxonomies of remote sensing data processing algorithms (including hyperspectral analysis methods) have been developed in the literature [5]. It should be noted, however, that most available hyperspectral data processing techniques have focused on analyzing the data without incorporating information on the spatially adjacent data, i.e., hyperspectral data are usually not treated as images, but as unordered listings of spectral measurements where the spatial coordinates can be randomly shuffled without affecting the analysis [6]. However, one of the distinguishing properties of hyperspectral data, as collected by available imaging spectrometers, is the multivariate information coupled with a two-dimensional pictorial representation amenable to image interpretation. Subsequently, there is a need to incorporate the image representation of the data in the development of appropriate application-oriented techniques for the understanding of hyperspectral data [7].

Specifically, the importance of analyzing spatial and spectral patterns simultaneously has been identified as a desired goal by many scientists devoted to hyperspectral data analysis [6], [8]. This type of processing has been approached in the past from various points of view. For instance, techniques have discussed the refinement of results obtained by applying spectral-based techniques to multispectral images (with tens of spectral channels) through a second step based on spatial context [9]. Such contextual classification, extended also to hyperspectral images [6], accounts for the tendency of certain ground cover classes to occur more frequently in some contexts than in others. This approach consists of two parts: the definition of a pixel neighborhood (surrounding each pixel) and the performance of a local operation so that the pixel may be changed into the label mostly represented in the window that defines the neighborhood. This simple operation separates spatial from spectral information, and thus the two types of information are not treated simultaneously.

In previous work, we have explored the application of mathematical morphology operations to integrate both spatial and spectral responses in hyperspectral data analysis [10], [11]. Mathematical morphology is a non-linear spatial processing technique that provides a remarkable framework to achieve the desired integration [12]. It was originally defined for binary images [13] and soon extended to greyscale and color image cases, but it has been seldom used to process hyperspectral images. In this chapter, we provide detailed insight on the use of extended morphological operations for integrating spatial and spectral information in different steps comprising the entire hyperspectral image processing chain, from the moment the data is collected onboard the sensor to the time in which an application-oriented outcome is provided to end-users.

Specifically, the main contribution of this chapter is the development and validation of a set of highly innovative techniques for spatial-spectral analysis of hyperspectral image scenes in the following relevant areas:

1. **Dimensionality reduction.** A new method for spatial-spectral feature extraction, based on morphological concepts, is developed and compared to a standard spectral-based technique for dimensionality reduction, i.e., the principal component transform (PCT) [3].
2. **Unsupervised classification/segmentation.** A novel methodology for unsupervised image classification/segmentation is developed by extending the morphological watershed transformation (commonly used for unsupervised segmentation of grayscale images) to hyperspectral image processing. This technique is compared to a standard approach for unsupervised classification of multicomponent data, i.e., the ISODATA algorithm [3].
3. **Spectral mixture analysis.** In order to address the problem of mixed pixels, resulting from the limited spatial resolution and also from naturally-occurring mixing phenomena, we develop two separate contributions:
   - **Endmember extraction.** First, we develop a new iterative version of a previously developed morphological method for extraction of pure spectral signatures (called endmembers in hyperspectral analysis terminology).

  The algorithm allows propagation of pure pixels between subsequent iterations, as opposed to the previous version of the algorithm. This algorithm is compared to other standard endmember extraction algorithms such as Pixel Purity Index (PPI) [14] or N-FINDR [15].
- **Unmixing.** Then, we develop a new spatial-spectral unmixing procedure based on the endmembers derived by the proposed extended morphological endmember extraction method. This technique is compared to other unmixing techniques in the literature such as unconstrained and fully constrained linear spectral unmixing [16], [4].

4. **Data compression.** In order to address the problems introduced by the extremely large volume of hyperspectral image data sets, we also develop a new spatial-spectral compression algorithm based on morphological concepts and compare the algorithms with standard techniques for data compression such as Set Partitioning in Hierarchical Trees (3D-SPIHT) [17] and JPEG 2000 multicomponent [18].
5. **Parallel implementations.** Last but not least, we develop high performance computing implementations [19] for all the spatial-spectral algorithms introduced in this chapter and thoroughly assess their performance on the 568-processor Thunderhead Beowulf cluster of computers at NASA's Goddard Space Flight Center in Maryland. The detailed cross-validation of parallel hyperspectral processing algorithms in different application domains, conducted in this chapter using a massively parallel computing platform, may help image analysts in selection of parallel algorithms for specific applications.

The remainder of the chapter is structured as follows. Section 7.2 presents the proposed framework to extend mathematical morphology to hyperspectral images. Section 7.3 provides a framework for parallel implementation of morphological spatial-spectral algorithms for hyperspectral image processing. Section 7.4 develops a collection of parallel morphological algorithms for dimensionality reduction (P-MORPHDIM), watershed-based classification (P-WSHED), morphological endmember extraction (P-MORPHEE), morphological spectral unmixing (P-MORPHSU), and morphological data compression (P-MORPHCOMP). These algorithms have been designed to be efficiently run on commodity cluster computing platforms. Section 7.5 presents experimental results and comparisons to other standardized algorithms in the literature, using two well-known, publicly available hyperspectral scenes: the AVIRIS Indian Pines scene and the AVIRIS Cuprite scene, both widely used as benchmark data to validate classification and spectral unmixing accuracy in hyperspectral imaging applications. This section also includes performance results on the Thunderhead massively parallel cluster, along with a summary of the main observations and lessons learned after the detailed quantitative and comparative assessment of algorithms conducted in this chapter. Finally, Section 7.6 concludes with some remarks and hints at plausible future research.

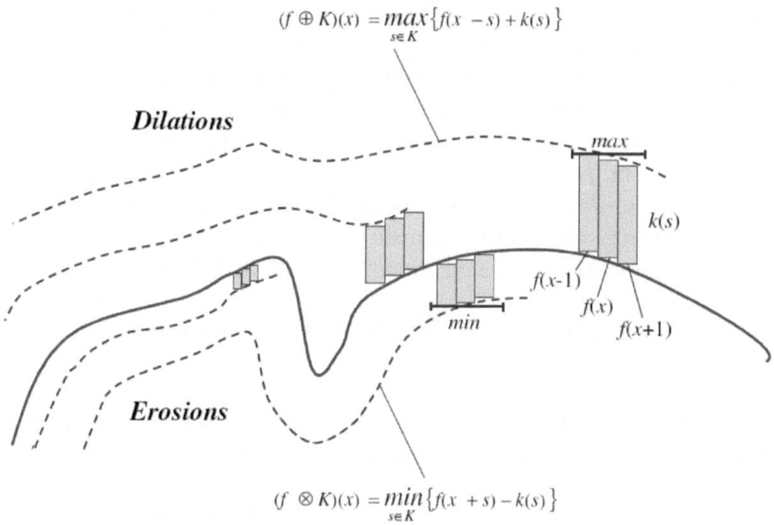

Fig. 7.2. Graphical interpretation of morphological erosion and dilation operations

7.2 Extended Mathematical Morphology

Mathematical morphology is a theory for spatial structure analysis that was established by introducing fundamental operators applied to two sets [13]. A set is processed by another one having a carefully selected shape and size, known as the structuring element (SE). The two basic operations of mathematical morphology are erosion and dilation. These operators can be graphically illustrated (in the context of greyscale morphology) by viewing the image data as an imaginary topographic relief in which the brighter the gray tone, the higher the corresponding elevation. With this assumption in mind, morphological operations can be interpreted as the result of sliding a SE over the topographical relief, so that the SE defines the new (dilated or eroded) scene based on its spatial properties such as height or width (see Fig. 7.2).

Extension of morphological operators to multichannel data such as hyperspectral imagery with hundreds of spectral channels is not straightforward. A simple approach consists in applying grayscale morphology techniques to each channel separately, an approach that has been called *marginal morphology* in the literature [20]. However, the marginal approach is often unacceptable in remote sensing applications because, when morphological techniques are applied independently to each image channel, analysis techniques are subject to the well-known problem of *false colors*; that is, it is very likely that new spectral constituents (not present in the original hyperspectral image) may be created as a result of processing the channels separately. An alternative (and perhaps more appropriate) way to approach the problem of multichannel morphology is to treat the data at each pixel as a vector [12]. Unfortunately, there is no unambiguous means of defining the minimum and maximum values between two vectors of more than one dimension, and thus it is important to define an appropriate arrangement of vectors in the selected vector space.

In this chapter, we develop an application-driven vector ordering technique based on a spectral purity-based criterion [10], where each pixel vector is ordered according to its spectral distance to other neighboring pixel vectors in the N-dimensional data set $\boldsymbol{f}$. More specifically, we adopt a distance-based technique which utilizes a cumulative distance between one particular pixel vector $\boldsymbol{f}(x, y)$, where (x, y) indicates the spatial coordinates, and all the pixel vectors in the spatial neighborhood given by a SE denoted by K as follows [7]:

$$C_K(\boldsymbol{f}(x,y)) = \sum_{(s,t)\in K} \text{SAD}(\boldsymbol{f}(x,y), \boldsymbol{f}(s,t)), \tag{7.1}$$

where SAD is the spectral angle distance [4]. The SAD between two pixel vectors $\boldsymbol{f}(x, y)$ and $\boldsymbol{f}(s, t)$ is given by the following expression:

$$\text{SAD}(\boldsymbol{f}(x,y), \boldsymbol{f}(s,t)) = cos^{-1}\left(\frac{\boldsymbol{f}(x,y)\cdot \boldsymbol{f}(s,t)}{\|\boldsymbol{f}(x,y)\| \cdot \|\boldsymbol{f}(s,t)\|}\right). \tag{7.2}$$

As a result,$C_K(\boldsymbol{f}(x, y))$ is given by the sum of SAD scores between $\boldsymbol{f}(x, y)$ and every other pixel vector in the K-neighborhood. At this point, we need to be able to define a maximum and an minimum given an arbitrary set of vectors $\mathbf{S} = \{\mathbf{v}_1, \mathbf{v}_2, \cdots, \mathbf{v}_p\}$, where k is the number of vectors in the set. This can be done by computing $C_K(\mathbf{S}) = \{C_K(\mathbf{v}_1), C_K(\mathbf{v}_2), \cdots, C_K(\mathbf{v}_k)\}$ and selecting $\mathbf{v}_i$ such that $C_K(\mathbf{v}_i)$ is the minimum of $C_K(\mathbf{S})$, with $1 \leq i \leq k$. In similar fashion, we can select $\mathbf{v}_j$ such that $C_K(\mathbf{v}_j)$ is the maximum of $C_K(\mathbf{S})$, with $1 \leq j \leq p$. Based on the definitions above, the extended erosion $\boldsymbol{f} \ominus K$ consists of selecting the K-neighborhood pixel vector that produces the minimum C_K value as follows [10]:

$$(\boldsymbol{f} \ominus K)(x,y) = argmin_{(s,t)\in K}\{C_K(\boldsymbol{f}(x+s, y+t)\}. \tag{7.3}$$

On the other hand, the extended dilation $\boldsymbol{f} \oplus K$ selects the K-neighborhood pixel that produces the maximum value for C_K as follows [10]:

$$(\boldsymbol{f} \oplus K)(x,y) = argmax_{(s,t)\in K}\{C_K(\boldsymbol{f}(x-s, y-t)\}. \tag{7.4}$$

For illustrative purposes, Fig. 7.3 shows a graphical representation of the performance of these two basic operators using a toy example in which a synthetic hyperspectral image is used for demonstration. As can be seen in Fig. 7.3, morphological dilation expands the spatial regions made up of pure pixel vectors in accordance with the spatial neighborhood defined by a 3×3 SE, while morphological erosion expands the regions made up of highly mixed pixel vectors in accordance with the same spatial neighborhood. In order to avoid changing the size and shape of the features in the image, a desirable feature for spatial filtering, extended morphological opening and closing operations have also been defined, respectively, as follows: $(\boldsymbol{f} \circ K)(x, y) = [(\boldsymbol{f} \ominus K) \oplus K](x, y)$, i.e., erosion followed by dilation, and $(\boldsymbol{f} \bullet K)(x, y) = [(\boldsymbol{f} \oplus K) \ominus K](x, y)$, i.e., dilation followed

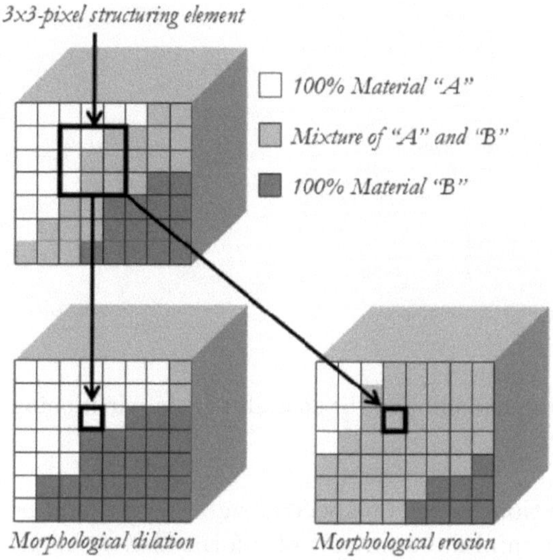

Fig. 7.3. Toy example showing the performance of extended morphological operations

by erosion [21]. With the above definitions in mind, the morphological gradient at $\boldsymbol{f}(x, y)$ using K can be defined as:

$$G_K(\boldsymbol{f}(x, y)) = \text{SAD}((\boldsymbol{f} \oplus K)(x, y), (\boldsymbol{f} \ominus K)(x, y)). \tag{7.5}$$

7.3 Parallel Implementation of Morphological Operations

This section discusses the development of parallel implementations for spatial-spectral hyperspectral image processing algorithms. An important consideration for this kind of algorithms is that, as shown by previous sections, these combined techniques make use of the information provided by the entire pixel vector *as a whole* and the spectral signature at a specific spatial location (x, y) is the minimum unit of information processed by the considered distance metric (i.e., the SAD distance). This source of information (spectral) is potentially complemented with that provided by other neighboring signatures to incorporate spatial context. The above consideration has a significant impact on the design of data partitioning strategies for parallelization purposes [22]. In particular, it has been shown in the literature that domain decomposition techniques provide flexibility and scalability in parallel multichannel image processing [23]. In parallel morphological processing for hyperspectral imaging, two types of partitioning can be exploited: spectral-domain partitioning and spatial-domain partitioning [24]:

1. **Spectral-domain partitioning** subdivides the volume into sub-volumes made up of contiguous spectral bands [see Fig. 7.4(a)], and assigns one or more sub-volumes to each processor. With this model, each pixel vector may be split amongst several processors and the communication cost associated

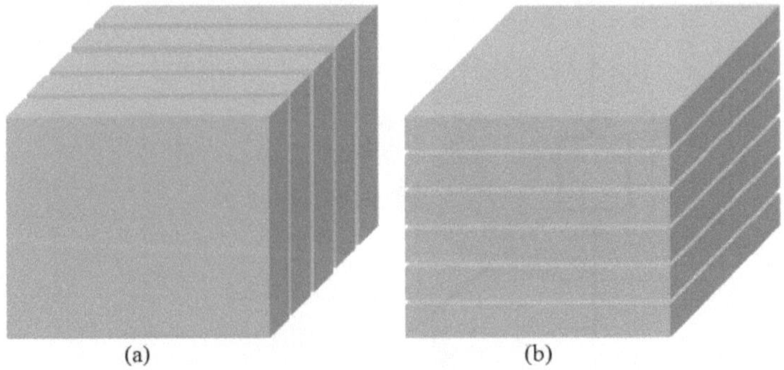

Fig. 7.4. (a) Spectral-domain, and (b) spatial-domain partitioning of a sample hyperspectral image cube

to the computations based on spectral signatures would be increased [22]. In this option, the minimum amount of information to be processed in parallel corresponds to a portion of the entire spectral signature allocated to a given specific location (x, y).

2. **Spatial-domain partitioning** subdivides the original data volume into slabs [see Fig. 7.4(b)] which are made up of (contiguous) pixel vector rows, thus retaining the full spectral information associated to each pixel vector at the same processor. In this option, the minimum amount of information to be processed in parallel corresponds to the entire spectral signature allocated to a given spatial location (x, y).

In order to exploit parallelism as much as possible, we have adopted a standard master-slave parallel processing paradigm combined with spatial-domain partitioning for the parallel implementations developed in this chapter [25]. The master-slave paradigm has been shown in previous work to perform effectively in the context of parallel image processing applications. On the other hand, our selection of spatial-domain instead of spectral-domain data partitioning is due to the following major reasons:

- First and foremost, spatial-domain partitioning is a natural approach for low-level image processing as many image processing operations require the same function to be applied to a small set of elements around each entire pixel vector in the image volume.
- A second major reason for our decision of adopting a spatial-domain decomposition framework in our application is that, in spectral-domain partitioning, the calculations made for each pixel vector need to originate from several processors and thus require intensive inter-processor communication. This is generally perceived as a shortcoming for parallel design, because the overhead introduced by inter-processor communication would increase linearly with the increase in the number of processing elements, thus complicating the design of scalable parallel algorithms.

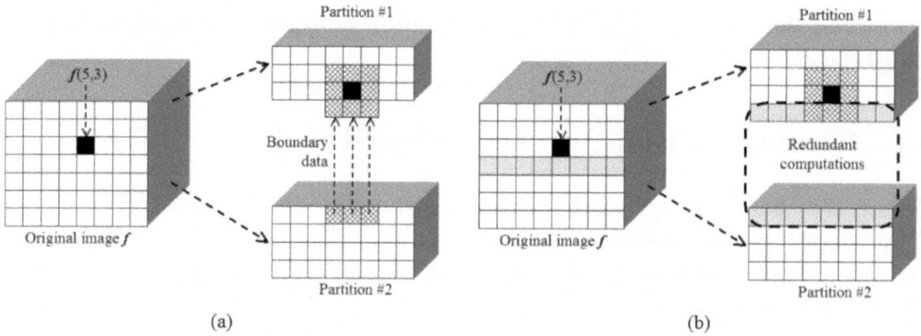

Fig. 7.5. (a) 3×3 SE computation split among two processing elements. (b) Introduction of an overlap border to minimize inter-processor communication in a 3×3 SE computation.

- A final major issue is code reusability; to reduce code redundancy and enhance portability, it is desirable to reuse much of the code for the sequential algorithm in the design of its correspondent parallel version and the spatial-domain decomposition approach greatly enhances code reuse.

A final consideration regarding hyperspectral data partitioning for morphological processing is noteworthy. In this type of processing, additional inter-processor communications will be required when the SE-based computation needs to be split amongst several different processing when the SE is centered around a pixel vector located in the border of one of the local partitions resulting after spatial-domain decomposition, as illustrated by Fig. 7.5(a), in which a small SE with 3×3 pixels in size is used for demonstration. In this case, the computations for the pixel vector at spatial coordinates $(5, 3)$ in the original image, denoted by $\boldsymbol{f}(5, 3)$, will need to originate from two processing elements since this pixel becomes a border pixel after spatial-domain partitioning. As a result, a communication overhead involving three N-dimensional pixel vectors (located in partition #2) is required in order to complete the SE-based computation for the pixel vector $\boldsymbol{f}(5, 3)$ in partition #1 (the same comment applies to all pixels in the third row of partition #1).

However, if an overlap border is carefully added to partition #1 (consisting of the entire first row of pixels allocated to partition #2), as illustrated in Fig. 7.5(b), then boundary data no longer need to be exchanged between neighboring processors in order to complete all calculations at partition #1, and therefore this data partition can be processed locally without the need for inter-processor communications. It is clear that such an overlap border would introduce redundant computations since the intersection between the two involved partitions would be non-empty. As a result, the solution above may be prohibitive for large SE sizes. Subsequently, for a given parallel platform, there is an application-dependent threshold to decide whether a redundant-information-based or data-exchange-based strategy should be adopted. This issue has been recently explored in [24] concluding that, for small SE sizes, a redundant-computation strategy is generally preferred.

7.4 Parallel Spatial-Spectral Algorithms

This section describes several parallel algorithms for spatial/spectral processing of hyperspectral images. Four types of parallel algorithms are considered: 1) dimensionality reduction; 2) unsupervised classification; 3) spectral unmixing; and 4) data compression.

7.4.1 Parallel Dimensionality Reduction (P-MORPHDIM)

One of the most widely used dimension reduction techniques in remote sensing is the PCT [3], which computes orthogonal projections that maximize the amount of data variance, and yields a dataset in a new uncorrelated coordinate system. This rotational transform is characterized by its global nature and, therefore, it might not preserve all the information useful to obtain a good classification [11]. In addition, the PCT relies on spectral properties of the data alone, thus neglecting the information related to the spatial arrangement of the pixels in the scene.

In the following, we develop a new parallel algorithm for morphological feature extraction which integrates the spatial and spectral information in simultaneous fashion. To achieve this goal, we apply sequences of extended opening by reconstruction operations using SE's of varying width (called morphological profiles) [21]. This type of morphological sequences have been applied in the past to characterize image structures in grayscale remotely sensed image data [26]. In this chapter, we extend this concept to feature extraction from hyperspectral image data, with the goal to capture the spatial and spectral information around each pixel vector through the combination of spatial-spectral morphological operations for increasingly larger spatial neighborhoods. It is important to emphasize that the use of opening and closing operations is essential for spatial-spectral filtering [11]. With this type of operators, the image features are either completely retained or completely removed in accordance with the size and shape of the structuring element, thus allowing us to perform accurate image filtering based on spatial-spectral content. The inputs to the parallel algorithm, called P-MORPHDIM, are an N-dimensional hyperspectral image cube, $\boldsymbol{f}$, a maximum number of filtering iterations, t, and a structuring element K with constant size of 3×3 pixels. The output is a transformed image cube, denoted by $\boldsymbol{g}$. The parallel algorithm is given by the following steps:

1. The master divides the original image cube $\boldsymbol{f}$ into P spatial-domain partitions, where P is the number of processors in the system. In order to avoid excessive inter-processor communication during morphological processing, the partitions are formed using overlap borders, as described in Fig. 7.5(b).
2. Each worker performs spatial-spectral filtering on its local spatial-domain partition using the following sequence of steps:

a) Compute an extended opening by reconstruction for each local pixel $f(x,y)$ as $(f \circ K)^t(x,y) = min_{t \leq 1}\{\delta_K^t(f \circ K|f)(x,y)\}$, with the basic operation $\delta_K^t(f \circ K|f)(x,y) = \delta_B\delta_B \cdots \delta_B(f \circ K|f)(x,y)$, i.e., δ_B is applied t times, and $\delta_B(f \circ K|f)(x,y) = min\{[(f \circ K) \oplus K](x,y), f(x,y)\}$.
b) Compute an extended closing by reconstruction for each local pixel $f(x,y)$ as $(f \bullet K)^t(x,y) = min_{t \leq 1}\{\delta_K^t(f \bullet K|f)(x,y)\}$, with the basic operation $\delta_K^t(f \bullet K|f)(x,y) = \delta_B\delta_B \cdots \delta_B(f \bullet K|f)(x,y)$, i.e., δ_B is applied t times, and $\delta_B(f \bullet K|f)(x,y) = min\{[(f \bullet K) \ominus K](x,y), f(x,y)\}$.
c) Compute the derivative of the extended opening profile as follows: $p_t^\circ = \{\text{SAD}[(f \circ K)^\lambda(x,y), (f \circ K)^{\lambda-1}(x,y)]\}$, with $\lambda = \{1, 2, \cdots, t\}$. Here, $f(x,y) = (f \circ K)^0(x,y)$ for $\lambda = 0$ by the definition of extended opening by reconstruction.
d) Compute the derivative of the extended closing profile as shown below: $p_t^\bullet = \{\text{SAD}[(f \bullet K)^\lambda(x,y), (f \bullet K)^{\lambda-1}(x,y)]\}$, with $\lambda = \{1, 2, \cdots, t\}$. Here, $f(x,y) = (f \bullet K)^0(x,y)$ for $\lambda = 0$ by the definition of extended closing by reconstruction.
e) Form a $(2t-1)$-dimensional morphological profile for each local pixel $f(x,y)$ by combining the derivatives of the extended opening and closing profiles as follows: $\text{MP}(x,y) = \{p_t^\circ(x,y), p_t^\bullet(x,y)\}$. The resulting morphological profile can be seen as a spatial-spectral feature vector on which a subsequent classification procedure may be applied.

3. The master processor gathers the individual $(2t-1)$-dimensional profiles provided by the workers and merges them into a new data cube g with $2t-1$ components. As shown by the parallel description above, this approach requires minimal coordination between the master and the workers, namely, at the beginning and ending of the parallel process, although it is subject to a redundant computation overhead introduced by the overlap borders used by the proposed data partitioning strategy.

7.4.2 Parallel Unsupervised Classification (P-WSHED)

This subsection provides a new parallel algorithm for unsupervised classification of hyperspectral images based on the morphological watershed transformation [27], originally introduced for grayscale images. This technique relies on a marker-controlled approach that considers the grayscale image as imaginary topographic relief in which, the brighter the intensity, the higher the corresponding elevation (see Fig. 7.2). Let us assume that a drop of water falls on such a topographic surface. The drop will flow down along the steepest slope path until it reaches a minimum. The set of points of the surface whose steepest slope path reach a given minimum constitutes the catchment basin associated with that minimum, while the watersheds are the zones dividing adjacent catchment basins [12]. Despite its encouraging results in several applications, the watershed transformation has not been fully exploited in multichannel image analysis, mainly due to the lack of well-defined vector ordering strategies required by basic morphological operations. In this chapter, we develop a multichannel watershed classification algorithm which consists of three main stages:

1. **Minima selection.** In order to select 'markers' or minima from which the watershed transform is started, we hierarchically order all minima according to their deepness, and then select only those above a threshold. The deepness of a basin is the level the water would reach, coming in through the minimum of the basin, before the water would overflow into a neighbor basin. Deepness can be computed using morphological reconstruction [12] applied to the multichannel gradient (reconstruction is a class of morphological transformation that does not introduce discontinuities). Given the multichannel gradient $G_K(\boldsymbol{f})$ of an N-dimensional image $\boldsymbol{f}$ processed using a SE denoted by K, the morphological reconstruction of $G_K(\boldsymbol{f})$ from $G_K(\boldsymbol{f}) \ominus K$ has a watershed transform in which the regions with deepness lower than a certain value v have been joined to the neighbor region with closer spectral properties.
2. **Flooding.** Let the set $\mathbf{S} = \{\mathbf{v}_1, \mathbf{v}_2, \cdots, \mathbf{v}_p\}$ denote the set of p minimum pixel vectors resulting from minima selection. Let the catchment basin associated with a minimum pixel $\mathbf{p}_i$ be denoted by $\mathrm{CB}(\mathbf{p}_i)$. The catchment basins are progressively created by simulating the flooding process. The first pixel vectors reached by water are the points of highest deepness score. From now on, the water either expands the region of the catchment basin already reached by water, or starts to flood the catchment basin whose minima have a deepness equal to $C_K(\mathbf{v}_l)$, where $\mathbf{v}_l$ is the deepest pixel in $\mathbf{S} - \{\mathbf{v}_j\}$. This operation is repeated until $\mathbf{S} = \emptyset$.
3. **Region merging.** To obtain the final classification, some of the regions $\{\mathrm{CB}(\mathbf{v}_l)\}_{l=1}^{p}$ resulting from the watershed can be merged to reduce the number of regions. First, all regions are ordered into a region adjacency graph (RAG) [28]. Each edge in the RAG is assigned a weight, so that the weight of an edge $e(\mathbf{v}_i, \mathbf{v}_j)$ is the value of $\mathrm{SAD}(\mathbf{v}_i, \mathbf{v}_j)$. Subsequently, regions $\mathrm{CB}(\mathbf{v}_i)$ and $\mathrm{CB}(\mathbf{v}_j)$ can be merged attending to spatial properties in the case of adjacent regions, and also according to pixel vector similarity criteria for non-adjacent regions. The output of the algorithm above is a 2-dimensional classification map based on the set of $\mathbf{S} = \{\mathbf{v}_1, \mathbf{v}_2, \cdots, \mathbf{v}_p\}$ class prototypes.

Our parallel implementation of the multichannel watershed algorithm above, called P-WSHED, uses a simple master-slave model [29], in which the master processor reads the whole multichannel image $\boldsymbol{f}$ and divides it into a set of P spatial-domain partitions, which are sent to different processors. The slave processors run the watershed algorithm on the respective partitions and also exchange data among themselves for uniform classification. After the classified regions become stable, the slaves send the output to the master, which combines all of them in a proper way and provides the final classification. If we assume that the parallel system has P processors available, then one of the processors is reserved to act as the master, while each of the $P-1$ remaining processors create a local queue Q_i with $1 \leq i \leq P-1$. The minima selection algorithm is run locally at each processor to obtain a set of minima pixels surrounded by non-minima, which are then used to initialize each queue Q_i. Flooding is then performed locally in each processor as in the serial algorithm.

It should be noted, however, that due to the image division, flooding is confined only to the local subdomain. There may exist parts of the subimage that cannot be reached by flooding since they are contained in other subimages. Our approach to deal with this problem is to first flood locally at every deepness score in the subimage. Once the local flooding is finished, each processor exchanges classification labels of pixels in the boundary with appropriate neighboring processors. Subsequently, a processor may receive classification labels corresponding to pixels in the extended subdomain. The processor must now *reflood* the local subdomain from those pixels, a procedure that may introduce changes in classification labels at the local subdomain. Communication and reflooding are again repeated until stabilization (i.e. no more changes occur).

7.4.3 Parallel Spectral Unmixing

While our focus in previous subsections has been on classification approaches, spectral unmixing has been an alluring exploitation goal since the earliest days of hyperspectral imaging [16]. No matter the spatial resolution, in natural environments the spectral signature for a nominal pixel is invariably a mixture of the signatures of the various materials found within the spatial extent of the ground instantaneous field view. In hyperspectral imagery, the number of spectral bands usually exceeds the number of pure spectral components, called endmembers in hyperspectral analysis terminology [30], and the unmixing problem is cast in terms of an over-determined system of equations in which, given the correct set of endmembers allows determination of the actual endmember abundance fractions through a numerical inversion process. Since each observed spectral signal is the result of an actual mixing process, the driving abundances must obey two constraints [4]. First, all abundances must be non-negative. Second, the sum of abundances for a given pixel must be unity. However, it is the derivation and validation of the correct suite of endmembers that has remained a challenging and goal for the past years [31], [32]. In the following subsections, we develop two novel parallel algorithms for endmember extraction and fractional abundance estimation in hyperspectral images which introduce an innovative component with regards to standard techniques in the literature: the incorporation of spatial information to the (spectrally-guided) problems of finding endmembers and approximating their abundance fractions in mixed pixels.

Parallel endmember extraction (P-MORPHEE)

Most available endmember extraction approaches, including popular and successful algorithms such as the PPI (available in Research Systems ENVI software) [14] or the N-FINDR (distributed by Technical Research Associates, Inc.) [15] have been designed from a spectroscopic viewpoint and, thus, tend to neglect the existing spatial correlation between pixels. In the following, we develop a novel parallel algorithm which integrates the spatial and spectral information in the endmember searching process. The algorithm, based on our previously developed automated morphological endmember extraction (AMEE) algorithm

[10], allows propagation of pure pixels between subsequent iterations, as opposed to the previous version of the algorithm. The inputs to the parallel algorithm, called P-MORPHEE, are the full hyperspectral data cube $\boldsymbol{f}$, a structuring element K, a maximum number of algorithm iterations I_{max}, and a number of endmembers to be extracted, p. The output is an endmember set, $\{\mathbf{e}_i\}_{i=1}^{q}$, with $q \leq p$. A step-by-step description of the algorithm follows:

1. The master divides the original image cube $\boldsymbol{f}$ into P spatial-domain partitions, where P is the number of processors in the system. In order to avoid excessive inter-processor communication during morphological processing, the partitions are formed using overlap borders, as described in Fig. 7.5(b).
2. Each worker performs an iterative spatial-spectral endmember extraction process on its local spatial-domain partition using the following sequence of steps:
   a) Set $i = 1$ and initialize a morphological eccentricity index [10], denoted by $\text{MEI}(x, y) = 0$, for each pixel $\boldsymbol{f}(x, y)$ in the local partition.
   b) Move K through all the pixels of the local partition data, defining a local spatial search area around each pixel $\boldsymbol{f}(x, y)$, and calculate the maximum and minimum pixel vectors at each K-neighborhood using extended morphological erosion and dilation. Then, update the MEI at each spatial location (x, y) using the following expression:

$$\text{MEI}(x, y) = \text{SAD}[(\boldsymbol{f} \ominus K)(x, y), (\boldsymbol{f} \oplus K)(x, y)] \tag{7.6}$$

   c) Set $i = i + 1$. If $i = I_{max}$, then return the MEI scores for all the pixels in the local partition to the master processor. Otherwise, replace the local partition with its dilation using K. This represents an optimization of the algorithm that propagates only the purest pixels at the local neighborhood to the following algorithm iteration. Then, go to step 2.
3. The master gathers the individual MEI scores provided by the workers and selects the set of p pixel vectors with higher associated MEI values (called endmember candidates). Then, the master forms a unique spectral set of $\{\mathbf{e}_i\}_{i=1}^{q}$ pixels, with $q \leq p$, by calculating the SAD for all pixel vector pairs.

Parallel fractional abundance estimation (P-MORPHSU)

To conclude this subsection on parallel spectral unmixing methods we outline a new parallel algorithm for fractional abundance estimation of a set of input endmembers. This method integrates the spatial and the spectral information by considering a spatial neighborhood (defined by a morphological SE denoted by K) around each mixed pixel. This method is similar to traditional approaches, in the sense that it makes use of the standard fully constrained least squares technique [4] to estimate abundance fractions. But it differs from traditional methods in the fact that the endmember set used for each pixel is adaptively calculated based on the spatial context. The inputs to the parallel method are the full hyperspectral data cube $\mathbf{f}$, a structuring element K, a tolerance threshold t_{SU}, and a set of endmembers $\{\mathbf{e}_i\}_{i=1}^q$. The output is an abundance fraction estimation for each endmember in each pixel of the input data set. The parallel algorithm, called P-MORPHSU, is based on the following steps:

1. The master divides the original image cube $\mathbf{f}$ into P spatial-domain partitions, using overlap borders as described in Fig. 7.5(b). The master also broadcasts the endmember set $\{\mathbf{e}_i\}_{i=1}^q$ to all the workers.
2. Each worker estimates the fractional abundance of each endmember in each pixel of its local partition using the following procedure:
   a) Before unmixing a certain local pixel, say $\mathbf{f}(x, y)$, a weight is assigned to the pixels in the K-neighborhood centered at spatial coordinates (x, y). This is done by first calculating, for each pixel in the K-neighborhood, the SAD distance to each one of the endmembers in the set $\{\mathbf{e}_i\}_{i=1}^q$, and labeling the pixel as an instance of a certain endmember (candidate) by using the minimum SAD score.
   b) Then, a weight is assigned to each endmember candidate (the weight is inversely proportional to the minimum SAD score reported for that candidate).
   c) Finally, all endmember candidates in the K-neighborhood are sorted by weight, and only those with associated weights above tolerance threshold t_{SU} are incorporated to the *local* endmember set which is finally used to unmix the pixel $\mathbf{f}(x, y)$ using standard FCLSU. In other words, the abundance estimation is still performed by using a fully constrained least squares technique [4], but the actual composition of the endmember set used to perform the least squares estimation may vary (for each particular pixel) depending on the spatial-spectral context around the pixel, as opposed to the traditional approach, in which the entire set of spectral endmembers is always used to unmix each pixel. Let us assume that the set of endmembers, obtained by the above procedure, is denoted by $\{\mathbf{e}_i\}_{i=1}^l$, with $1 \leq l \leq q$. The goal is to achieve a decomposition of the pixel $\mathbf{f}(x, y)$ using the set of l endmembers above as follows:

$$\mathbf{f}(x, y) = \mathbf{e}_1 \cdot a_1(x, y) + \mathbf{e}_2 \cdot a_2(x, y) + \cdots + \mathbf{e}_l \cdot a_l(x, y). \qquad (7.7)$$

   To achieve this goal, the pixel is multiplied by $(\mathbf{M}^{\mathrm{T}}\mathbf{M})^{-1}\mathbf{M}^{\mathrm{T}}$, where $\mathbf{M} = \{\mathbf{e}_i\}_{i=1}^l$ and the superscript 'T' denotes the matrix transpose operation. In the expression above, abundance sum-to-one and non-negativity constraints are imposed, i.e., $\sum_{i=1}^l a_i(x, y) = 1$ and $a_i(x, y) \geq 0$ for all (x, y).
3. The master gathers the individual fractional abundance estimations provided by the workers and merges them together to generate a set of endmember fractional abundance estimations for all the pixels in the original hyperspectral image.

7.4.4 Parallel Data Compression (P-MORPHCOMP)

In this subsection, we describe a lossy compression algorithm for hyperspectral imagery which relies on the concept of spectral unmixing introduced in the previous subsection. The main advantage of the proposed technique with regards to other standard compression algorithms in the literature, such as 3D-SPIHT or JPEG 2000 multicomponent, is the ability to deal with mixed pixels and subpixel targets [33]. A mixed pixel is a mixture of two or more different substances present in the same pixel, as outlined before in this chapter. A subpixel target is a mixed pixel with size smaller than the available pixel size (spatial resolution). When hyperspectral image compression is performed, it is critical and crucial to take into account these two issues, which have been generally overlooked in the development of lossy compression techniques in the literature. The idea of the proposed data compression algorithm is to represent each pixel vector $\boldsymbol{f}(x,y)$ with N spectral bands by a set of $l \leq N$ fractional abundance estimations corresponding to the set of $\{\mathbf{e}_j(x,y)\}_{j=1}^{l}$ endmembers that contribute to the mixture in that pixel [34]. More precisely, for each N-dimensional pixel vector $\boldsymbol{f}(x,y)$, its associated l-dimensional abundance vector $\mathbf{a}(x,y) = a_1(x,y), a_2(x,y), \cdots, a_l(x,y)$, estimated using the P-MORPHSU algorithm, is used as a fingerprint of $\boldsymbol{f}(x,y)$ with regards to l endmembers obtained as a subset of the full endmember set produced by the P-MORPHEE algorithm, i.e., $\{\mathbf{e}_j(x,y)\}_{j=1}^{l} \subseteq \{\mathbf{e}_i\}_{i=1}^{q}$. The proposed parallel data compression algorithm receives as input parameters a hyperspectral image cube, $\boldsymbol{f}$, and a maximum number of endmembers to be retained per pixel, q, and can be summarized by the following steps:

1. Use the P-MORPHEE parallel algorithm to adaptively obtain a set of q endmembers, $\{\mathbf{e}_i\}_{i=1}^{q}$ from the input hyperspectral image $\boldsymbol{f}$.
2. Use the P-MORPHSU parallel algorithm to adaptively estimate the corresponding abundance fractions $\mathbf{a}(x,y) = a_1(x,y), a_2(x,y), \cdots, a_l(x,y)$ of a subset of l endmembers, with $\{\mathbf{e}_j(x,y)\}_{j=1}^{l} \subseteq \{\mathbf{e}_i\}_{i=1}^{q}$, for each pixel vector in the input scene, thus approximating the pixel signature by the following expression: $\boldsymbol{f}(x,y) = \mathbf{e}_1 \cdot a_1(x,y), \mathbf{e}_2 \cdot a_2(x,y), \cdots, \mathbf{e}_l \cdot a_l(x,y)$. Note that this is a reconstruction of the spectral signature at $\boldsymbol{f}(x,y)$.
3. The master processor applies lossless predictive coding to reduce spatial redundancy within each of the fractional abundance estimates for each pixel in the input scene, using Huffman coding to encode predictive errors [34].

7.5 Experimental Results

This section provides an assessment of the effectiveness of the proposed parallel algorithms in the task of providing significant performance gains, without loss of accuracy, in the analysis of real hyperspectral data sets. Two different application domains: 1) land-cover classification in agriculture; and 2) mapping of geological features, are used to provide a detailed cross-validation of parallel algorithms for hyperspectral image analysis, addressing the current need for application-oriented inter-comparisons of parallel in this emerging and fast growing research area. The section is organized as follows. First, we provide a brief outline of Thunderhead, a Beowulf cluster available at NASA's Goddard Space Flight Center that has been used as our baseline parallel platform. Then, we provide an overview of the hyperspectral image data sets used in this study. A detailed computational cost-performance analysis of the parallel algorithms in the context of two highly representative application domains follows. The section concludes with a summary and detailed discussion on the results obtained.

7.5.1 Parallel Computing Platfom

The parallel computing platform used for experimental validation in this work is the Thunderhead system at NASA's Goddard Space Flight Center in Maryland. This Beowulf cluster can be seen as an evolution of the HIVE (Highly Parallel Virtual Environment) project, started in spring of 1997 to build a commodity cluster intended to be exploited by different users in a wide range of scientific applications. The idea was to have workstations distributed among many offices and a large number of compute nodes (the compute core) concentrated in one area. The workstations would share the compute core as though it was apart of each. Thunderhead is currently composed of 268 dual 2.4 Ghz Intel 4 Xeon nodes, each with 1 GB of memory and 80 GB of hard disk (see http://thunderhead.gsfc.nasa.gov for additional details). The total disk space available in the system is 21.44 Tbyte, and the theoretical peak performance of the system is 2.5728 Tflops (1.2 Tflops on the Linpack benchmark). The current estimated cost of the Thunderhead system is 1.25M U.S. dollars. Along with the 568-processor computer core (out of which 256 were used for experiments), Thunderhead has several nodes attached to the core with Myrinet 2000 connectivity. Our parallel algorithms were run from one of such nodes, called thunder1. The operating system is Linux Fedora Core, and MPICH was the message-passing library used.

7.5.2 Hyperspectral Data Sets

Two different hyperspectral data sets collected by the NASA's Jet Propulsion Laboratory AVIRIS instrument have been selected for experimental validation in this study. All the considered scenes have been geometrically corrected by JPL and have extensive ground-truth information available, thus allowing us to validate the performance of parallel algorithms in several different application domains.

AVIRIS Hyperspectral Data over Indiana's Indian Pines Region

This scene was gathered by AVIRIS over the Indian Pines test site in Northwestern Indiana, a mixed agricultural/forested area, early in the growing season, and consists of 1939×677 pixels and 224 spectral bands in the wavelength range 0.4–2.5 μm (574 MB in size). The AVIRIS Indian Pines data set represents a very challenging classification problem dominated by similar spectral classes and

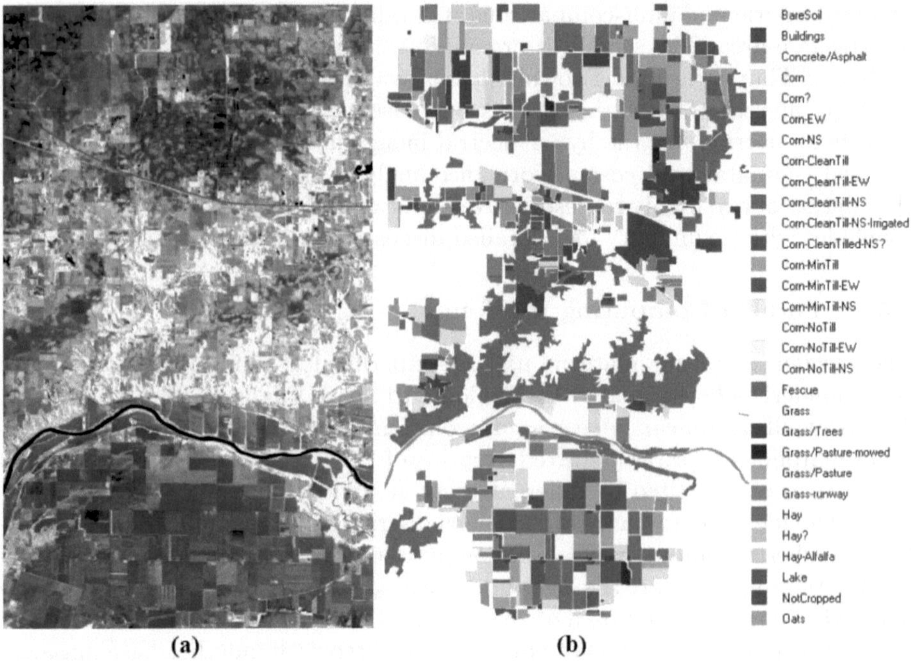

Fig. 7.6. (a) Spectral band at 587 nm wavelength of an AVIRIS scene comprising agricultural and forest features at Indian Pines region. (b) Ground-truth map with 30 mutually-exclusive land-cover classes.

mixed pixels. Specifically, the primary crops of the area, mainly corn and soybeans, were very early in their growth cycle with only about 5% canopy cover. This fact makes most of the scene pixels highly mixed in nature. Discriminating among the major crops under this circumstances can be very difficult, a fact that has made this scene an extensively used benchmark to validate classification accuracy of hyperspectral imaging algorithms. For illustrative purposes, Fig. 7.6(a) shows the spectral band at 587 nm of the original scene and Fig. 7.6(b) shows the corresponding ground-truth map, displayed in the form of a class assignment for each labeled pixel, with 30 mutually exclusive ground-truth classes. Part of these data, including ground-truth, are available online from Purdue University (for details, see http://cobweb.ecn.purdue.edu/ biehl/MultiSpec/).

AVIRIS Hyperspectral Data over Cuprite Mining District, Nevada

Another AVIRIS scene collected over the Cuprite mining district in Nevada was also used in experiments to evaluate the proposed parallel algorithms in the context of a mineral mapping application. The data set (available from http://aviris.jpl.nasa.gov/html/aviris.freedata.html) consists of 614 × 512 pixels and 224 bands in the wavelength range 0.4–2.5 μm (137 MB in size). As opposed to the previously described Indian Pines data set, the Cuprite data set

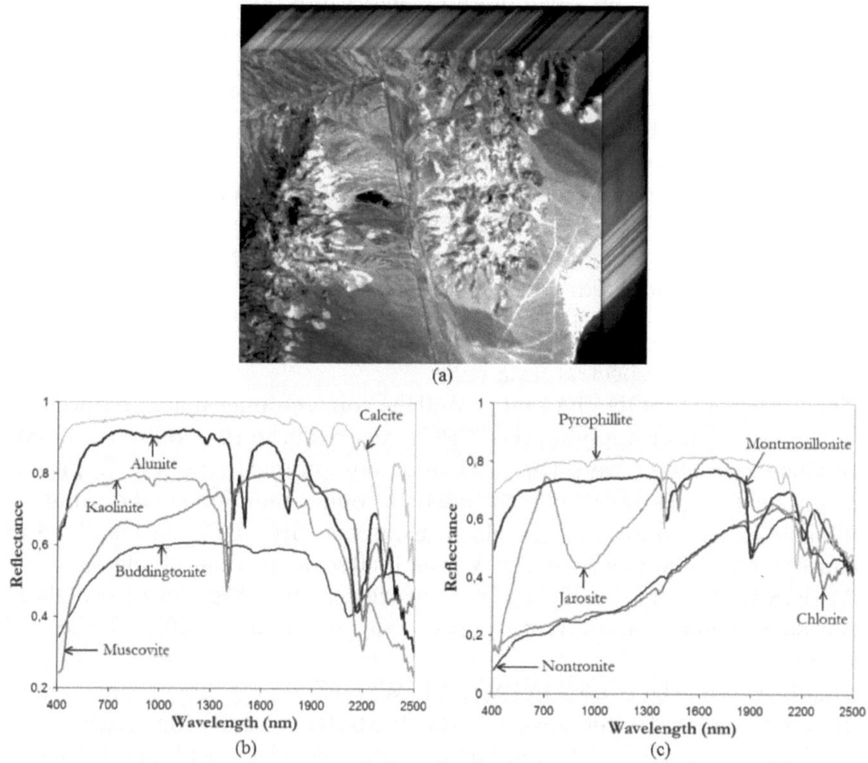

Fig. 7.7. (a) AVIRIS scene over Cuprite mining district. (b-c) Ground-truth mineral spectra provided by USGS.

is atmospherically corrected and available in reflectance units (not in at-sensor radiance), thus allowing direct comparison of pixel vectors to ground spectral signatures. The Cuprite site has been extensively mapped by the U.S. Geological Survey (USGS) in the last twenty years, and there is extensive ground-truth information available, including a library of mineral signatures collected on the field (see http://speclab.cr.usgs.gov/spectral-lib.html). Fig. 7.7(a) shows the spectral band at 587 nm wavelength of the AVIRIS scene. The spectra of USGS ground minerals: alunite, buddingtonite, calcite, kaolinite, muscovite [Fig. 7.7(b)], chlorite, jarosite, montmorillonite, nontronite, pyrophilite [Fig. 7.7(c)] are also displayed. These selected spectral signatures will be used in this work to evaluate endmember extraction accuracy.

7.5.3 Experimental Assessment of Parallel Algorithms

Experiment 1: Land-Cover Classification in Agriculture

In this subsection, we discuss the performance of the proposed parallel algorithms in the context of an agricultural classification application. In order to validate the experimental results provided by those algorithms, we resort to ground-truth measurements collected by Purdue University over the Indian Pines region [see Fig. 7.6(b)]. In this experiment, we assessed the feature extraction performance of the proposed P-MORPH dimensionality reduction technique in comparison with that of a parallel PCT (P-PCT) algorithm proposed in [35]. Also, we evaluated the classification performance of the proposed P-WSHED technique in comparison with a parallel version of the ISODATA algorithm (P-ISODATA) [36]. It should be noted that ISODATA is widely regarded as a benchmark for validation of unsupervised classification algorithms [3]. Specifically, the following experiments were conducted:

1. First, we run the P-ISODATA and the P-WSHED on the original N-dimensional hyperspectral data cube.
2. Then, we run P-ISODATA and P-WSHED on a q-dimensional, reduced data set, obtained after applying the P-PCT algorithm to the original data cube. In this experiment, we set $q = 41$ and thus retained the first 41 principal components. This decision was based on our estimation of the number of distinct signal sources in the data using the virtual dimensionality (VD) concept [37]. The fact that the VD estimated more than 30 classes for the AVIRIS Indian Pines scene was not surprising, since Fig. 7.6(b) reveals that the 30 available ground-truth classes only cover about 60% of the entire scene.
3. Finally, we run the P-ISODATA and P-WSHED on a q-dimensional, reduced data set obtained after applying the P-MORPH algorithm (with $t = 21$, resulting in $2t - 1 = 41$ components after morphological, spatial-spectral dimensionality reduction).

In all cases, we carefully adjusted algorithm parameters based on our experience with those algorithms in different application domains [22], [30]. With the above experimental settings in mind, Table 7.1 reports the overall and individual classification accuracies (with regards to some selected ground-truth classes) produced by the parallel algorithms after using different algorithm configurations (it should be noted that the overall accuracies refer to the entire set of 30 ground-truth classes available, not displayed here for space considerations). As shown by Table 7.1, when P-ISODATA and P-WSHED were combined with P-MORPH for dimensionality reduction, the classification accuracies increased significantly with respect to the cases in which the same algorithms were applied to the original image, or to a reduced version of the image using P-PCT. Overall, the P-WSHED provided significantly better classification results than P-ISODATA, with the combination P-MORPH+P-WSHED clearly resulting in the highest individual and overall classification accuracies. This indicates that the incorporation of spatial information in both feature extraction and spectral classification may offer important advantages, in particular, when the spatial information is an added value for the analysis, as it seems to be the case in the considered application environment given the spatially correlated distribution of land-cover classes in the Indian Pines region.

Table 7.1. Individual (per-class) and overall percentages of correctly classified pixels in the AVIRIS Indian Pines scene by different combinations of parallel algorithms. Sequential execution times (in seconds) measured in a single node of NASA's Thunderhead cluster are reported for each algorithm combination.

	(Original image)		P-PCT+		P-MORPH+	
	P-ISODATA	P-WSHED	P-ISODATA	P-WSHED	P-ISODATA	P-WSHED
Bare soil	60.32	71.48	62.34	72.71	77.81	87.34
Buildings	52.45	64.56	52.89	73.28	65.23	79.12
Concrete	58.33	69.95	61.12	62.01	75.94	85.99
Corn	51.23	63.56	54.27	63.48	67.01	80.31
Fescue	56.03	67.39	55.03	64.34	69.23	79.98
Grass	62.37	74.01	60.48	72.90	73.91	86.00
Hay	55.15	67.03	53.49	65.23	72.16	85.43
Lake	58.23	69.50	59.25	73.42	73.45	87.22
Oats	53.12	62.89	56.78	66.58	69.32	83.91
Overall accuracy	55.48	66.17	57.21	68.56	72.26	84.58
Processing time	49912	60356	48007	50096	50221	52310

For illustrative purposes, Table 7.1 also reports the execution times (in seconds) measured for the different algorithms executed on a single processor of the Thunderhead system. We emphasize that these times correspond to real sequential versions of the considered algorithms, not to parallel versions using one master and one worker. As shown by the table, the two considered dimensionality reduction techniques were computationally expensive but led to significant reductions in the processing times of P-ISODATA and P-WSHED algorithms. We note that, in this example, the number of bands was reduced from more than 200 to 41 according to the VD concept. In the case of P-MORPH, the spatial-spectral feature extraction accomplished by the algorithm always helped increase the classification scores despite the significant reduction in the number of bands, thus indicating that this approach was able to retain the information relevant to the separation of the classes in all considered experiments.

To empirically investigate the scaling properties of the considered parallel algorithms, Fig. 7.8(a) plots their speedup factors as a function of the number of available processors on Thunderhead. Results in Fig. 7.8(a) reveal that the performance drop from linear speedup in both P-PCT and P-ISODATA algorithms increases significantly as the number of processors increase. This is due to the data dependencies and sequential calculations involved in those algorithms. A similar effect was observed for the P-WSHED algorithm, in which the region-growing process at each local partition may result in different processing times, thus affecting the overall scalability of the algorithm. On the other hand, we can observe in Fig. 7.8(a) that the P-MORPH algorithm was clearly the algorithm which achieved better scalability on Thunderhead. This result comes as no surprise since P-MORPH was identified as *pleasingly parallel* (despite the use of overlap borders in the data partitioning procedure), and involves very few data dependencies. This observation also resulted in faster execution times for the parallel algorithm combinations using P-MORPH for dimensionality reduction, as reported in Fig. 7.8(b). For illustrative purposes, Table 7.2 displays the execution times in seconds measured for different processors in this application case study.

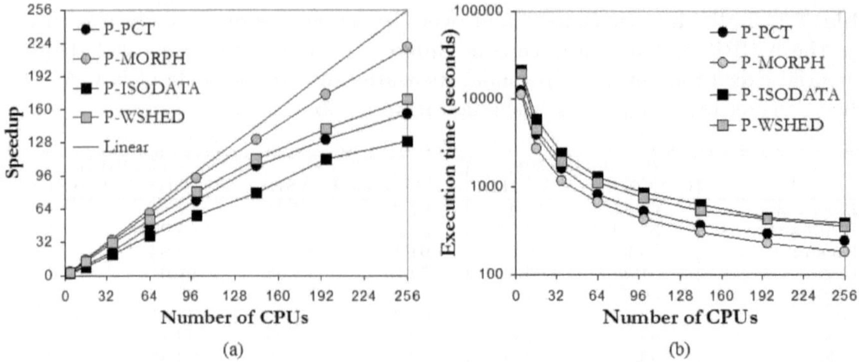

Fig. 7.8. Performance results after processing the AVRIS Indian Pines scene with the proposed parallel algorithms on Thunderhead: (a) Parallel algorithm scalability; (b) Graphical representation of execution times as a function of the number of CPUs

Table 7.2. Execution times (in seconds) measured on Thunderhead for the parallel algorithms using the AVIRIS Indian Pines scene

CPUs	1	4	16	36	64	100	144	196	256
P-PCT	38025	12467	3978	1622	815	527	361	290	245
P-MORPH	40239	11178	2728	1183	668	430	307	231	183
P-ISODATA	49912	21330	5907	2428	1299	865	630	444	386
P-WSHED	60356	19345	4569	1933	1129	752	536	426	355

Summarizing, experimental results in this subsection reveal that significant improvements can be obtained, both from the viewpoint of classification accuracy and computational performance, by using spatial-spectral parallel algorithms in the framework of our land-cover classification application. For instance, the combination of P-MORPH for dimensionality reduction followed by P-WSHED for classification was able to provide a relatively accurate result (taking into account the high complexity of the scene and the unsupervised nature of the algorithm) in only 9 minutes using 256 processors, as opposed to several hours of computation required by the sequential version of the same algorithm combination executed on a single Thunderhead processor. Despite these encouraging results, further experimentation in other application areas is required to rigorously establish the statistics obtained in this case study.

Experiment 2: Mapping of Geological Features

In this second experiment, we have conducted a cross-validation of parallel spectral unmixing algorithms in the context of a mineral mapping application, using the well-known AVIRIS Cuprite data set for demonstration purposes. It should be noted that ground-truth information for this scene is only available in the form of a collection of USGS mineral signatures, and therefore the parallel algorithms cannot be evaluated in terms of their classification accuracy as in the previous examples. For comparative purposes, we used the parallel versions of PPI (P-PPI) and N-FINDR (P-FINDR) developed in previous work [25] to evaluate the performance of the proposed P-MORPHEE endmember extraction algorithm. Specifically, our experimentation in this subsection comprised the following steps:

1. First, we run P-PPI and P-FINDR using their original configurations, i.e., using a dimension reduction technique (the P-PCT in our experiments) to reduce the dimensionality of the input data from N to q (with $q = 15$), obtaining a set of 15 spectral endmembers in both cases. This value was obtained using the VD concept in [37]. In order to establish a fair comparison of our P-MORPHEE algorithm with the above two parallel algorithms, we also used a reduced version of the original data cube (obtained by the P-PCT in [35]) when running our algorithm.
2. Then, we repeated the previous experiment but this time using P-MORPH instead of P-PCT to perform feature extraction from the input hyperspectral scene. Here, we used $t = 8$, resulting in 15 components, which is consistent with the dimensionality estimation provided by the VD concept.

Table 7.3 shows the SAD values between the endmembers in the final endmember set (extracted by different combinations of a parallel dimensionality reduction algorithm followed by a parallel endmember extraction algorithm) and the corresponding spectral signatures in the USGS library. In order to display the results in a more effective manner, we only report the SAD score associated to the most similar spectral endmember (out of 15 endmembers obtained for each algorithm combination) with regards to its corresponding USGS signature. It is important to emphasize that smaller SAD values indicate higher spectral similarity. As shown by Table 7.3, the P-PCT+P-FINDR combination resulted in the largest number of minimal SAD values (displayed in bold typeface in the table) among all considered combinations. Quite opposite, all the combinations which used P-MORPH for feature extraction prior to endmember extraction generally produced endmembers which were less similar, spectrally, with regards to reference USGS signatures. This is indeed a very interesting result, which indicates that spectral information is more important than spatial information in this particular application case study and, specifically, in the feature extraction task prior to endmember extraction. This results from the fact that geological features in the Cuprite mining district appear quite scattered, thus exhibiting little spatial correlation. Therefore, in this example spectral information is more important than spatial information in order to discriminate between subtle mineral signatures in this application. As a result, it is not surprising that the performance of P-PCT in this example was better than that exhibited by P-MORPH, which is particularly tuned for the integration of spatial and spectral information.

We would like to emphasize that the proposed combinations of parallel algorithms have also been evaluated from the viewpoint of their capacity to produce high-quality abundance estimations for geological features in the Cuprite mining district. This has been done by estimating the fractional abundance of

Table 7.3. SAD-based spectral similarity scores between the USGS mineral spectra and their corresponding endmember pixels produced by several combinations of a parallel dimensionality reduction algorithm followed by a parallel endmember extraction algorithm (the most similar endmember for each mineral is shown in bold typeface). Sequential execution times (in seconds), measured in a single node of NASA's Thunderhead cluster, are also given.

	P-PCT+			P-MORPH+		
	P-PPI	**P-FINDR**	**P-MORPHEE**	**P-PPI**	**P-FINDR**	**P-MORPHEE**
Alunite	0.084	**0.081**	0.084	0.106	0.103	0.095
Buddingtonite	0.106	**0.084**	0.094	0.122	0.109	0.108
Calcite	**0.105**	**0.105**	0.110	0.120	0.112	0.110
Kaolinite	**0.125**	0.136	0.136	0.144	0.136	0.131
Muscovite	**0.136**	**0.136**	**0.136**	0.153	0.150	0.145
Chlorite	0.112	**0.102**	0.108	0.135	0.118	0.116
Jarosite	0.106	**0.089**	0.096	0.122	0.115	0.109
Montmorillonite	0.108	**0.094**	0.106	0.126	0.119	0.117
Nontronite	0.102	**0.099**	**0.099**	0.124	0.120	0.113
Pyrophilite	0.094	**0.090**	**0.090**	0.115	0.112	0.105
Processing time	12251	10201	10573	12680	10630	11002

Table 7.4. Execution times (in seconds) measured on Thunderhead for the parallel algorithms using the AVIRIS Cuprite scene

CPUs	**1**	**4**	**16**	**36**	**64**	**100**	**144**	**196**	**256**
P-PCT	9506	3117	994	405	204	132	90	72	61
P-MORPH	9935	2760	674	292	165	106	76	57	45
P-PPI	2745	807	244	112	56	36	34	24	21
P-FINDR	695	302	73	25	18	12	10	8	7
P-MORPHEE	1067	333	76	33	19	12	9	6	5
P-MORPHSU	2308	615	152	66	37	24	17	12	10

endmembers provided by P-PPI, P-FINDR and P-MORPHEE using a straightforward parallel linear spectral unmixing (P-LSU) algorithm and the proposed P-MORPHSU algorithm for spatial-spectral abundance estimation. In both cases, we tested unconstrained and fully constrained versions (i.e., with sum-to-one and non-negativity restrictions) of the algorithms. Although ground-truth information on endmember fractional abundances at sub-pixel levels is not available for the Cuprite data set (this type of reference information is very difficult to be obtained in real-world scenarios), our quantitative experiments demonstrated that the use of the unconstrained P-MORPHSU generally resulted in very few negative abundance estimations, while the constrained P-MORPSHU provided very similar results to those reported by the unconstrained version of the same algorithm. In contrast, a more significant fraction of negative abundances was obtained by the unconstrained P-LSU with regards to the constrained P-LSU. It should be noted that a common indicator of poor model fitting and/or inappropriate selection of endmembers is estimation of negative abundance fractions by unconstrained linear models.

Fig. 7.9(a) shows the scalability of the considered parallel endmember extraction and spectral unmixing algorithms on Thunderhead. As shown by Fig. 7.9(a), the P-MORPHSU algorithm scaled nicely due to the lack of data dependencies

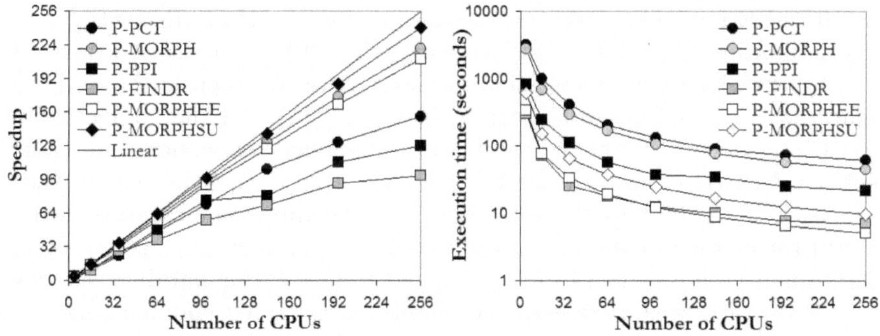

Fig. 7.9. Performance results after processing the AVRIS Cuprite scene with the proposed parallel algorithms on Thunderhead: (a) Parallel algorithm scalability; (b) Graphical representation of execution times as a function of the number of CPUs

Table 7.5. Spectral similarity scores among USGS spectra and the endmembers extracted from the original image, and from several reconstructed versions of the original image after applying the P-MORPHCOMP, JPEG2000 and 3D-SPIHT algorithms with different compression ratios

Mineral signature	Original image	P-MORPHCOMP 20:1	40:1	80:1	JPEG 2000 20:1	40:1	80:1	3D-SPIHT 20:1	40:1	80:1
Alunite	0.084	0.096	0.105	0.112	0.128	0.134	0.143	0.115	0.121	0.124
Buddingtonite	0.106	0.124	0.131	0.142	0.168	0.174	0.179	0.147	0.152	0.163
Calcite	0.105	0.118	0.129	0.138	0.166	0.175	0.184	0.131	0.148	0.165
Kaolinite	0.125	0.134	0.142	0.150	0.189	0.194	0.199	0.156	0.161	0.169
Muscovite	0.136	0.140	0.147	0.158	0.178	0.196	0.205	0.160	0.169	0.178
Chlorite	0.112	0.121	0.128	0.139	0.170	0.176	0.181	0.154	0.161	0.169
Jarosite	0.106	0.121	0.128	0.135	0.156	0.165	0.172	0.143	0.149	0.157
Montmorillonite	0.108	0.125	0.136	0.140	0.163	0.169	0.175	0.148	0.154	0.162
Nontronite	0.102	0.112	0.120	0.131	0.157	0.172	0.179	0.134	0.145	0.156
Pyrophilite	0.094	0.103	0.110	0.116	0.130	0.138	0.146	0.119	0.123	0.133

and inter-processor communications in its parallel implementation. A similar comment applies to the P-MORPHEE algorithm, which scaled almost as effectively. Quite opposite, both the P-PPI and P-FINDR resulted in lower speedups due to the less *pleasingly parallel* nature of their implementation, as reported in previous work [25]. Although the sequential versions of these algorithms took less time that P-MORPHEE to produce their output in a single processor (see Table 7.3), the better scalability of P-MORPHEE as the number of processors was increased resulted in lower execution times in Fig. 7.9(b). Finally, both P-PCT and P-MORPH behaved in similar terms as in the previous experiment, with significantly reduced execution times (by a factor of approximately 4) resulting from the smaller size in MB of the scene. For illustrative purposes, Table 7.4 displays the execution times in seconds measured for different processors in this application case study.

To conclude this subsection, we evaluate the data compression capabilities of the proposed P-MORPHCOMP method, which comprises the combination of

P-MORPHEE for endmember extraction followed by P-MORPHSU for abundance estimation. In order to explore the quality of the compressed images produced by the proposed compression method, Table 7.5 reports the SAD scores among the USGS reference signatures in Fig. 7.7(b) and the endmembers extracted by the best endmember extraction combination reported in Table 7.3 (P-PCT+P-FINDR) from the resulting images after data compression (the lowest the scores, the highest the similarity), using compression ratios of 20:1, 40:1 and 80:1 (given by different tested values of input parameter q). As expected, the highest-quality endmembers were extracted from the original data set. As the compression ratio was increased, the quality of extracted endmembers was decreased. For illustrative purposes, we have also included the results provided by two standard methods in our comparison, i.e., the 3-D SPIHT method [17] and the wavelet-based JPEG2000 multi-component method [18]). The JPEG2000 implementation used for our experiments was the one available in kakadu software (http://www.kakadusoftware.com). Both techniques are 3-D compression algorithms that treat the hyperspectral data as a 3-D volume, where the spectral information is the third dimension. Results in Table 7.5 show that such 3-D techniques, which enjoy great success in classical image processing, may not necessarily find equal success in hyperspectral image compression. Specifically, for the same compression ratio, a 3-D lossy compression may result in significant loss of spectral information which can be preserved much better, in turn, by a spectral unmixing-oriented lossy compression algorithm such as the proposed P-MORPHCOMP.

7.5.4 Summary

In this section we have thoroughly analyzed the performance of a suite of parallel algorithms for spatial-spectral processing of hyperspectral images in the context of two different application areas. Here, we intend to provide a quick reference of the main observations and lessons learned after each experiment. For that purpose, Table 7.6 summarizes the outcome of processing experiments conducted using parallel algorithms, including the algorithm combinations that performed best in each area, the best sequential and parallel processing times measured, and the use of spatial-spectral or just spectral information in the design of parallel algorithms. As Table 7.6 shows, the incorporation of spatial information to the traditionally spectrally-guided approach used in hyperspectral imaging can be very beneficial from the viewpoint of both algorithm accuracy and parallel algorithm design. We believe that the compendium of parallel spatial-spectral processing techniques and their detailed cross-validation in the context of real application domains, summarized on Table 7.6, may help hyperspectral image analysts and practitioners in this field in the task of selecting advanced data processing techniques and strategies for specific applications. Our experimental assessment of parallel algorithms also revealed important considerations which have not been previously addressed in the hyperspectral imaging literature to the authors' best knowledge:

Table 7.6. Summary of experiments using parallel spatial-spectral techniques

	Dimensionality reduction	Unsupervised classification	Endmember extraction	Abundance estimation	Data compression
Algorithms compared	P-PCT P-MORPH	P-ISODATA P-WSHED	P-PPI,P-FINDR P-MORPH	P-LSU P-MORPHSU	SPIHT,JPEG 2000 P-MORPHCOMP
Best method:	P-MORPH	P-WSHED	P-FINDR	P-MORPHSU	P-MORPHCOMP
Spatial info?	Yes	Yes	No	Yes	Yes
Time (seq.)	9935	60356	695	2308	3375
Time (par.)	45	355	7	10	15
Speedup	220.7	170.1	99.2	230.8	225.0

- Contrary to the common perception that spatial-spectral algorithms involve more complex operations than traditional, spectral-based techniques, results in this chapter indicate that spatial-spectral techniques, when carefully designed and implemented, can indeed be more *pleasingly parallel* than spectral-based techniques, mainly because they can reduce sequential computations at the master and only involve minimal communication between the parallel tasks, namely, at the beginning and ending of such tasks.
- Another important issue confirmed by experimental results is that the performance of spatial-spectral techniques is directly linked to the application domain. These techniques generally performed accurately in applications involving data sets with high spatial auto-correlation. On the other hand, applications in which the relevant image features are not spatially correlated generally benefit from the use of more spectrally-guided approaches.
- A relevant observation is that abundance estimation can be greatly improved by the incorporation of spatial context into the estimation. Standard techniques for fractional abundance determination in the literature have only resorted to the spectral-based techniques, and therefore the use of spatial-spectral information in the unmixing process (after a set of endmembers has been extracted) is a novel contribution first presented in this chapter.
- It is also important to emphasize that when the application considered requires high-quality preservation of the rich spectral information present in the original hyperspectral image, compression techniques based on spectral unmixing concepts are generally more effective than traditional 3-D compression techniques which have enjoyed great success in the image processing literature but may not be as successful in exploitation-based hyperspectral data compression.
- As a final major remark, this chapter has shown that spatial-spectral techniques are very appealing for the design of efficient parallel implementations, thus allowing their rapid execution in commodity cluster-based parallel systems.

7.6 Conclusions and Future Research

In this chapter, we have discussed the role of joint spatial-spectral information (via specialized morphological processing) in the analysis of hyperspectral images.

Specifically, we have explored the performance of five types of spatial-spectral algorithms for dimensionality reduction, unsupervised classification, endmember extraction, abundance estimation and data compression in the context of two different application domains, i.e., land-cover classification in agricultural applications, and mapping of geological features. Our experimental assessment of parallel spatial-spectral algorithms revealed important considerations about the properties and nature of such algorithms. On the other hand, performance results measured on the Thunderhead system at NASA's Goddard Space Flight Center indicate that our parallel implementations were able to provide adequate results in both the quality of the solutions and the time to obtain them, in particular, when they are implemented on commodity Beowulf clusters. The compendium of parallel spatial-spectral techniques presented in this chapter reflects the increasing sophistication of a field that is rapidly maturing at the intersection of many different disciplines, including image and signal processing, sensor design and instrumentation, parallel and distributed computing, and environmental applications. As future work, we plan to implement the full suite of parallel spatial-spectral algorithms discussed in this chapter on alternative high performance computing architectures, such Grid computing environments and specialized hardware platforms, including field programmable gate arrays (FPGAs) and general-purpose graphic processing units (GPUs). These platforms may allow us to fully accomplish the goal of real-time processing of hyperspectral image data, with potential applications in on-board hyperspectral image data compression and analysis.

Acknowledgement

The author would like to acknowledge J. Dorband, J.C. Tilton and J.A. Gualtieri for their support and collaboration in the experiments developed on the NASA Thunderhead Beowulf cluster. The author also gratefully thanks A. Paz for generating some of the illustrations displayed in this chapter.

References

1. Goetz, A.F.H., Vane, G., Solomon, J.E., Rock, B.N.: Imaging spectrometry for Earth remote sensing. Science 22, 1147–1153 (1985)
2. Green, R.O., et al.: Imaging spectroscopy and the airborne visible/infrared imaging spectrometer (AVIRIS). Remote Sensing of Environment 65, 227–248 (1998)
3. Richards, J.A., Xia, J.: Remote sensing digital image analysis: an introduction, 4th edn. Springer, Heidelberg (2006)
4. Chang, C.I.: Hyperspectral imaging: techniques for spectral detection and classification. Kluwer, New York (2003)
5. Richards, J.A.: Analysis of remotely sensed data: the formative decades and the future. IEEE Trans. Geoscience and Remote Sensing 43, 422–432 (2005)
6. Jimenez, L.O., Rivera-Medina, J.L., Rodriguez-Diaz, E., Arzuaga-Cruz, E., Ramirez-Velez, M.: Integration of spatial and spectral information by means of unsupervised extraction and classification for homogenous objects applied to multispectral and hyperspectral data. IEEE Trans. Geoscience and Remote Sensing 43, 844–851 (2005)

7. Gamba, P., Dell'Acqua, F., Ferrari, A., Palmason, J.A., Benediktsson, J.A., Arnasson, J.: Exploiting spectral and spatial information in hyperspectral urban data with high resolution. IEEE Geoscience and Remote Sensing Letters 1, 322–326 (2004)
8. Chanussot, J., Benediktsson, J.A., Fauvel, M.: Classification of remote sensing images from urban areas using a fuzzy probabilistic model. IEEE Geoscience and Remote Sensing Letters 3, 40–44 (2006)
9. Landgrebe, D.A.: Signal theory methods in multispectral remote sensing. Wiley, Hoboken (2003)
10. Plaza, A., Martinez, P., Perez, R., Plaza, J.: Spatial/spectral endmember extraction by multi-dimensional morphological operations. IEEE Trans. Geoscience and Remote Sensing 40, 2025–2041 (2002)
11. Plaza, A., Martinez, P., Plaza, J., Perez, R.: Dimensionality reduction and classification of hyperspectral image data using sequences of extended morphological transformations. IEEE Trans. Geoscience and Remote Sensing 43, 466–479 (2005)
12. Soille, P.: Morphological image analysis: principles and applications. Springer, Heidelberg (2003)
13. Serra, J.: Image analysis and mathematical morphology. Academic, New York (1982)
14. Boardman, J.: Automating spectral unmixing of aviris data using convex geometry concepts. In: Green, R.O. (ed.) Summaries of Airborne Earth Science Workshop, pp. 111–114. JPL Publication 93-26, Pasadena, CA (1993)
15. Winter, M.: Algorithm for fast autonomous spectral endmember determination in hyperspectral data. In: Descour, M.R., Shen, S.S. (eds.) Imaging Spectrometry V, Proceedings of SPIE, pp. 266–275 (1999)
16. Adams, J.B., Smith, M.O., Johnson, P.E.: Spectral mixture modeling: a new analysis of rock and soil types at the Viking Lander 1 site. Journal of Geophysical Research 91, 8098–8112 (1986)
17. Said, A., Pearlman, W.A.: A New, fast, and efficient image codec based on set partitioning in hierarchical trees. IEEE Trans. Circuits and Systems 6, 243–350 (1996)
18. Taubman, D.S., Marcellin, M.W.: JPEG2000: Image compression fundamentals, standards and practice. Kluwer, Boston (2002)
19. Plaza, A., Chang, C.I.: High performance computing in remote sensing. Chapman & Hall/CRC Press, Boca Raton (2007)
20. Comer, M., Delp, E.: Morphological operations for color image processing. Journal of Electronic Imaging 8, 279–289 (1999)
21. Plaza, A., Martinez, P., Perez, R., Plaza, J.: A new approach to mixed pixel classification of hyperspectral imagery based on extended morphological profiles. Pattern Recognition 37, 1097–1116 (2004)
22. Plaza, A., Valencia, D., Plaza, J., Martinez, P.: Commodity cluster-based parallel processing of hyperspectral imagery. Journal of Parallel and Distributed Computing 66, 345–358 (2006)
23. Seinstra, F.J., Koelma, D.: User transparency: a fully sequential programming model for efficient data parallel image processing. Concurrency and Computation: Practice and Experience 16, 611–644 (2004)
24. Plaza, A., Plaza, J., Valencia, D.: Impact of platform heterogeneity on the design of parallel algorithms for morphological processing of high-dimensional image data. Journal of Supercomputing 40, 81–107 (2007)

25. Plaza, A., Valencia, D., Plaza, J., Chang, C.I.: Parallel implementation of endmember extraction algorithms from hyperspectral data. IEEE Geoscience and Remote Sensing Letters 3, 334–338 (2006)
26. Benediktsson, J.A., Palmason, J.A., Sveinsson, J.R.: Classification of hyperspectral data from urban areas based on extended morphological profiles. IEEE Trans. Geoscience and Remote Sensing 42, 480–491 (2005)
27. Beucher, S.: Watershed, hierarchical segmentation and waterfall algorithm. In: Dougherty, E. (ed.) Mathematical morphology and its applications to image processing. Kluwer, Boston (1994)
28. Malpica, N., Ortuno, J.E., Santos, A.: A multichannel watershed-based algorithm for supervised texture segmentation. Pattern Recognition Letters 24, 1545–1554 (2003)
29. Moga, A.N., Gabbouj, M.: Parallel marker-based image segmentation with watershed transformation. Journal of Parallel and Distributed Computing 51, 27–45 (1998)
30. Plaza, A., Martinez, P., Perez, R., Plaza, J.: A quantitative and comparative analysis of endmember extraction algorithms from hyperspectral data. IEEE Trans. Geoscience and Remote Sensing 42, 650–663 (2004)
31. Plaza, A., Chang, C.I.: Impact of initialization on design of endmember extraction algorithms. IEEE Trans. Geoscience and Remote Sensing 44, 3397–3407 (2006)
32. Graña, M., Hernandez, C., Gallego, J.: A single individual evolutionary strategy for endmember search in hyperspectral images. Information Sciences 161, 181–197 (2004)
33. Ramakhrishna, B., Plaza, A., Chang, C.I., Ren, H., Du, Q., Chang, C.C.: Spectral/spatial hyperspectral image compression. In: Storer, J.A., Motta, G. (eds.) Hyperspectral data compression. Springer, Heidelberg (2006)
34. Valencia, D., Plaza, A.: FPGA-based compression of hyperspectral imagery using spectral unmixing and the pixel purity index algorithm. LNCS, vol. 3993, pp. 24–31. Springer, Heidelberg (2006)
35. Achalakul, T., Taylor, S.: A distributed spectral-screening PCT algorithm. Journal of Parallel and Distributed Computing 63, 373–384 (2003)
36. Dhodhi, M.K., Saghri, J.A., Ahmad, I., Ul-Mustafa, R.: D-ISODATA: A distributed algorithm for unsupervised classification of remotely sensed data on networks of workstations. Journal of Parallel and Distributed Computing 59, 280–301 (1999)
37. Du, Q., Chang, C.I.: Estimation of number of spectrally distinct signal sources in hyperspectral imagery. IEEE Trans. Geoscience and Remote Sensing 42, 608–619 (2004)

8

Parallel Classification of Hyperspectral Images Using Neural Networks

Javier Plaza, Antonio Plaza, Rosa Pérez, and Pablo Martínez

Department of Technology of Computers and Communications
University of Extremadura, Avda. de la Universidad s/n
E-10071 Cáceres, Spain
{jplaza,aplaza,rosapere,pablomar}@unex.es

Summary. Neural networks represent a widely used alternative to deal with remotely sensed image data. The improvement of spatial and spectral resolution in latest-generation Earth observation instruments is expected to introduce extremely high computational requirements in neural network-based algorithms for classification of high-dimensional data sets such as hyperspectral images, with hundreds of spectral channels and very fine spatial resolution. A significant advantage of neural networks versus other types of processing algorithms for hyperspectral imaging is that they are inherently amenable for parallel implementation. As a result, they can benefit from advances in low cost parallel computing architectures such as heterogeneous networks of computers, which have soon become a standard tool of choice for dealing with the massive amount of image data sets. In this chapter, several techniques for classification of hyperspectral imagery using neural networks are presented and discussed. Experimental results are provided from the viewpoint of both classification accuracy and parallel performance on a variety of parallel computing platforms, including two networks of workstations at University of Maryland and a massively parallel Beowulf cluster at NASA's Goddard Space Flight Center in Maryland. Two different application areas are addressed for demonstration: land-cover classification using hyperspectral data collected by NASA over the valley of Salinas, California, and urban city classification using data collected by the German Aerospace Agency (DLR) over the city of Pavia, Italy.

8.1 Introduction

Many international agencies and research organizations are currently devoted to the analysis and interpretation of high-dimensional image data collected over the surface of the Earth [1]. For instance, NASA is continuously gathering hyperspectral images (at different wavelength channels) using Jet Propulsion Laboratory's Airborne Visible-Infrared Imaging Spectrometer (AVIRIS) [2].

The incorporation of hyperspectral instruments aboard satellite platforms is now producing a near-continual stream of high-dimensional remotely sensed data, and computationally efficient data processing techniques are required in a variety of time-critical applications, including wildland fire monitoring, detection of chemical and biological agents in waters and atmosphere, or target detection for military purposes.

M. Graña and R.J. Duro (Eds.): Comput. Intel. for Remote Sensing, SCI 133, pp. 193–216, 2008.
springerlink.com

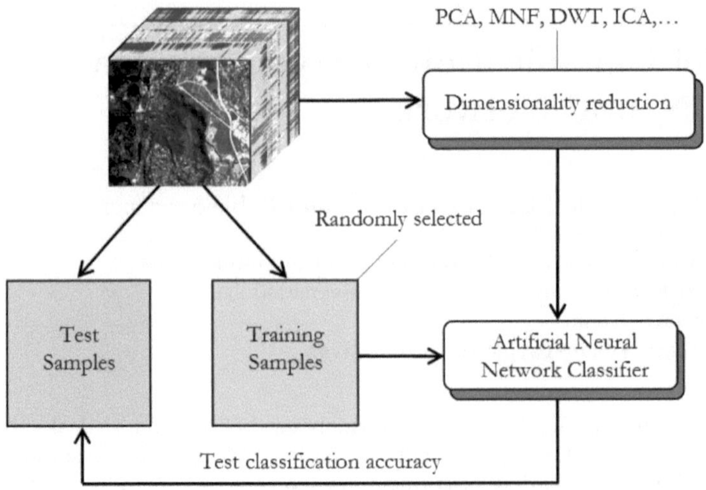

Fig. 8.1. Neural network-based hyperspectral data classification

Neural networks have been widely used in previous work to analyze hyperspectral images [3]. In particular, neural architectures have demonstrated great potential to model mixed pixels, which result from limited spatial resolution in certain application domains and also from the mixed nature of the interaction between photons and particles of observed land-cover surfaces. Since this process is inherently nonlinear, neural networks are an appropriate tool for mixed pixel classification due to their capacity to approximate complex nonlinear functions [4].

The standard classification approach adopted for neural networks is illustrated in Fig. 8.1, in which the original input data is first reduced in its dimensionality to avoid the Hughes effect. Previous research has demonstrated that the high-dimensional data space spanned by hyperspectral data sets is usually empty [5], indicating that the data structure involved exists primarily in a subspace. Commonly used techniques to reduce the dimensionality of the data have been the principal component transform (PCT) or the minimum noise fraction (MNF) [6]. The discrete wavelet transform (DWT) or independent component analysis (ICA) have also been proposed for this task [7]. However, these approaches rely on spectral properties of the data alone, thus neglecting the information related to the spatial arrangement of the pixels in the scene. Usually, we need to manage very high-dimensional data volumes in which spatial correlation between spectral responses of neighboring pixels can be potentially high [8]. As a result, there is a need for feature extraction techniques able to integrate the spatial and spectral information available from the data simultaneously [7]. Once a set of relevant features have been extracted from the input data, artificial neural network-based classification usually follows a supervised strategy, in which a few randomly selected training samples are selected from available labeled data and used to train the neural classifier. The trained classifier is then tested using the remaining samples.

Although many neural network architectures have been explored in the literature, feedforward networks of various layers, such as the multi-layer perceptron (MLP), have been widely used in hyperspectral imaging applications [9]. The MLP is typically trained using the error back-propagation algorithm, a supervised technique of training with three phases. In the first one, an initial vector is presented to the network, which leads to the activation of the network as a whole. The second phase computes an error between the output vector and a vector of desired values for each output unit, and propagates it successively back through the network. The last phase computes the changes for the connection weights, which are randomly generated at the beginning of the process. It has been shown in the literature that MLP-based neural models, when trained accordingly, generally outperform other nonlinear models such as regression trees or fuzzy classifiers [9].

Despite the success of neural network models for classification of remotely sensed hyperspectral images, several challenges still remain open in order to incorporate such models to real applications. An important limitation of these models is the fact that their computational complexity can be quite high [10], in particular, when they are used to analyze large hyperspectral data sets (or data repositories) comprising hundreds of spectral channels per scene [4].

To address the computational needs introduced by neural network-based algorithms in hyperspectral imaging applications, several efforts have been recently directed towards the incorporation of parallel computing models in remote sensing, specially with the advent of relatively cheap Beoulf clusters [11], [12]. The goal is to create parallel computing systems from commodity components to satisfy specific requirements for the Earth and space sciences community [13]. Although most dedicated parallel machines employed by NASA and other institutions during the last decade have been chiefly homogeneous in nature [14], a current trend is to utilize heterogeneous and distributed parallel computing platforms [15]. In particular, computing on heterogeneous networks of computers (HNOCs) is an economical alternative which can benefit from local (user) computing resources while, at the same time, achieve high communication speed at lower prices. These properties have led HNOCs to become a standard tool for high-performance computing in many ongoing and planned remote sensing missions [16], [17], thus taking advantage of from the considerable amount of work done in dynamic, resource-aware static and dynamic task scheduling and load balancing in distributed platforms, including distributed systems and Grid computing environments.

Despite the increasing importance of HNOCs in remote sensing data interpretation, there is a lack of consolidated neural network-based algorithms specifically developed for heterogeneous computing platforms. To address the need for cost-effective and innovative parallel neural network classification algorithms, this chapter develops a new morphological/neural parallel algorithm for classification of hyperspectral imagery. The algorithm is inspired by previous work on morphological neural networks, such as autoassociative morphological memories and morphological perceptrons [18], although it is based on different concepts.

Most importantly, the algorithm can be tuned for very efficient execution on both HNOCs and massively parallel, Beowulf-type commodity clusters.

The remainder of the chapter is structured as follows. Section 8.2 describes the proposed heterogeneous parallel algorithm, which consists of two main processing steps: 1) parallel morphological feature extraction taking into account the spatial and spectral information, and 2) robust classification using a parallel multi-layer neural network with back-propagation learning. Section 8.3 describes the algorithm's accuracy and parallel performance. Classification accuracy is discussed in the context of a real applications, including land-cover classification of agricultural fields in the valley of Salinas, California and urban classification using hyperspectral data collected over the city of Pavia, Italy. Parallel performance in the context of the above-mentioned application is also provided by comparing the efficiency achieved by an heterogeneous parallel version of the proposed algorithm, executed on a fully heterogeneous network, with the efficiency achieved by its equivalent homogeneous version, executed on a fully homogeneous network with the same aggregate performance as the heterogeneous one. For comparative purposes, performance data on Thunderhead, a massively parallel Beowulf cluster at NASA's Goddard Space Flight Center, are also given. Finally, Section 8.4 concludes with some remarks and hints at plausible future research.

8.2 Parallel Morphological/Neural Classification Algorithm

In this section we present a new parallel approach for neural network-based classification of hyperspectral data, which has been specifically tuned for efficient execution in heterogeneous parallel platforms. The section is structured as follows. First, we formulate a general optimization problem in the context of HNOCs. Then, we separately describe the two main steps of the proposed parallel neural algorithm. The first stage (morphological feature extraction) relies on a feature selection stage performed by using extended morphological operations specifically tuned to deal with hyperspectral images. The second stage (neural network classification) is a supervised technique in which the extracted features are fed to a MLP-based neural architecture and used to classify the data according to a gradient descent algorithm.

8.2.1 Optimization Problem

Heterogeneous networks are composed of different-speed processors that communicate through links at different capacities [15]. This type of platform can be simply modeled as a complete graph $G = (P, E)$ where each node models a computing resource p_i weighted by its relative cycle-time w_i . Each edge in the graph models a communication link weighted by its relative capacity, where c_{ij} denotes the maximum capacity of the slowest link in the path of physical communication links from p_i to p_j. We also assume that the system has symmetric costs, i.e., $c_{ij} = c_{ji}$. Under the above assumptions, processor p_i will accomplish a share of $\alpha_i \times W$ of the total workload W, with $\alpha_i \geq 0$ for $1 \leq i \leq P$ and $\sum_{i=1}^{P} \alpha_i = 1$. With the above assumptions in mind, an abstract view of our problem can be simply stated in the form of a client-server architecture, in which the server is responsible for the efficient distribution of work among the P nodes, and the clients operate with the spatial and spectral information contained in a local partition. The partitions are then updated locally and the resulting calculations may also be exchanged between the clients, or between the server and the clients. Below, we describe the two main steps of our parallel algorithm.

8.2.2 Parallel Morphological Feature Extraction

This section develops a parallel morphological feature extraction algorithm for hyperspectral image analysis. First, we describe the morphological algorithm and its principles, then we study several possibilities for its parallel implementation for HNOCs.

Morphological feature extraction algorithm

The feature extraction algorithm presented in this section is based on concepts from mathematical morphology theory [19], which provides a remarkable framework to achieve the desired integration of the complementary nature of spatial and spectral information in simultaneous fashion, thus alleviating the problems related to each of them taken separately. In order to extend classical morphological operations from gray-scale image analysis to hyperspectral image scenarios, the main goal is to impose an ordering relation (in terms of spectral purity) in the set of pixel vectors lying within a spatial search window (called structuring element) designed by B [7]. This is done by defining a cumulative distance between a pixel vector $\boldsymbol{f}(x, y)$ and all the pixel vectors in the spatial neighborhood given by B (B-neighborhood) as follows: $D_B[\boldsymbol{f}(x,y)] = \sum_i \sum_j \text{SAD}[\boldsymbol{f}(x,y), \boldsymbol{f}(i,j)]$, where (x, y) refers to the spatial coordinates in the B-neighborhood and SAD is the spectral angle distance, given by the following expression:

$$\text{SAD}(\boldsymbol{f}(x,y), \boldsymbol{f}(i,j)) = \arccos\left(\frac{\boldsymbol{f}(x,y) \cdot \boldsymbol{f}(i,j)}{\|\boldsymbol{f}(x,y)\| \cdot \|\boldsymbol{f}(i,j)\|}\right). \tag{8.1}$$

From the above definitions, two standard morphological operations called extended erosion and dilation can be respectively defined as follows:

$$(\boldsymbol{f} \otimes B)(x,y) = \text{argmin}_{(s,t) \in Z^2(B)} \sum_s \sum_t \text{SAD}(f(x,y), f(x+s, y+t)), \tag{8.2}$$

$$(\boldsymbol{f} \oplus B)(x,y) = \text{argmax}_{(s,t) \in Z^2(B)} \sum_s \sum_t \text{SAD}(\boldsymbol{f}(x,y), \boldsymbol{f}(x-s, y-t)). \tag{8.3}$$

Using the above operations, the opening filter is defined as $(\boldsymbol{f} \circ B)(x,y) = [(\boldsymbol{f} \otimes B) \oplus B](x,y)$ (erosion followed by dilation), while the closing filter is defined as

$(\boldsymbol{f}\bullet B)(x,y) = [(\boldsymbol{f}\oplus B)\otimes B](x,y)$ (dilation followed by erosion). The composition of opening and closing operations is called a spatial/spectral profile, which is defined as a vector which stores the relative spectral variation for every step of an increasing series. Let us denote by $\{(\boldsymbol{f}\circ B)^{\lambda}(x,y)\}, with\lambda = \{0,1,...,k\}$, the *opening series* at $\boldsymbol{f}(x,y)$, meaning that several consecutive opening filters are applied using the same window B. Similarly, let us denote by $\{(\boldsymbol{f}\bullet B)^{\lambda}(x,y)\}, \lambda = \{0,1,...,k\}$, the *closing series* at $\boldsymbol{f}(x,y)$. Then, the spatial/spectral profile at $\boldsymbol{f}(x,y)$ is given by the following vector:

$$\boldsymbol{p}(x,y) = \{\mathrm{SAM}((\boldsymbol{f}\circ B)^{\lambda}(x,y),(\boldsymbol{f}\circ B)^{\lambda-1}(x,y))\}\cup\cdots \\ \cdots\cup\{\mathrm{SAM}((\boldsymbol{f}\bullet B)^{\lambda}(x,y),(\boldsymbol{f}\bullet B)^{\lambda-1}(x,y))\}. \quad (8.4)$$

Here, the step of the opening/closing series iteration at which the spatial/spectral profile provides a maximum value gives an intuitive idea of both the spectral and spatial distribution in the B-neighborhood [7]. As a result, the morphological profile can be used as a feature vector on which the classification is performed using a spatial/spectral criterion.

Parallel implementations

Two types of data partitioning can be exploited in the parallelization of spatial/spectral algorithms such as the one addressed above [14]:

1. **Spectral-domain partitioning.** Subdivides the volume into small cells or sub-volumes made up of contiguous spectral bands, and assigns one or more sub-volumes to each processor. With this model, each pixel vector is split amongst several processors, which breaks the spectral identity of the data because the calculations for each pixel vector (e.g., for the SAD calculation) need to originate from several different processing units.
2. **Spatial-domain partitioning.** Provides data chunks in which the same pixel vector is never partitioned among several processors. As a result, each pixel vector is always entirely stored in the same processing unit, thus entirely preserving the full spectral signature associated to the pixel at the same processor.

In this chapter, we adopt a spatial-domain partitioning approach due to several reasons. First and foremost, the application of spatial-domain partitioning is a natural approach for morphological image processing, as many operations require the same function to be applied to a small set of elements around each data element present in the image data structure, as indicated in the previous subsection.

A second reason has to do with the cost of inter-processor communication. For instance, in spectral-domain partitioning, the structuring element-based calculations made for each hyperspectral pixel would need to originate from several processing elements, thus requiring intensive inter-processor communication. However, if redundant information such as an overlap border is added to each of the adjacent partitions to avoid accesses outside the image domain, then boundary data to be communicated between neighboring processors can be greatly minimized. Such an overlapping scatter would obviously introduce redundant computations, since the intersection between partitions would be non-empty. Our implementation makes use of a constant structuring element B (with size of 3×3 pixels) which is repeatedly iterated to increase the spatial context, and the total amount of redundant information is minimized. To do so, we have implemented three different approximations in order to handle with these so-called border pixels:

- **MP-1** implements a non-overlapping scatter operation followed by overlap communication for every hyperspectral pixel vector, thus communicating small sets of pixels very often.
- **MP-2** implements a standard non-overlapping scatter operation followed by a special overlap communication which sends all border pixels beforehand , but only once.
- **MP-3** performs a special 'overlapping scatter' operation that also sends out the overlap border data as part of the scatter operation itself (i.e., redundant computations replace communications).

Here, we make use of MPI *derived datatypes* to directly scatter hyperspectral data structures, which may be stored non-contiguously in memory, in a single communication step. A pseudo-code of the proposed parallel morphological feature extraction algorithm is given below. The inputs to the algorithm are an N-dimensional cube $\boldsymbol{f}$, and a structuring element B. The output is a set of morphological profiles for each pixel.

1. Obtain information about the heterogeneous system, including the number of processors, P, each processors identification number, $\{p_i\}_{i=1}^P$, and processor cycle-times, $\{w_i\}_{i=1}^P$.
2. Using B and the information obtained in step 1, determine the total volume of information, R, that needs to be replicated from the original data volume, V, according to the data communication strategies outlined above, and let the total workload W to be handled by the algorithm be given by $W = V+R$. Then, partition the data using one of the three considered partitioning strategies for morphological processing: MP-1, MP-2 or MP-3.
3. Set $\alpha_i = \lfloor \frac{(P/w_i)}{\sum_{i=1}^P (1/w_i)} \rfloor$ for all $i \in \{1, ..., P\}$.
4. For $m = \sum_{i=1}^P \alpha_i$ to $(V + R)$, find $k \in \{1, .., P\}$ so that $w_k \cdot (\alpha_k + 1) = \min\{w_i \cdot (\alpha_i + 1)\}_{i=1}^P$ and set $\alpha_k = \alpha_k + 1$.
5. Use the resulting $\{\alpha_i\}_{i=1}^P$ to obtain a set of P spatial-domain heterogeneous partitions of W in accordance with the selected strategy: MP-1, MP-2 or MP-3, and send each partition to processor p_i, along with B.
6. Calculate the morphological profiles $\boldsymbol{p}(x, y)$ for the pixels in the local data partitions (in parallel) at each heterogeneous processor.
7. Collect all the individual results and merge them together to produce the final output at the master processor.

It should be noted that a homogeneous version of the algorithm above can be simply obtained by replacing step 4 with $\alpha_i = P/w_i$ for all $i \in \{1, ..., P\}$, where w_i is the communication speed between processor pairs in the network, assumed to be homogeneous.

8.2.3 Parallel Neural Classification Algorithm

The second step of the proposed parallel algorithm consists of a supervised parallel classifier based on a MLP neural network with back-propagation learning. The neural network is trained with selected features from the previous morphological feature extraction. This approach has been shown in previous work to be very robust for classification of hyperspectral imagery [20]. However, the considered neural architecture and back-propagation-type learning algorithm introduce additional considerations for parallel implementation on HNOCs. This section first describes the neural network architecture and learning procedure, and then describes several strategies for its parallelization, resulting in two main approaches for efficient implementation in HNOCs, namely, Exemplar partitioning and Hybrid partitioning.

Network architecture and learning

The architecture adopted for the proposed MLP-based neural network classifier is shown in Fig. 8.2. The number of input neurons equals the number of spectral bands acquired by the sensor. In case either PCT-based pre-processing or morphological feature extraction are applied as a pre-processing steps, then the number of neurons at the input layer equals the dimensionality of feature vectors used for classification. The second layer is the hidden layer, where the number of nodes, M, is usually estimated empirically. Finally, the number of neurons at the output layer, C, equals the number of distinct classes to be identified in the input data. With the above architecture, the activation function for hidden and output nodes can be selected by the user. In this chapter, we have used sigmoid activation for all experiments. With the above architectural design in mind, the standard back-propagation learning algorithm can be outlined by the following steps:

1. *Forward phase.* Let the individual components of an input pattern be denoted by $f_j(x, y)$, with $j = 1, 2, ..., N$. The output of the neurons at the hidden layer are obtained as: $H_i = \varphi(\sum_{j=1}^{N} \omega_{ij} \cdot \boldsymbol{f}_j(x, y))$ with $i = 1, 2, ..., M$, where $\varphi(\cdot)$ is the activation function and ω_{ij} is the weight associated to the connection between the i-th input node and the j-th hidden node. The outputs of the MLP are obtained using $O_k = \varphi(\sum_{i=1}^{M} \omega_{ki} \cdot H_i)$, with $k = 1, 2, ..., C$. Here, ω_{ki} is the weight associated to the connection between the i-th hidden node and the k-th output node.
2. *Error back-propagation.* In this stage, the differences between the desired and obtained network outputs are calculated and back-propagated. The *delta* terms for every node in the output layer are calculated using $\delta_k^o = (O_k - d_k) \cdot$

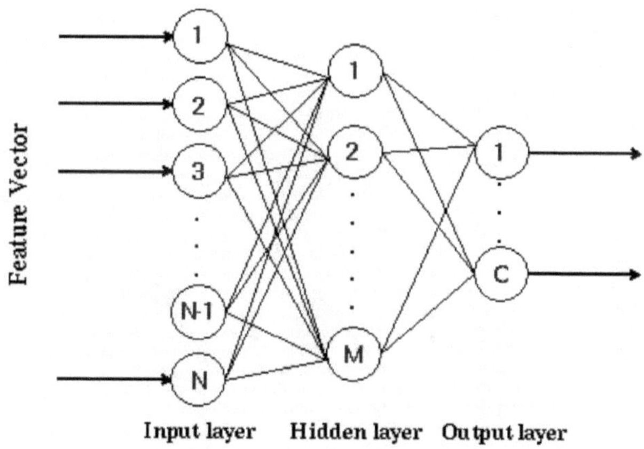

Fig. 8.2. MLP neural network topology

$\varphi'(\cdot)$, with $i = 1, 2, ..., C$. Here, $\varphi'(\cdot)$ is the first derivative of the activation function. Similarly, *delta* terms for the hidden nodes are obtained using $\delta_i^h = \sum_{k=1}^{C}(\omega_{ki} \cdot \delta_i^o) \cdot \varphi(\cdot))$, with $i = 1, 2, ..., M$.

3. *Weight update.* After the back-propagation step, all the weights of the network need to be updated according to the *delta* terms and to η, a learning rate parameter. This is done using $\omega_{ij} = \omega_{ij} + \eta \cdot \delta_i^h \cdot f_j(x, y)$ and $\omega_{ki} = \omega_{ki} + \eta \cdot \delta_k^o \cdot H_i$. Once this stage is accomplished, another training pattern is presented to the network and the procedure is repeated for all incoming training patterns.

Once the back-propagation learning algorithm is finalized, a classification stage follows, in which each test input pixel vector is classified using the weights obtained by the network during the training stage [20]. Diffferent strategies for parallel classification using the proposed MLP architecture are given in the following subsection.

Parallel MLP classification

Several partitioning schemes can be analyzed when mapping a MLP neural network on a cluster architecture [21]. The choice is application-dependent, and a key issue is the number of training patterns to be used during the learning stage. In this chapter, we present two different schemes for the multi-layer perceptron classifier partitioning:

1. **Exemplar partitioning.** This approach, also called training example parallelism, explores data level parallelism and can be easily obtained by simply partitioning the training pattern data set (see Fig. 8.3). Each process determines the weight changes for a disjoint subset of the training population and then changes are combined and applied to the neural network at the end of each epoch. As will be shown by experiments, this scheme requires

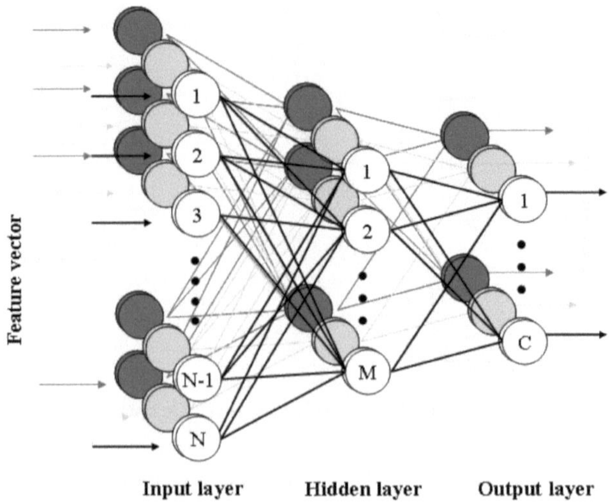

Fig. 8.3. Exemplar partitioning scheme over MLP topology. Training data is divided into different subsets which are used to train three different subnetworks, i.e., white, grey and black.

a large number of training patterns to produce significant speedups, which is the most common situation in most remote sensing applications due to the limited availability of training samples and the great difficulty to generate accurately labeled ground-truth samples prior to analyzing the collected data.

2. **Hybrid partitioning.** This approach is based on a combination of neuronal level as well as synaptic level parallelism [22] which allows us to reduce the processor intercommunications at each iteration (see Fig. 8.4). This approach results from a combination of neuronal level parallelism and synaptic level parallelism. In the former (also called vertical partitioning) all the incoming weights to the neurons local to the processor are computed by a single processor. In the latter, each workstation will compute only the outgoing weight connections of the nodes (neurons) local to the processor. The hybrid scheme combines those approaches, i.e., the hidden layer is partitioned using neuronal parallelism while weight connections adopt synaptic scheme. As a result, all inter-processor communications will be reduced to a cumulative sum at each epoch, thus significantly reducing processing time on parallel platforms.

Implementation of parallel Exemplar partitioning

As mentioned above, the exemplar parallelism strategy is relatively easy to implement. It only requires the addition of an initialization algorithm and a synchronization step on each iteration. The initialization procedure distribute the training patterns among the different processors according to their relative speed and memory characteristics. As shown in Fig. 8.3, each processor will implement a complete neural network topology which will compute only the previously related pattern subset. On each iteration, a synchronization step is required in order to integrate the weights matrices actualization obtained at each MLP architecture considering its respective training pattern subset. It can be easily achieved by having all the processors communicate their partial weights change matrices to the master node which computes the total sum of all partial matrices and applies the changes to the network. It then broadcasts the new weights matrix to the slave nodes. In our approximation, we improve the communication task by taking advantage of MPI predefined communication directives, which allow us to perform an all to all actualization and communication in a single MPI directive.

A pseudo-code of the Exemplar partitioning algorithm can be summarized as follows. The inputs to the algorithm are an N-dimensional cube $\boldsymbol{f}$, and a set of training patterns $\boldsymbol{f}_j(x,y)$. The output is a classification label for each image pixel.

1. Use steps 1-4 of the parallel morphological feature extraction algorithm to obtain a set of values $(\alpha_i)_{i=1}^{P}$ which will determine the share of the workload to be accomplished by each heterogeneous processor.
2. Use the resulting $(\alpha_i)_{i=1}^{P}$ to obtain a set of P heterogeneous partitions of the training patterns and map the resulting partitions $\boldsymbol{f}_j^P(x,y)$ among the P heterogeneous processors (which also store the full multi-layer perceptron architecture).
3. *Parallel training.* For each training pattern contained on each partition, the following three steps are executed in parallel for each processor:
   a) *Parallel forward phase.* In this phase, the activation value of the hidden neurons local to the processors are calculated. For each input pattern, the activation value for the hidden neurons is calculated using $H_i^P = \varphi(\sum_{j=1}^{N} \omega_{ij} \cdot \boldsymbol{f}_j^P(x,y))$. Here the activation values of output neurons are calculated according to $O_k^P = \varphi(\sum_{i=1}^{M} \omega_{ki}^P \cdot H_i^P)$, with $k = 1, 2, ..., C$.
   b) *Parallel error back-propagation.* In this phase, each processor calculates the error terms for the local training patters. To do so, *delta* terms for the output neurons are first calculated using $(\delta_k^o)^P = (O_k - d_k)^P \cdot \varphi'(\cdot)$, with $i = 1, 2, ..., C$. Then, error terms for the hidden layer are computed using $(\delta_i^h)^P = \sum_{k=1}^{P} (\omega_{ki}^P \cdot (\delta_k^o)^P) \cdot \varphi'(\cdot)$, with $i = 1, 2, ..., N$.
   c) *Parallel weight update.* In this phase, the weight connections between the input and hidden layers are updated by $\omega_{ij} = \omega_{ij} + \eta^P \cdot (\delta_i^h)^P \cdot \boldsymbol{f}_j^P(x,y)$. Similarly, the weight connections between the hidden and output layers are updated using $\omega_{ki}^P = \omega_{ki}^P + \eta^P \cdot (\delta_k^o)^P \cdot H_i^P$.
   d) *Broadcast and intialization of weight matrices.* In this phase, each node sends its partial weight matrices to its neighbor node, which sums it to its partial matrix and proceed to send it again to the neighbor. When all nodes have added their local matrices, then the resulting total weight matrices are broadcast to be used by all processors in the next iteration.

4. *Classification.* For each pixel vector in the input data cube $\boldsymbol{f}$, calculate (in parallel) $\sum_{j=1}^{P} O_k^j$, with $k = 1, 2, ..., C$. A classification label for each pixel can be obtained using the winner-take-all criterion commonly used in neural network applications by finding the cumulative sum with maximum value, say $\sum_{j=1}^{P} O_{k^*}^j$, with $k^* = \arg\{\max_{1 \leq k \leq C} \sum_{j=1}^{P} O_k^j\}$.

Implementation of parallel Hybrid partitioning

In the hybrid classifier, the hidden layer is partitioned so that each heterogeneous processor receives a number of hidden neurons which depends on its relative speed. Figure 8.4 shows how each processor stores the weight connections between the neurons local to the processor. Since the fully connected MLP network is partitioned into P partitions and then mapped onto P heterogeneous processors using the above framework, each processor is required to communicate with every other processor to simulate the complete network. For this purpose, each of the processors in the network executes the three phases of the back-propagation learning algorithm described above.

A pseudo-code of the Hybrid partitioning algorithm can be summarized as follows. The inputs to the algorithm are an N-dimensional data cube $\boldsymbol{f}$, and a set of training patterns $\boldsymbol{f}_j(x, y)$. The output is a classification label for each image pixel.

1. Use steps 1-4 of the parallel morphological feature extraction algorithm to obtain a set of values $(\alpha_i)_{i=1}^{P}$ which will determine the share of the workload to be accomplished by each heterogeneous processor.
2. Use the resulting $(\alpha_i)_{i=1}^{P}$ to obtain a set of P heterogeneous partitions of the hidden layer and map the resulting partitions among the P heterogeneous processors (which also store the full input and output layers along with all connections involving local neurons).
3. *Parallel training.* For each considered training pattern, the following three parallel steps are executed:
   a) *Parallel forward phase.* In this phase, the activation value of the hidden neurons local to the processors are calculated. For each input pattern, the activation value for the hidden neurons is calculated using $H_i^P = \varphi(\sum_{j=1}^{N} \omega_{ij} \cdot \boldsymbol{f}_j(x, y))$. Here, the activation values and weight connections of neurons present in other processors are required to calculate the activation values of output neurons according to $O_k^P = \varphi(\sum_{i=1}^{M/P} \omega_{ki}^P \cdot H_i^P)$, with $k = 1, 2, ..., C$. In our implementation, broadcasting the weights and activation values is circumvented by calculating the partial sum of the activation values of the output neurons.
   b) *Parallel error back-propagation.* In this phase, each processor calculates the error terms for the local hidden neurons. To do so, *delta* terms for the output neurons are first calculated using the formula $(\delta_k^o)^P = (O_k - d_k)^P \cdot \varphi'(\cdot)$, with $i = 1, 2, ..., C$. Then, error terms for the hidden layer are computed using $(\delta_i^h)^P = \sum_{k=1}^{P} (\omega_{ki}^P \cdot (\delta_k^o)^P) \cdot \varphi'(\cdot)$, with $i = 1, 2, ..., N$.

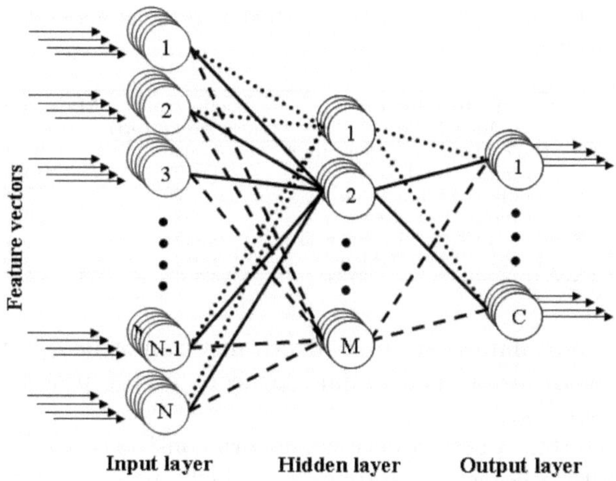

Fig. 8.4. Hybrid partitioning scheme over MLP topology. The input and output layers are common to all processors. The hidden nodes are distributed among different processors (lines, dotted-lines and dashed-lines denote weight connections corresponding to three different processors).

c) *Parallel weight update.* In this phase, the weight connections between the input and hidden layers are updated by $\omega_{ij} = \omega_{ij} + \eta^P \cdot (\delta_i^h)^P \cdot \boldsymbol{f}_j(x, y)$. Similarly, the weight connections between the hidden and output layers are updated using the expression: $\omega_{ki}^P = \omega_{ki}^P + \eta^P \cdot (\delta_k^o)^P \cdot H_i^P$.

4. *Classification.* For each pixel vector in the input data cube $\boldsymbol{f}$, calculate (in parallel) $\sum_{j=1}^{P} O_k^j$, with $k = 1, 2, ..., C$. A classification label for each pixel can be obtained using the winner-take-all criterion commonly used in neural networks by finding the cumulative sum with maximum value, say $\sum_{j=1}^{P} O_{k^*}^j$, with $k^* = \arg\{\max_{1 \leq k \leq C} \sum_{j=1}^{P} O_k^j\}$.

8.3 Experimental Results

This section provides an assessment of the effectiveness of the parallel algorithms described in section 2. The section is organized as follows. First, we describe the parallel computing platforms used in this chapter for evaluation purposes. Then, we provide performance data for the proposed parallel algorithms in the context of two real hyperspectral imaging applications.

8.3.1 Parallel Computing Platforms

Before describing the parallel systems used for evaluation in this chapter, we briefly outline a recent study [23] on performance analysis for heterogeneous algorithms that we have adopted in this work for evaluation purposes. Specifically, we assess the parallel performance of the proposed heterogeneous algorithms

Table 8.1. Specifications of processors in a heterogeneous network of computers at University of Maryland

Processor identification	Architecture description	Cycle-time (secs/megaflop)	Main memory (MB)	Cache (KB)
p_1	Free BSD – i386 Intel Pentium 4	0.0058	2048	1024
p_2, p_5, p_8	Linux – Intel Xeon	0.0102	1024	512
p_3	Linux – AMD Athlon	0.0026	7748	512
p_4, p_6, p_7, p_9	Linux – Intel Xeon	0.0072	1024	1024
p_{10}	SunOS – SUNW UltraSparc-5	0.0451	512	2048
$p_{11} - p_{16}$	Linux – AMD Athlon	0.0131	2048	1024

using the basic postulate that these algorithms cannot be executed on a heterogeneous network faster than its homogeneous prototypes on the equivalent homogeneous network.

Let us assume that a heterogeneous network consists of $\{p_i\}_i^P$ heterogeneous workstations with different cycle-times w_i, which span m communication segments $\{s_j\}_{j=1}^m$, where $c^{(j)}$ denotes the communication speed of segment s_j. Similarly, let $p^{(j)}$ be the number of processors that belong to s_j, and let $w_t^{(j)}$ be the speed of the t-th processor connected to s_j, where $t = 1, ..., p^{(j)}$. Finally, let $c^{(j,k)}$ be the speed of the communication link between segments s_j and s_k, with $j, k = 1, ..., m$. According to [23], the above network can be considered equivalent to a homogeneous one made up of $\{q_i\}_{i=1}^P$processors with constant cycle-time and interconnected through a homogeneous communication network with speed c if the following expressions are satisfied:

$$c = \frac{\sum_{j=1}^m c^{(j)} \cdot \left[\frac{p^{(j)}(p^{(j)}-1)}{2}\right]}{\frac{P(P-1)}{2}} + \frac{\sum_{j=1}^m \sum_{k=j+1}^m p^{(j)} \cdot p^{(k)} \cdot c^{(j,k)}}{\frac{P(P-1)}{2}}, \tag{8.5}$$

$$w = \frac{\sum_{j=1}^m \sum_{t=1}^{p^{(j)}} w_t^{(j)}}{P}, \tag{8.6}$$

where equation (8.5) states that the average speed of point-to-point communications between processors $\{p_i\}_{i=1}^P$ in the heterogeneous network should be equal to the speed of point-to-point communications between processors $\{q_i\}_{i=1}^P$ in the homogeneous network, with both networks having the same number of processors. On the other hand, equation (8.6) states that the aggregate performance of processors $\{p_i\}_{i=1}^P$ should be equal to the aggregate performance of processors $\{q_i\}_{i=1}^P$.

We have configured two networks of workstations at University of Maryland to serve as sample networks for testing the performance of the proposed heterogeneous hyperspectral imaging algorithm. The networks are considered approximately equivalent under the above framework. In addition, we have also used a massively parallel Beowulf cluster at NASA's Goddard Space Flight Center in Maryland to test the scalability of the proposed parallel algorithms in a

Table 8.2. Capacity of communication links (time in milliseconds to transfer a one-megabit message) between processors in the heterogeneous network at University of Maryland

Processor	$p_1 - p_4$	$p_5 - p_8$	$p_9 - p_{10}$	$p_{11} - p_{16}$
$p_1 - p_4$	19.26	48.31	96.62	154.76
$p_5 - p_8$	48.31	17.65	48.31	106.45
$p_9 - p_{10}$	96.62	48.31	16.38	58.14
$p_{11} - p_{16}$	154.76	106.45	58.14	14.05

large-scale parallel platform. A more detailed description of the parallel platforms used in this chapter follows:

1. **Fully heterogeneous network.** Consists of 16 different workstations, and four communication segments. Table 8.1 shows the properties of the 16 heterogeneous workstations, where processors $\{p_i\}_{i=1}^{4}$ are attached to communication segment s_1, processors $\{p_i\}_{i=5}^{8}$ communicate through s_2,processors $\{p_i\}_{i=9}^{10}$ are interconnected via s_3, and processors $\{p_i\}_{i=11}^{16}$ share the communication segment s_4. The communication links between the different segments $\{s_j\}_{j=1}^{4}$ only support serial communication. For illustrative purposes, Table 8.2 also shows the capacity of all point-to-point communications in the heterogeneous network, expressed as the time in milliseconds to transfer a one-megabit message between each processor pair (p_i, p_j) in the heterogeneous system. As noted, the communication network of the fully heterogeneous network consists of four relatively fast homogeneous communication segments, interconnected by three slower communication links with capacities $c^{(1,2)} = 29.05$, $c^{(2,3)} = 48.31$, $c^{(3,4)} = 58.14$ in milliseconds, respectively. Although this is a simple architecture, it is also a quite typical and realistic one as well.
2. **Fully homogeneous network.** Consists of 16 identical Linux workstations $\{q_i\}_{i=1}^{16}$ with processor cycle-time of $w = 0.0131$ seconds per megaflop, interconnected via a homogeneous communication network where the capacity of links is $c = 26.64$ milliseconds.
3. **Thunderhead Beowulf cluster.** Consists of 268 dual 2.4 Ghz Intel 4 Xeon nodes, each with 1 GB of memory and 80 GB of hard disk (see http://thunderhead.gsfc.nasa.gov for additional details). The total disk space available in the system is 21.44 Tbyte, and the theoretical peak performance of the system is 2.5728 Tflops (1.2 Tflops on the Linpack benchmark). Along with the 568-processor computer core, Thunderhead has several nodes attached to the core with Myrinet 2000 connectivity (see Fig. 8.5). Our parallel algorithms were run from one of such nodes, called thunder1. The operating system is Linux Fedora Core, and MPICH was the message-passing library used (see http://www-unix.mcs.anl.gov/mpi/mpich).

Fig. 8.5. Thunderhead Beowulf cluster at NASA's Goddard Space Flight Center

8.3.2 Performance Evaluation

Before empirically investigating the performance of the proposed parallel hyperspectral imaging algorithms in the five considered platforms, we describe two hyperspectral image scenes that will be used in experiments.

1. **AVIRIS data.** The first scene was collected by the 224-band AVIRIS sensor over Salinas Valley, California, and is characterized by high spatial resolution (3.7-meter pixels). The area covered comprises 512 lines by 217 samples. 20 bands were discarded previous to the analysis due to low signal-to-noise ratio in those bands. Fig. 6(a) shows the spectral band at 587 nm wavelength, and Fig. 8.6(b) shows the ground-truth map, in the form of a class assignment for each labeled pixel with 15 mutually exclusive ground-truth classes. As shown by Fig. 8.6(b), ground truth is available for nearly half of Salinas scene. The data set above represents a very challenging classification problem (due to the spectral similarity of most classes, discriminating among them is very difficult). This fact has made the scene a universal and widely used benchmark test site to validate classification accuracy of hyperspectral algorithms [7].
2. **DAIS 7915 data.** The second hyperspectral data set used in experiments was collected by the DAIS 7915 airbone imaging spectrometer of the German Aerospace Agency (DLR). It was collected by a flight at 1500 meters altitude over the city of Pavia, Italy. The resulting scene has spatial resolution of 5 meters and size of 400 lines by 400 samples. Fig. 8.7(a) shows the image collected at 639 nm by DAIS 7915 imaging spectrometer, which reveals a dense residential area on one side of the river, as well as open areas and

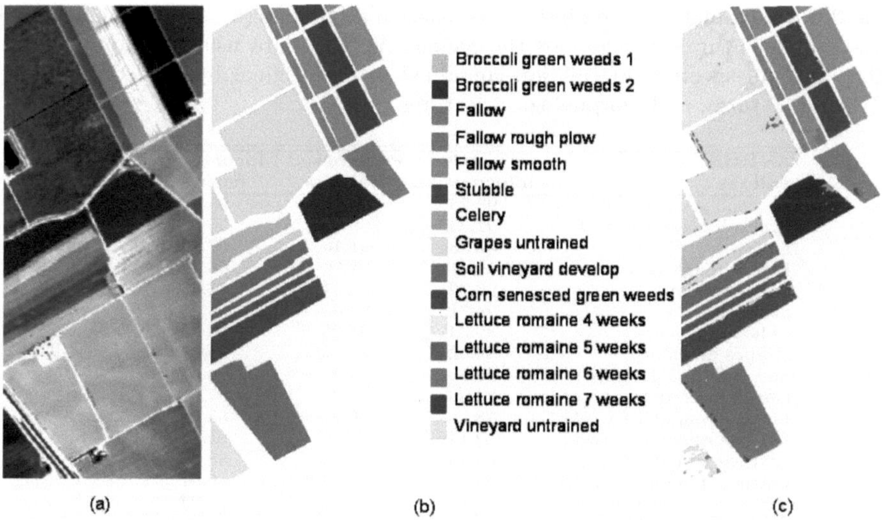

Fig. 8.6. (a) AVIRIS data collected over Salinas Valley, California, at 587 nm. (b) Salinas land-cover ground classes for several areas of the scene. (c) Classification results using the proposed morphological/neural algorithm.

meadows on the other side. Ground-truth is available for several areas of the scene (see Fig. 8.7(b)). Following a previous research study on this scene [7], we take into account only 40 spectral bands of reflective energy, and thus skip thermal infrared and middle infrared bands above 1958 nm because of low signal to noise ratio in those bands.

In order to test the accuracy of the proposed parallel morphological/neural classifier in the context of a land-cover classification of agricultural crop fields, a random sample of less than 2% of the pixels was chosen from the known ground truth of the two images described above. Morphological profiles were then constructed in parallel for the selected training samples using 10 iterations, which resulted in feature vectors with dimensionality of 20 (i.e., 10 structuring element iterations for the *opening series* and 10 iterations for the *closing series*). The resulting features were then used to train the parallel back-propagation neural network classifier with one hidden layer, where the number of hidden neurons was selected empirically as the square root of the product of the number of input features and information classes (several configurations of the hidden layer were tested and the one that gave the highest overall accuracies was reported). The trained classifier was then applied to the remaining 98% of labeled pixels in each considered scene, yielding the classification results shown qualitatively for visual evaluation in Fig. 8.6(c) and quantitatively in Table 8.3.

For comparative purposes, the accuracies obtained using the full spectral information and PCT-reduced features as input to the neural classifier are also reported in Table 8.3. As shown by the table, morphological input features

Table 8.3. Individual and overall test classification accuracies (in percentage) achieved by the parallel neural classifier for the Salinas AVIRIS data using as input feature vectors: the full spectral information contained in the original hyperspectral image, PCT-based features, and morphological features

AVIRIS Salinas	Spectral information	PCT-based features	Morphological features
Broccoli green weeds 1	93.28	90.03	95.69
Broccoli green weeds 2	92.33	89.27	95.02
Fallow	96.23	91.16	97.10
Fallow rough plow	96.51	91.90	96.78
Fallow smooth	93.72	93.21	97.63
Stubble	94.71	95.43	98.96
Celery	89.34	94.28	98.03
Grapes untrained	88.02	86.38	95.34
Soil vineyard develop	88.55	84.21	90.45
Corn senesced green weeds	82.46	75.33	87.54
Lettuce romaine 4 weeks	78.86	76.34	83.21
Lettuce romaine 5 weeks	82.14	77.80	91.35
Lettuce romaine 6 weeks	84.53	78.03	88.56
Lettuce romaine 7 weeks	84.85	81.54	86.57
Vineyard untrained	87.14	84.63	92.93
Overall accuracy	87.25	86.21	95.08

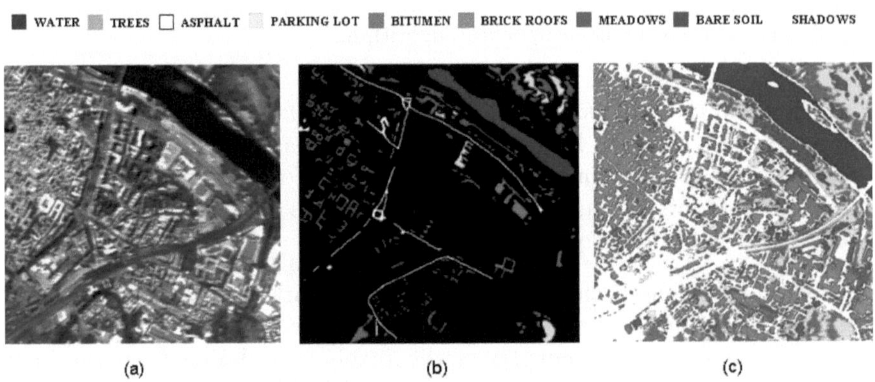

Fig. 8.7. (a) DAIS 1795 data collected at 1500m. over the city of Pavia, Italy, at 639 nm. (b) Pavia land-cover ground classes for several areas of the scene. (c) Classification results using the proposed morphological/neural algorithm.

substantially improve individual and overall classification accuracies with regards to PCT-based features and the full spectral information. This is not surprising since morphological operations use both spatial and spectral information as opposed to the other methods which rely on spectral information alone.

In order to substantiate the performance of the parallel algorithms in a different (and perhaps more challenging) application domain, we repeated the experiments above using the DAIS 7915 hyperspectral scene collected over downtown Pavia, Italy. The classification results achieved by the proposed morphological/neural classification algorithm are visually displayed in Fig. 8.7(c), and

Table 8.4. Individual and overall test classification accuracies (in percentage) achieved by the morphological/neural classifier for the DAIS 7915 Pavia scene using as input feature vectors: the full spectral information contained in the original hyperspectral image, PCT-based features, and morphological features

DAIS 7915 Pavia	Spectral information	PCT-based features	Morphological features
Water	87.30	86.17	100.00
Trees	94.64	97.62	98.72
Asphalt	97.79	84.48	98.88
Parking lot	83.82	81.93	71.77
Bitumen	86.11	75.48	98.68
Brick roofs	83.69	82.36	99.37
Meadow	88.88	89.86	92.61
Bare soil	79.85	84.68	95.11
Shadows	89.64	92.81	96.19
Overall accuracy	88.65	89.75	96.16

Table 8.5. Execution times and load balancing rates (in the parentheses) for the different alternatives tested in the implementation of the morphological and neural stages of the proposed parallel classifier

	Morphological stage			Neural stage	
Heterogeneous versions:	**MP-1**	**MP-2**	**MP-3**	**Exemplar**	**Hybrid**
Heterogeneous network:	267 (1.13)	211 (1.02)	214 (1.03)	156 (1.04)	125 (1.02)
Homogeneous network:	279 (1.15)	216 (1.03)	221 (1.04)	178 (1.03)	141 (1.01)
Homogeneous versions:	**MP-1**	**MP-2**	**MP-3**	**Exemplar**	**Hybrid**
Heterogeneous network:	2871 (1.89)	2535 (1.75)	2261 (1.68)	1261 (1.69)	1123 (1.61)
Homogeneous network:	265 (1.13)	209 (1.02)	212 (1.03)	152 (1.04)	121 (1.02)

quantitatively reported in the form of individual and overall test classification accuracies in Table 8.4. The results obtained in this example confirm our introspection that the use of spatial/spectral filtering prior to classification can substantially increase the results obtained using the full original spectral information or a reduced version of this spectral-based information using standard transformations such as the PCT. Although these results are promising from the viewpoint of classification accuracy, further analyses in terms of computational complexity are required to fully substantiate the suitability of the proposed parallel morphological/neural algorithm for being used in real application with time-critical constraints.

To investigate the properties of the parallel morphological/neural classification algorithm developed in this chapter, the performance of its two main modules (morphological feature extraction and neural classification) was first tested by timing the program using the heterogeneous network and its equivalent homogeneous one. Since the proposed parallel morphological/neural classification algorithm is dominated by regular computations, and given the satisfactory classification results reported in two completely different application scenarios, only parallel performance results for the Pavia (DAIS 7195 data) scene will be provided in the following subsection for the sake of simplicity.

For illustrative purposes, Table 8.5 shows the execution times measured for morphological and neural heterogeneous algorithms and their respective homogeneous versions on the two HNOCs (homogeneous and heterogeneous) at University of Maryland. Three alternative implementations of the parallel morphological feature extraction algorithm (denoted as MP-1, MP-2 and MP-3) and two different implementations of the parallel neural classification algorithm (Exemplar and Hybrid) were tested. As expected, the execution times reported on Table 8.5 for morphological and neural heterogeneous algorithms and their respective homogeneous versions indicate that the heterogeneous implementations were able to adapt much better to the heterogeneous computing environment than the homogeneous ones, which were only able to perform satisfactorily on the homogeneous network. From Table 8.5, one can also see that the heterogeneous algorithms were always several times faster than their homogeneous counterparts in the heterogeneous network, while the homogeneous algorithms only slightly outperformed their heterogeneous counterparts in the homogeneous network. Interestingly, Table 8.5 also reveals that the performance of the heterogeneous algorithms on the heterogeneous network was almost the same as that evidenced by the equivalent homogeneous algorithms on the homogeneous network (i.e., the algorithms achieved essentially the same speed, but each on its network). This seems to indicate that the heterogeneous algorithms are very close to the optimal heterogeneous modification of the basic homogeneous ones.

In order to measure load balance, Table 8.5 also shows (in the parentheses) the imbalance scores achieved by the different stages of the parallel algorithm on the two considered networks of workstations. The imbalance is simply defined as $D = R_{max}/R_{min}$, where R_{max} and R_{min} are the maxima and minima processor run times across all processors, respectively. Therefore, perfect balance is achieved when $D = 1$. As we can see from Table 8.5, the proposed heterogeneous implementations were all effective in terms of load balance in all cases except for MP-1 which communicates pixels too often. On the other hand, the homogeneous algorithms executed on the heterogeneous platforms achieved the worst load-balancing ratios as expected.

For the sake of quantitative comparison, Table 8.6 reports the measured execution times achieved by all tested algorithms on Thunderhead, using different numbers of processors. It should be noted that the processing time measured for the sequential version of morphological feature extraction on one single Thunderhead node was 2057 seconds, while the processing time measured for the sequential version of the neural network classification stage was 1638 seconds. The parallel times reported on Table 8.6 reveal that the combination of MP-2 for spatial/spectral feature extraction, followed by the Hybrid neural classifier for robust classification is able to provide highly accurate hyperspectral classification results (in light of Tables 8.3 and 8.4), but also quickly enough for practical use. For instance, using 256 Thunderhead processors, the proposed classifier was able to provide a highly accurate classification for the Salinas AVIRIS scene in less than 20 seconds. In this regard, the measured processing times represent a significant improvement over commonly used processing strategies for this

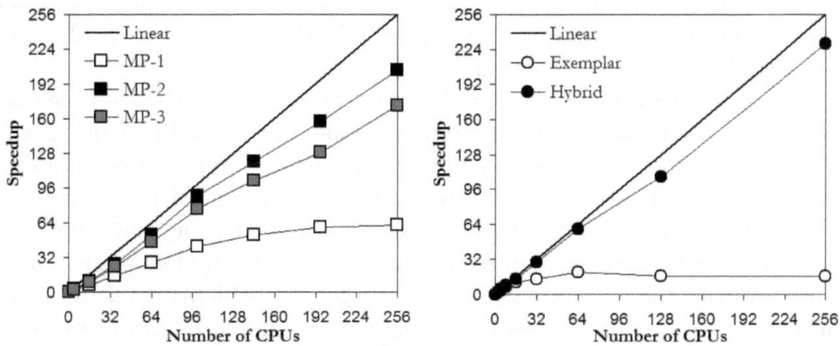

Fig. 8.8. Scalability of parallel morphological feature extraction algorithms (MP-1, MP-2 and MP-3) and parallel neural classifiers (Exemplar and Hybrid) on Thunderhead.

Table 8.6. Processing times in seconds and speedups (in the parentheses) achieved by multi-processor runs of the considered parallel algorithms on the Thunderhead Beowulf cluster at NASA's Goddard Space Flight Center

	4	**16**	**36**	**64**	**100**	**144**	**196**	**256**
MP-1	1177 (1.8)	339 (6.5)	146 (15.0)	81 (27.2)	53 (41.5)	42 (52.4)	37 (59.5)	36 (61.2)
MP-2	797 (2.5)	203 (10.0)	79 (25.8)	39 (52.3)	23 (88.73)	17 (120.0)	13 (157.0)	10 (204.1)
MP-3	826 (2.4)	215 (9.5)	88 (23.3)	45 (45.7)	27 (76.2)	20 (102.8)	16 (128.5)	12 (171.5)
	2	**4**	**8**	**16**	**32**	**64**	**128**	**256**
Exmp.	1041 (1.9)	414 (4.8)	248 (8.1)	174 (11.5)	142 (14.1)	99 (20.2)	120 (16.7)	120 (16.7)
Hyb.	973 (1.6)	458 (3.5)	222 (7.2)	114 (14.0)	55 (29.2)	27 (59.5)	15 (107.1)	7 (229.5)

kind of high-dimensional data sets, which can take up to more than one hour of computation for the considered problem size as evidenced by the sequential computation times reported above.

Taking into account the results presented above, and with the ultimate goal of exploring issues of scalability (considered to be a highly desirable property in the design of heterogeneous parallel algorithms), we have also compared the speedups achieved by the heterogeneous algorithms on the Thunderhead Beowulf cluster. Fig. 8.8 plots the speedups achieved by multi-processor runs of the heterogeneous parallel implementations of the three developed versions of morphological feature extraction algorithms (MP-1, MP-2 and MP-3) over the corresponding single-processor runs of each considered algorithm on Thunderhead. Fig. 8.8 also shows similar results for the two presented parallel neural network classifiers (Exemplar and Hybrid). Although not displayed in Fig. 8.8, the scalability of homogeneous algorithms was essentially the same as that evidenced by their heterogeneous versions, with MP-2 and MP-3 showing scalability results close to linear and the expected slow down of MP-1 due to the great amount of small communications introduced, indicating that the ratio of communications to computations is progressively more significant as the number of processors is increased, and parallel performance is significantly degraded. The above results clearly indicate that the proposed data-replication strategy is more appropriate than the tested data-communication strategy in the design of a parallel version of morphological feature extraction.

As for parallel neural classifiers, the scalability of the Hybrid approach is very close to linear. As mentioned in previous sections, the Hybrid parallel method is based in the distribution of weight connections among different processors, which does not affect the basis of back-propagation learning algorithm. However, the speedup of Exemplar parallel method saturates for 64 processors and slightly declines afterwards. This is mainly due to the limited number of training patterns used during the training and the convergence problems resulting from the execution of several neural classifiers in parallel. It should be noticed that back-propagation algorithm is strongly dependent of the number of selected training patterns. Since the Exemplar method divides the total training pattern set into as many subsets as CPUs available, each local processing will likely have different local minima and convergence challenges, thus introducing load imbalance problems. According to our experimentation, the Exemplar approach is suitable for applications in which the number of available training patters is very high, which is often not the case in the context of remote sensing applications such as those discussed in this chapter.

8.4 Conclusions and Future Research

In this chapter, we have presented several innovative parallel neural network-based algorithms for hyperspectral image classification and implemented them on high performance computing platforms such as heterogeneous and homogeneous networks of workstations and commodity Beowulf clusters. As a case study of specific issues involved in the exploitation of heterogeneous algorithms for hyperspectral image information extraction, this work provided a detailed discussion on the effects that platform heterogeneity has on degrading parallel performance of a new morphological/neural classification algorithm, able to exploit the spatial and spectral information in simultaneous fashion.

The proposed parallel classification approach was tested using hyperspectral data sets corresponding to real applications, such as a complex urban mapping scenario and an agricultural land-cover classification problem, both of which have been commonly used as benchmark test sites in the remote sensing community. The parallel performance evaluation strategy conducted in this work for assessing heterogeneous parallel algorithms was based on experimentally assessing each heterogeneous algorithm by comparing its efficiency on a fully heterogeneous network (made up of processing units with different speeds and highly heterogeneous communication links) with the efficiency achieved by its equivalent homogeneous version on an equally powerful homogeneous network. Scalability results on a massively parallel commodity cluster are also provided.

Experimental results in this work anticipate that the combination of the (readily available) computational power offered by heterogeneous parallel platforms and the recent advances in the design of advanced parallel algorithms for hyperspectral data classification algorithms (such as those presented in this work) may introduce substantial changes in the systems currently used for exploiting the sheer volume of hyperspectral data which is now being collected worldwide, on a daily basis.

Although the experimental results presented in this chapter are encouraging, further work is still needed to arrive to optimal parallel design and implementations for the considered parallel algorithms. We also plan to implement the proposed parallel techniques on other massively parallel computing architectures, such as NASA's Project Columbia and Grid computing environments. We are also developing implementations of the proposed algorithms on field programmable gate arrays (FPGAs) and commodity graphics processing units (GPUs), which represent a very appealing type of high performance hardware architecture for onboard hyperspectral image classification and compression.

Acknowledgement

The authors would like to thank J. Dorband, J. C. Tilton and J. A. Gualtieri for many helpful discussions, and also for their support with experiments on NASA's Thunderhead system. They also state their appreciation for Profs. M. Valero and F. Tirado.

References

1. Chang, C.I.: Hyperspectral imaging: techniques for spectral detection and classification. Kluwer, New York (2003)
2. Green, R.O., et al.: Imaging spectroscopy and the airborne visible/infrared imaging spectrometer (AVIRIS). Remote Sensing of Environment 65, 227–248 (1998)
3. Benediktsson, J.A., Swain, P.H., Ersoy, O.K.: Neural network approaches versus statistical methods in classification of multisource remote-sensing data. IEEE Trans. Geoscience and and Remote Sensing 28, 540–552 (1990)
4. Vilmann, T., Merenyi, E., Hammer, B.: Neural maps in remote sensing image analysis. Neural Networks 16, 389–403 (2003)
5. Landgrebe, D.A.: Signal theory methods in multispectral remote sensing. Wiley, Hoboken (2003)
6. Richards, J.A., Xia, J.: Remote sensing digital image analysis: an introduction, 4th edn. Springer, Berlin (2006)
7. Plaza, A., Martinez, P., Plaza, J., Perez, R.: Dimensionality reduction and classification of hyperspectral image data using sequences of extended morphological transformations. IEEE Trans. Geoscience and Remote Sensing 43, 466–479 (2005)
8. Plaza, A., Benediktsson, J.A., Boardman, J., Brazile, J., Bruzzone, L., Camps-Valls, G., Chanussot, J., Fauvel, M., Gamba, P., Gualtieri, J.A., Tilton, J.C., Trianni, G.: Advanced processing of hyperspectral images. In: Proceedings of the IEEE International Geoscience and Remote Sensing Symposium, pp. 1348–1351 (2006)
9. Baraldi, A., Binaghi, E., Blonda, P., Brivio, P.A., Rampini, A.: Comparison of the multilayer perceptron with neuro-fuzzy techniques in the estimation of cover class mixture in remotely sensed data. IEEE Trans. Geoscience and Remote Sensing 39, 994–1005 (2001)

10. Zhang, D., Pal, S.K.: Neural Networks and Systolic Array Design. World Scientific, Singapore (2001)
11. Sterling, T.: Beowulf cluster computing with Linux. MIT Press, Cambridge (2002)
12. Dorband, J., Palencia, J., Ranawake, U.: Commodity clusters at Goddard Space Flight Center. Journal of Space Communication 1, 227–248 (2003)
13. El-Ghazawi, T., Kaewpijit, S., Moigne, J.L.: Parallel and adaptive reduction of hyperspectral data to intrinsic dimensionality. In: Proceedings of the IEEE International Conference on Cluster Computing, pp. 102–109 (2001)
14. Plaza, A., Valencia, D., Plaza, J., Martinez, P.: Commodity cluster-based parallel processing of hyperspectral imagery. Journal of Parallel and Distributed Computing 66, 345–358 (2006)
15. Lastovetsky, A.: Parallel computing on heterogeneous networks. Wiley-Interscience, Hoboken (2003)
16. Aloisio, G., Cafaro, M.: A dynamic earth observation system. Parallel Computing 29, 1357–1362 (2003)
17. Wang, P., Liu, K.Y., Cwik, T., Green, R.O.: MODTRAN on supercomputers and parallel computers. Parallel Computing 28, 53–64 (2002)
18. Ritter, G.X., Sussner, P., Diaz, J.L.: Morphological associative memories. IEEE Transactions on Neural Networks 9, 281–293 (2004)
19. Soille, P.: Morphological image analysis: principles and applications. Springer, Berlin (2003)
20. Plaza, J., Plaza, A., Perez, R.M., Martinez, P.: Automated generation of semi-labeled training samples for nonlinear neural network-based abundance estimation in hyperspectral data. In: Proceedings of the IEEE International Geoscience and Remote Sensing Symposium, pp. 345–350 (2005)
21. Plaza, J., Perez, R.M., Plaza, A., Martinez, P.: Parallel morphological/neural classification of remote sensing images using fully heterogeneous and homogeneous commodity clusters. In: Proceedings of the IEEE International Conference on Cluster Computing, pp. 328–337 (2006)
22. Suresh, S., Omkar, S.N., Mani, V.: Parallel implementation of back-propagation algorithm in networks of workstations. IEEE Transactions on Parallel and Distributed Systems 16, 24–34 (2005)
23. Lastovetsky, A., Reddy, R.: On performance analysis of heterogeneous parallel algorithms. Parallel Computing 30, 1195–1216 (2004)

9

Positioning Weather Systems from Remote Sensing Data Using Genetic Algorithms

Wong Ka Yan[1] and Yip Chi Lap[2]

[1] Department of Computer Science, The University of Hong Kong, Pokfulam, Hong Kong
kywong@cs.hku.hk
[2] Department of Computer Science, The University of Hong Kong, Pokfulam, Hong Kong
clyip@cs.hku.hk

Summary. Remote sensing technology is widely used in meteorology for weather system positioning. Yet, these remote sensing data are often analyzed manually based on forecasters' experience, and results may vary among forecasters. In this chapter, we briefly introduce the problem of weather system positioning, and discuss how evolutionary algorithms can be used to solve the problem. A genetic algorithm-based framework for automatic weather system positioning is introduced. Examples on positioning tropical cyclones and line-shaped weather systems on radar data are used to demonstrate its practical use.

9.1 Introduction

Weather systems[3] of different scales, from hailstorms and tornadoes to typhoons, cold fronts and pressure systems, affect our daily lives. Some of them are destructive and life-threatening. It is of prime importance to have them identified, tracked and forecast. With the extensive deployment of modern remote sensing technologies such as radars and optical sensors mounted in airplanes and satellites, terabytes of meteorological data are being collected every day for analysis. Yet, the job of identifying and positioning weather systems is still largely done manually or in a semiautomated way. Forecasters manually identify and estimate the position of weather system by tracing radar echoes or cloud movements from remote sensing data. Such manual identification and positioning process often requires subjective and fuzzy matching by experienced forecasters. Thus, they are error-prone, and results may vary between forecasters. The positioning tasks are challenging due to a number of reasons. First, there is the data complexity problem[61]. Meteorological data are often incomplete, noisy, indirect, and spatially and temporally scarce. Second, the processing of the huge volumes of observations for nowcasting and forecasting is a task with tough real-time constraint.

Third, there is a lack of well-established and fast rules to determine the positions of weather systems under concern. There is no universally accepted methods and techniques for automatic weather system identification and positioning.

M. Graña and R.J. Duro (Eds.): Comput. Intel. for Remote Sensing, SCI 133, pp. 217–243, 2008.
springerlink.com

These discrepancies are reflected in the disparity between the analyses produced by different forecasters or weather centers from the same meteorological data.

In contrast to manual approaches, automated weather system positioning methods employ objective similarity measures. Major approaches include pattern matching and motion vector field analysis. The existing approaches require computationally expensive operations such as wind field or motion vector field construction, parameter estimation using searching algorithms, and extensive block or object matching. Because of the large volume and rate of data, this problem can only be solved using high performance computing systems to generate timely results.

Computational Intelligence techniques, specifically evolutionary algorithms (EA), are proposed in this chapter to position weather systems. The properties of weather systems can be modeled, and the way that human forecasters identify and position weather systems can be formalized in the framework of EA. A number of research work has shown that EA have been applied successfully to real-world problems[6, 11, 39, 47]. Yet, interestingly, their application to weather forecasting has received little attention. The only example we know is [23] on bounded weak echo region detection. Can Computational Intelligence techniques be applied to the weather system positioning problem? How can this be done using remote sensing data? In this chapter, a genetic algorithm-based framework for automatic weather system positioning from remote sensing data is introduced, with case studies on several severe weather systems.

The rest of the chapter is organized as follows. Data sources that are often used for weather system positioning are first introduced in Section 9.2. Then, a brief review of the work of computer scientists, meteorologists and forecasters on weather system identification is given in Section 9.3. The genetic algorithm-based framework for weather system positioning is then introduced in Section 9.4, followed by case studies on tropical cyclones and line-shaped weather systems in Sections 9.5 and 9.6 respectively. Finally, Section 9.7 summarizes this chapter.

9.2 Data Sources

There are three major types of remote sensing data used for automatic weather system positioning: radar reflectivity data, Doppler velocity data, and satellite images. Since these data types are collected or generated by different sensors and focus on different aspects of the atmosphere, specialized algorithms for processing and analysis are needed to handle each of them. Here are brief descriptions of each data type.

Radar reflectivity and Doppler velocity data come from weather radars, mostly grounded on earth. Weather radars send out microwave signals to the atmosphere to do a volume scan by monitoring the reflected signals. Radar reflectivity values show the strengths of echoes reflected from rain, snow, ice, or hail aloft. Doppler velocity values show the radial velocities of these water species with respect to the radar station. The reflected signals allow the construction of a volumetric map of the atmosphere, which is preprocessed to extract the 2D slices relevant

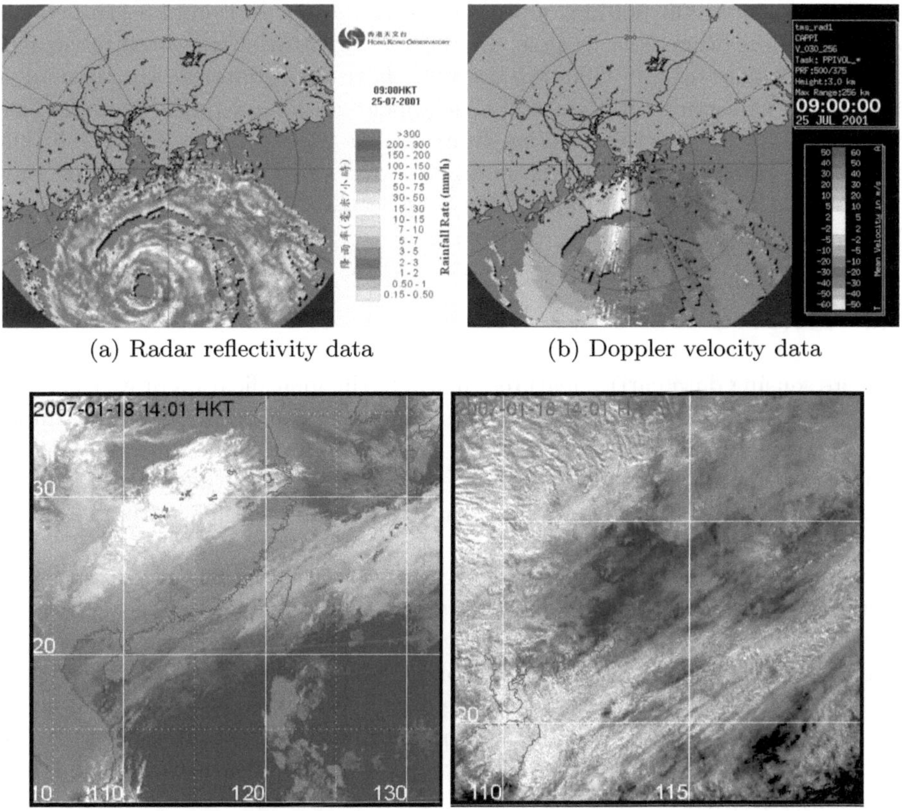

(a) Radar reflectivity data

(b) Doppler velocity data

(c) Infra-red data

(d) Visible light data

Fig. 9.1. Types of remote sensing data

Data sources: 1(a),1(b): Tai Mo Shan Doppler weather radar of the Hong Kong Observatory (HKO)[16].
1(c),1(d): Multi-functional Transport Satellite-1R(MTSAT-1R) of Japan Meteorological Agency (JMA)[2].

for the analysis. Radar data at 3 km from sea level, called 3 km Constant Altitude Plan Position Indicator (CAPPI), is often used for weather system positioning. The range of weather radars is limited to a few hundred kilometers and their data are updated at intervals of a few minutes. With a relatively high resolution of tens of meters, they are ideal for observing weather situations near the station site. Fig. 1(a) and 1(b) are a radar reflectivity and a Doppler velocity image respectively.

Meteorological satellites carry optical sensors that point to the ground to obtain a bird's view of the earth. Images of the earth at different wavelengths from infrared (IR) to visible light (VIS) are captured in different channels. Satellites can be polar-orbiting or geostationary. Polar-orbiting satellites have relatively low-flying orbits (400–900 km) and take images of the portion of earth below it while traveling from pole to pole. In contrast, geostationary satellites usually take much higher orbits (35800 km) and fly above the equator at the same angular velocity as earth's rotation, and thus appear stationary to the ground. They transmit data in intervals ranging from one to twelve hours. The resolutions of weather satellites typically range from 1 to 5 km. Fig.s 1(c) and 1(d) show images from the Multi-functional Transport Satellite-1R (MTSAT-1R) of Japan Meteorological Agency (JMA).

9.3 Related Work

Remote sensing data carries features that help the identification of weather phenomena. Time sequences of remote sensing images can be presented as animations, showing the dynamics of weather system evolution and allowing analysis by forecasters and meteorologists. We review some studies that have been done on feature identification, positioning and tracking of weather systems from remote sensing data.

According to the shape, texture, organizational patterns and thickness[34], clouds are classified into types such as stratus, cumulus and cirrus. The types of cloud form a basis for understanding and identifying weather systems from satellite images. Techniques used for cloud classification include the use of principal component analysis[30], independent component analysis, artificial neural networks[42, 44, 45], and support vector machines[4]. Contour analysis[10, 62] and data mining techniques[19, 20] are used to classify hurricane and non-hurricane patterns.

For weather system positioning, the primary interest is on severe weather systems that span over a period of time, such as tropical cyclones (TCs). A TC is a low pressure system, and has a rain-free circulation center where spiral rainbands whirls into: the eye. The localization of the eye position is of outmost importance to predict the TC evolution. Traditional TC eye fix methods are manual or semiautomated. Forecasters estimate the TC center location either by tracing the movement of spiral rainbands on an animation loop of remote sensing images, or by overlaying spiral templates on a display or printout of remote sensing images looking for the best match [41]. These manual techniques require trained and experienced forecasters. In recent years, they are supplemented by computer-assisted objective analysis tools.

Methods to automate or semiautomate the TC eye fix process mostly fall into two categories: motion vector field analysis and pattern matching. In motion vector field analysis, motion fields are formed either from wind data or the analysis of sequences of remote sensing images. Techniques for motion field construction include the Tracking Radar Echoes by Correlation (TREC) algorithm [35], motion compensated prediction [50, 54], the construction of cloud motion winds (CMW) by cross correlation [31, 37], and estimating cloud motion by semi-fluid motion model [14]. The motion field built is then considered as a map of the horizontal winds of the TC[46], it is used for tropical cyclone observation [21, 25, 26] and to fix the TC center[32, 33]. By decomposing the motion field [7, 8], critical points such as swirl, vortices, sinks, or sources can be extracted and characterized by examining the corresponding linear phase portrait matrices [7, 8, 40], or vector field pattern matching [36]. Vortex centers can then be regarded as TC centers. However, these methods can be computationally expensive, and well-structured motion fields fitting the mathematical model are often required to obtain reasonable results. To overcome the computational cost limitations, a motion field structure analysis approach is introduced in[50, 54]. A simple voting technique is proposed and the TC center is found by finding the point covered by the largest number of sectors generated by rotated vectors of the motion vector fields. The motion field analysis techniques mentioned above require the use of a sequence of images. Hence, they may not be suitable if the image is sampled infrequently, partly missing, of low quality, or when the TC moves fast.

Another major approach for TC eye fix, namely pattern matching, can be performed using only a single image or temporally discontinuous data. In a pattern matching approach, a TC eye is fixed by finding the best match between a predefined TC model and the remote sensing data. Examples of TC models include logarithmic spiral model [49, 57] and vortex flow model[58]. Templates are generated for matching. An automatic template generation method is introduced in [57]. In this method, the spiral parameters are estimated at different latitude and longitude positions by a number of image processing and transformation operations . Templates generated by the estimated parameters are then used to match against radar images at plausible latitude-longitude positions. An average error of about 0.16 degrees on a Mercator projected map with respect to best track data (manually found center locations using all available data after the event) was reported, which is well within the relative error of about 0.3 degrees[24] given by different TC warning centers. The World Meteorological Organization (WMO) considers an error within 0.5 degrees as "good" for forecast[38].

Another issue on weather system positioning is to locate the line-shaped weather systems, such as frontal systems, troughs, ridges, and squall lines. In practice, they are located by identifying the sharp edges of clouds on satellite images[1]. To locate frontal systems, besides remote sensing data, forecasters also make use of local observations, following specific strategies and guidelines such as [22] for data analysis.

Unlike TCs, research studies on automating the process of identifying line-shaped weather systems are few and far between. Most of the methods are based on physical analysis on wind profile data[27] or numerical weather prediction (NWP) data. Techniques include the use of thermal front parameter (TFP)[15], pattern matching on grid points of geopotential height and other data[18], spatial aggregation techniques[17], and Bayesian classification[12]. The use of remote sensing data is only addressed by [5] on the detection of synoptic-scale fronts using Lagrangian Scalar Integration, and [13] on thin-line detection using wavelet transformation.

In our recent work [51, 52, 53, 55, 56, 59, 60], a framework for automatic weather system identification and positioning is proposed. The framework allows the use of both remote sensing and NWP data. Genetic algorithms (GA) are used in the work. GA makes use of Darwin's idea of "survival for the fittest" [9], in which a population of randomly generated individuals (each individual corresponding to a set of model parameter values) goes through a simulated process of evolution. Through the employment of a fitness function (a matching quality measure), each individual is evaluated and ranked according to its relative strength within the population. The fitter individuals that maximize the fitness function are then selected for reproduction by mutation or crossover in an iterative stage. Because fitter individuals tend to produce fitter offsprings, the population would evolve towards the desired solution with time. The use of GA makes the search more efficient and helps to escape from local maxima values of the fitness functions. Different fitness functions can be designed for positioning different weather systems from meteorological data. With the use of radar data, [51, 56, 59, 60] and [52] suggested the fitness functions for positioning tropical cyclones and squall lines respectively. To locate pressure systems, frontal systems and troughs from NWP data, fitness functions suggested in [53, 55] can be used. This GA-based weather system identification framework is introduced in the following sections.

9.4 System Framework

Our weather system positioning framework adopts the pattern matching approach. Rather than using traditional gradient ascend algorithms to search for the best match location, genetic algorithm is used to speed up the search and to escape from local maxima. According to the properties of the weather system under concern, the fitness function, i.e. the matching criteria, can then be formulated.

To handle huge volumes of remote sensing data, the search space needs to be reduced as much as possible so that the time needed for positioning the weather system can be reduced. Furthermore, noise reduction techniques are required to handle noisy results produced by incomplete or missing remote sensing data. All these can be done through a number of operations in successive order:

- Data preprocessing;
- Search space limitation;
- Genetic algorithm-based pattern matching; and
- Result post-processing.

A general architecture that incorporates all these operations is shown in Fig. 9.2. The system is organized in a modular manner to support operations such as meteorological data acquisition, data preprocessing, Region Of Interest (ROI) generation, genetic algorithm-based positioning, and result post-processing. A past result database is used to store previous weather system locations and associated information such as ROI, to assist the weather system positioning process.

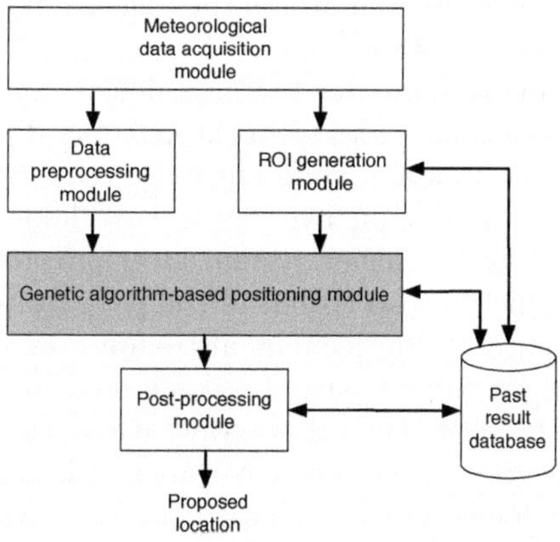

Fig. 9.2. GA-based framework for weather system positioning

The computational steps of the weather system positioning process are as follows. First, remote sensing images are obtained from the meteorological data acquisition module. They are then preprocessed in the data preprocessing module.

In the proposed framework, an ROI is an area defined using domain knowledge such as the type of weather systems and topographical information. It can be found by some ROI estimation techniques, such as zero isodop analysis suggested in [57], or motion estimation techniques using the ROI of the previous image as a reference, which can be obtained from the past result database. To allow interaction with human forecasters, the module also supports a user-specified ROI.

Next, pattern matching is performed in the genetic algorithm-based positioning module to find an initial estimation of the weather system location. Unlike traditional pattern matching approaches that treat every image or data independently, temporal information can be used in the framework if it is available. The pattern matching algorithm is as follows. Initially, *nInit* template candidate parameter sets are generated randomly inside an ROI. To profit from the use of temporal information, at least *nBest* of them are the fittest candidates of the last iteration of previous time frame, if available. After the initial set of candidate is generated, the algorithm enters an iterative phase. According to the fitness function, the fitness value for each candidate is calculated and the fittest *nBest* candidates are retained as parents. *nChild* children templates are then generated from the parents using crossover and mutation probabilities of p_x and p_m respectively, with at least *nRetain* of them being verbatim copies of the parents. The crossover genetic operator splices and exchanges parameter values (genes) of the parent after a randomly selected crossover point, whereas mutation alters only one of the genes randomly. The iterative phase ends when the best score does not improve for *boreCount* iterations after the algorithm runs for at least *minLoop* iterations, or when the score is over an user-defined threshold *minScore*. This threshold allows the algorithm to give an answer once the fitness score satisfies the user.

Finally, to smooth out the noise due to image defects and/or motion prediction errors, the estimated locations found from the GA-based positioning module are fed to the post-processing module. Different noise reduction techniques, such as Kalman filtering[28, 48] and moving average, are available in the post-processing module. The post-processed weather system location provides the final output of the whole system, which is also stored in the past result database.

To improve the accuracy, the system also supports user input of weather system locations. This provides a basic feedback step in which the forecaster is given the opportunity to verify the proposed weather system locations whenever necessary. The practicality of this system framework is demonstrated by case studies on tropical cyclones and line-shaped weather systems in the following sections.

9.5 Case Study: Tropical Cyclones

The application of the GA-based framework for weather system positioning is demonstrated in this section for fixing the eye, or the center, of one of the most destructive weather systems, namely tropical cyclones. The data used are radar images, which usually have a higher spatial and temporal resolutions than satellite images. A simple model of TC[51, 56, 59, 60] is introduced in Section 9.5.1, and the parameters used for fixing the TC center is described in Section 9.5.2. The performance of the framework is then evaluated in Section 9.5.3.

9.5.1 A Model of Tropical Cyclone

A TC has a center (point C in Fig. 9.3) at longitude lon and latitude lat where a spiral rainband (curve EDBA) with the polar equation $r = ae^{\theta \cot \alpha}$ whirls into. For TCs in the Northern Hemisphere, $\cot \alpha$ is negative, giving the spiral shape as shown in the figure. TCs in the Southern Hemisphere have positive $\cot \alpha$ and rainbands swirl in clockwisely. A TC has an eyewall (inner circle of Fig. 9.3) with radius R (distance BC), which is the boundary between rainy and no-rain areas. Places with a distance slightly larger than R from the center (the outer circle with radius $R + d$) would be rainy. The spiral rainband outside the eyewall (curve BDE) has a length of l, which relates to the distance of influence of the TC. With this model, the six parameters lat, lon, a, α, l and R define the template for matching stage.

9.5.2 Positioning Tropical Cyclone Using Radar Data

The GA-based framework is used for eye fix as follows. A radar reflectivity image is preprocessed in the data preprocessing module by thresholding and contrast enhancement to make spiral rainbands stand out (Fig. 9.4). Thresholding with histogram cumulative frequency of 82–86% is used, so that the brightest pixels

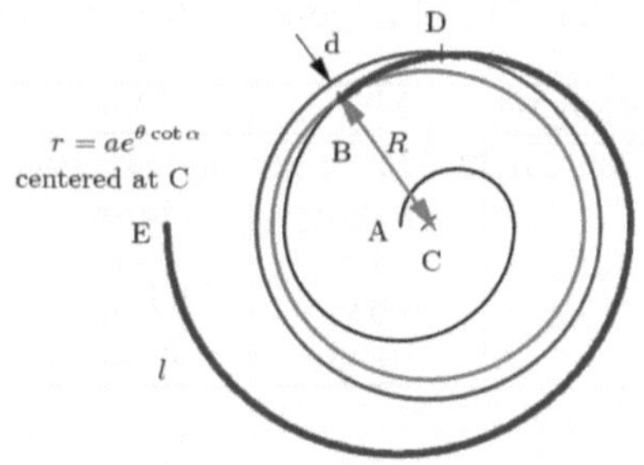

Fig. 9.3. A model of tropical cyclone

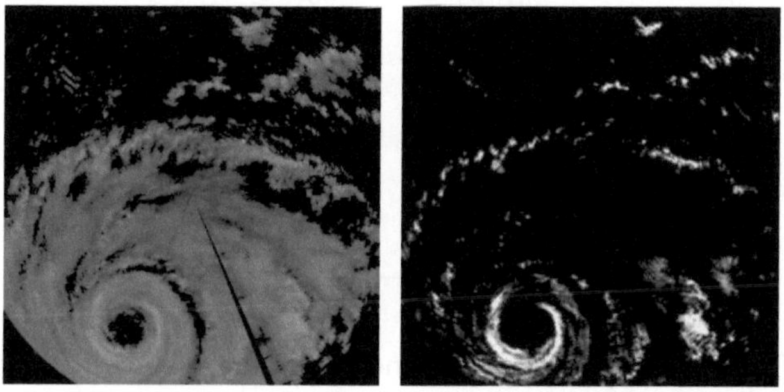

(a) Original (b) Thresholded at 34 dBZ

Fig. 9.4. Original and preprocessed image for TC eye fix

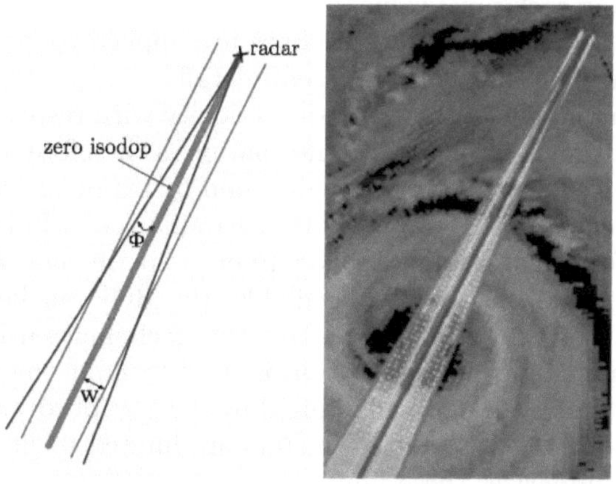

(a) ROI parameters (b) Radar reflectivity data with ROI

Fig. 9.5. Region of Interest (ROI) for TC eye fix

Table 9.1. Model parameters for TC eye fix

Parameter	Min	Max
(lon, lat)	256 km from radar station	
a	10 km	45 km
α	-1.5	-1
l	180 km	720 km
R	10 km	45 km

Table 9.2. GA parameters for TC eye fix

Parameter	Experiment setting
$nInit$	50
$nBest$	10
$nChild$	50
$nRetain$	2
$boreCount$	10
$minLoop$	20
p_x	80%
p_m	20%

are retained. This percentage is determined by domain knowledge in meteorology and corresponds to a reflectivity threshold of about 34 dBZ. To enhance the contrast, equalization is applied to the retained pixels. The image is further preprocessed by Gaussian smoothing and max filtering. Gaussian smoothing is done using a Gaussian kernel of size 7×7 with $\sigma = 1.0$, while 3×3 max filtering is done by replacing each pixel value in an image with the maximum value of its eight neighbors and itself. These techniques are applied to smooth and reduce noise in images, while preserving the useful details.

Zero isodop, or the line with zero radial velocity with respect to the radar, is where the TC center should lie. This information is therefore used to determine the ROI for TC eye fix. The zero isodop is found by log-polar transformation[63] using Doppler velocity data. The area within Φ radian and w km from the straight zero isodop line forms the ROI (Fig. 9.5). In our experiments, $\Phi = \pi/45$, $w = 3$. If the zero isodop information is unavailable, the ROI can be an user-defined rectangular area, or found by motion estimation techniques using previous ROI information. Domain-specific information is used to limit the values of the six template parameters. lat and lon are limited by the area of coverage of radar, and the values of the other template parameters are limited to fit values of typical TCs. Table 9.1 summarizes these limits.

The quality of match is calculated using a fitness function, which is a modified correlation function designed so that high reflectivity areas match the spiral segment BDE and the outer circle of the template, and low reflectivity areas match the inner circle. A more detailed formula with explanation can be found in Appendix A. Genetic algorithm described in Section 9.4 is used to find the set of parameters for the best match. The setting of GA parameters for TC eye fix is listed in Table 9.2.

The best candidate found by genetic algorithm gives the (lon, lat) location. It is then fed to the post-processing module in which a Kalman filter smooths out noises caused by bad matches and image defects. Historical TC data, such as average TC speed, are used to determine Kalman filter parameters such as system noise variance. Latitude and longitude values are separately smoothed, with the assumption that they are statistically independent. The Kalman filtered TC center location gives the final output.

9.5.3 Evaluation

A Java-based system prototype is built for evaluation. Data used for evaluation are sequences of 3 km CAPPI radar reflectivity images with a range of 256 km captured every 6 minutes, along with their Doppler velocity counterparts. Table 9.3 gives the details of the experimental data set. The data set covers different types of TCs as well as different stages of a TC life cycle from strengthening to dissipation.

The effectiveness of the algorithm is evaluated by finding the average Euclidean distance between the center found by the algorithm and the corresponding linearly interpolated best track location from the Hong Kong Observatory. Best tracks are the hourly TC locations determined after the event by a TC warning center using all available data.

Using typical parameter values (Table 9.2), the algorithm gives average error (Table 9.4) well within the relative error of about 0.3 degrees given by different TC warning centers [24]. The comparison between the interpolated best track and estimated track are shown in Fig. 9.6, and Fig.s 9.7, 9.8 and 9.9 show some of the eye fix results of Severe Tropical Storm (STS) Maria, Typhoon Yutu and STS Hagupit respectively. In general, the TC tracks proposed by our system are close to the best tracks, especially for Typhoon Yutu. An inspection of radar images show that Yutu has a better spiral structure (Fig. 9.8) which matches the

Table 9.3. Experimental tropical cyclone data set

Name	Time (HKT)	No. of images	Characteristics
STS Maria	2000-08-31 22:00 to 2000-09-01 14:54	180	Spiral rainbands, dissipated at the end
Typhoon Yutu	2001-07-24 20:00 to 2001-07-25 19:54	240	Spiral rainbands, well defined TC eye
STS Hagupit	2002-09-11 12:00 to 2002-09-11 23:54	120	Spiral rainbands

Table 9.4. Experimental results on accuracy

TC name	Error (average) (in degrees)	Error (s.d.) (in degrees)
STS Maria	0.270	0.152
Typhoon Yutu	0.158	0.075
STS Hagupit	0.178	0.075

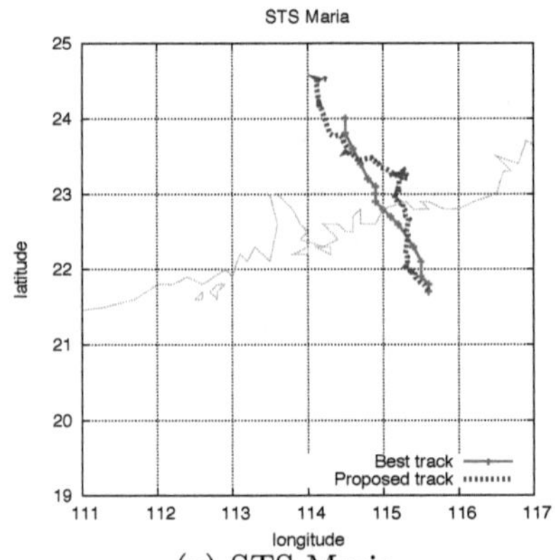

(a) STS Maria

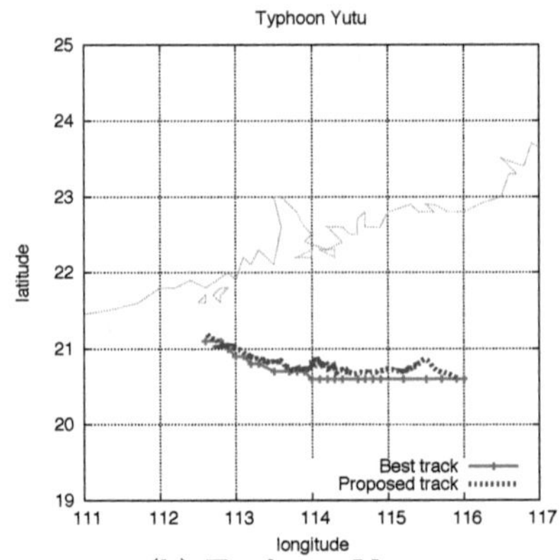

(b) Typhoon Yutu

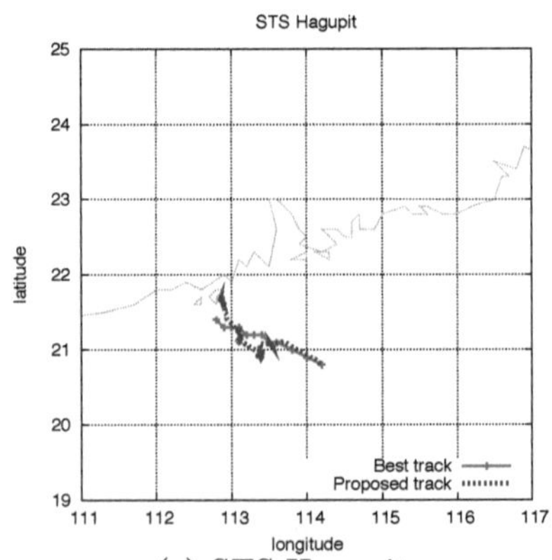

(c) STS Hagupit

Fig. 9.6. Comparison of estimated track and best track

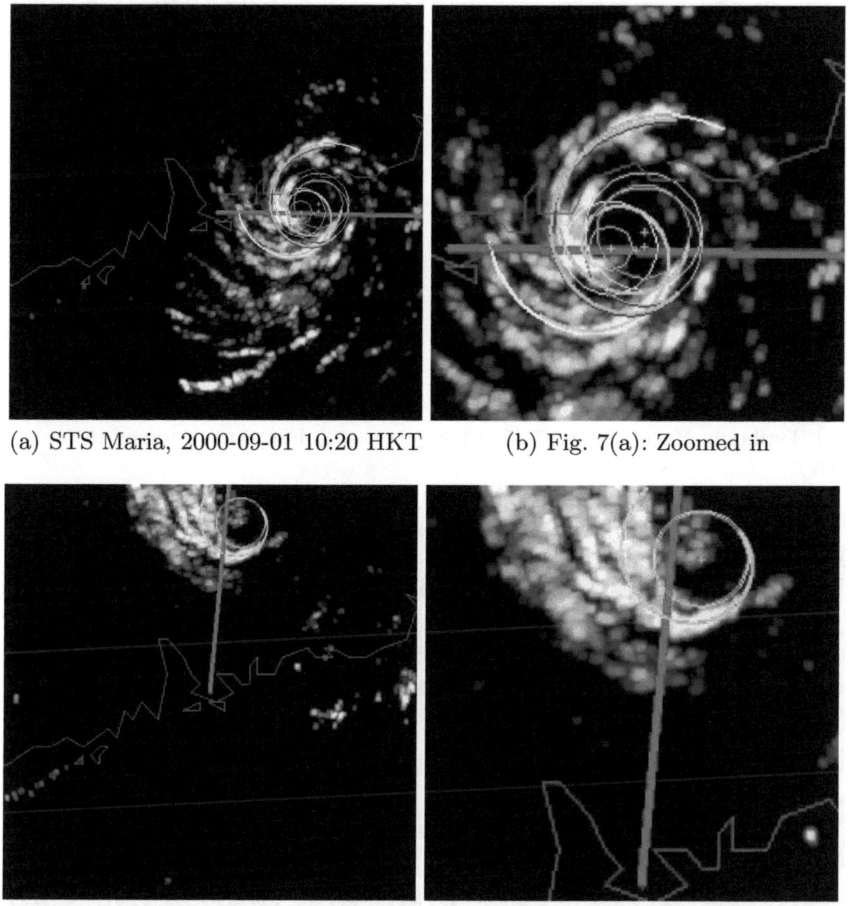

(a) STS Maria, 2000-09-01 10:20 HKT (b) Fig. 7(a): Zoomed in

(c) STS Maria, 2000-09-01 14:36 HKT (d) Fig. 7(c): Zoomed in

Fig. 9.7. Results of eye fix: STS Maria. Preprocessed radar reflectivity images overlaid with results of eye fix. The straight line indicates the zero isodop, the polyline indicates the coastline, the spirals are the top 5 GA candidates, and the cross indicates the proposed TC center. The cross in cyan color (the most accurate one) is the final output of the system.

model well. For STS Maria, a significant portion of the error with our proposed eye fix method can be attributed to its rapid dissipation after its landfall. The TC center vanished (Fig. 7(c)), resulting the large average error. For STS Hagupit, the algorithm found center positions that are close to the best tracks. During its later stage, its spiral rainbands close to the circulation center weakened. This made the model matching less effective, and contributes to the deviation of the proposed center locations from the best track.

The efficiency of the algorithm is measured by the sum of system time and user time in timing tests on a Macintosh PowerMac with Dual 2.5GHz PowerPC

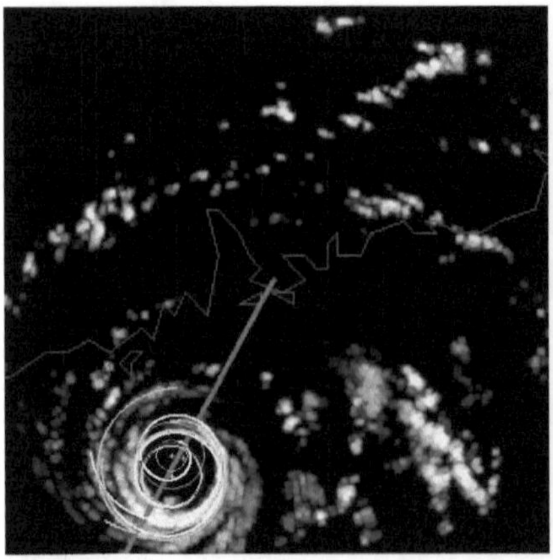

(a) Typhoon Yutu, 2001-07-25 14:30 HKT

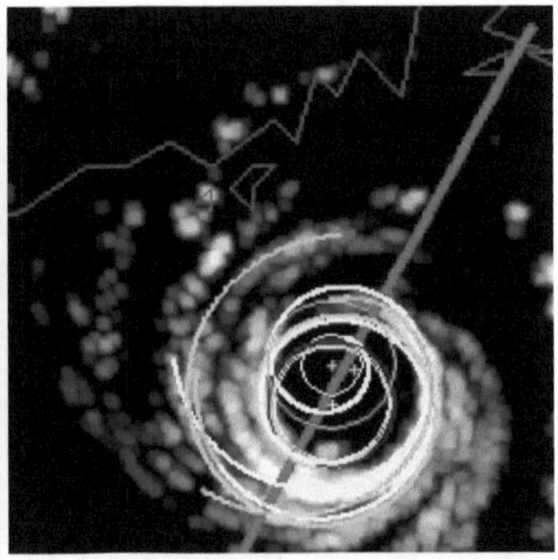

(b) Fig. 8(a): Zoomed in

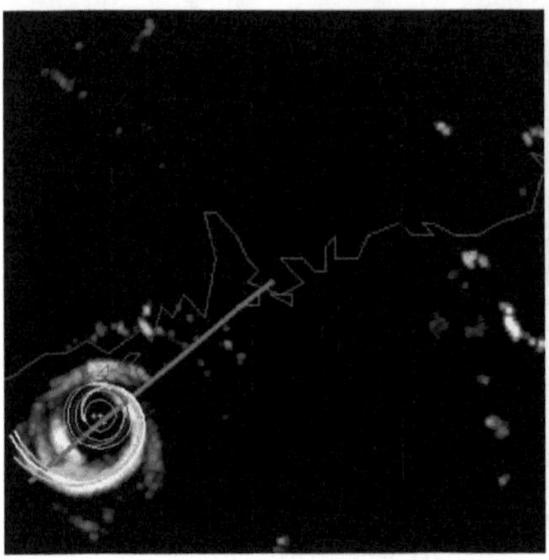

(c) Typhoon Yutu, 2001-07-25 19:54 HKT

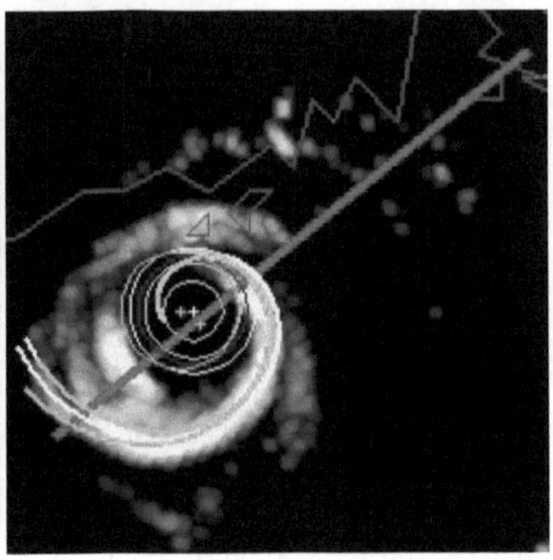

(d) Fig. 8(c): Zoomed in

Fig. 9.8. Results of eye fix: Typhoon Yutu Preprocessed radar reflectivity images overlaid with results of eye fix. The straight line indicates the zero isodop, the polyline indicates the coastline, the spirals are the top 5 GA candidates, and the cross indicates the proposed TC center. The cross in cyan color (the most accurate one) is the final output of the system.

G5 processor and 3GB RAM running Mac OS 10.4.8, with only one CPU running the Java program. Using the typical parameter set in Table 9.2, on average, the algorithm can complete an eye fix exercise within 16 seconds, including all I/O operations. This is well within the six minute-deadline, when the next set of radar data is available and needs to be processed, leaving ample time for other analyses.

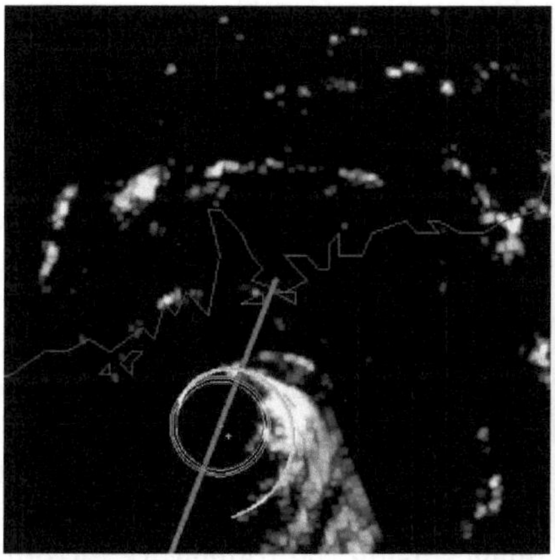

(a) STS Hagupit, 2002-09-11 13:36 HKT

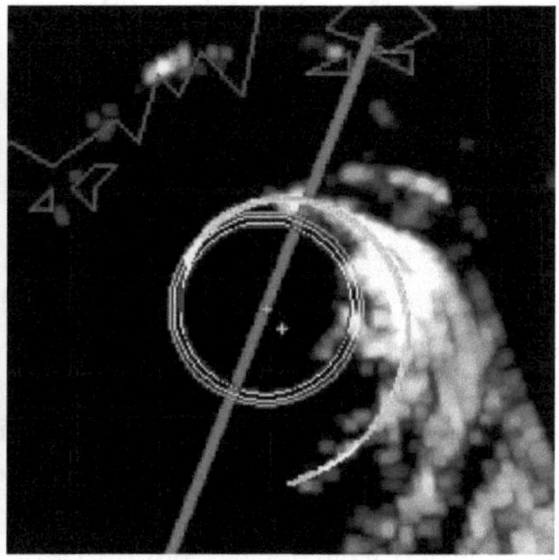

(b) Fig. 9(a): Zoomed in

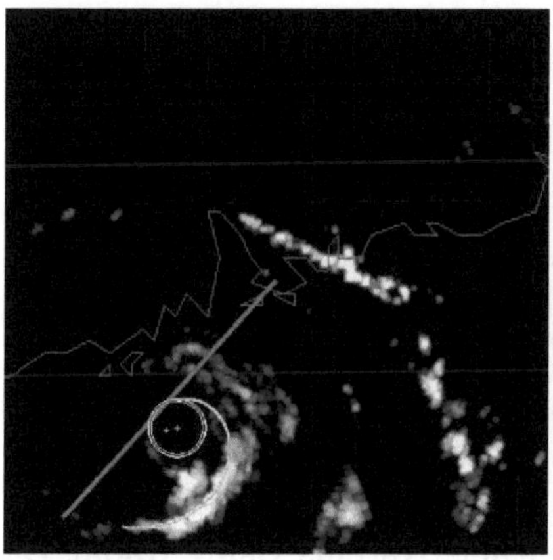

(c) STS Hagupit, 2002-09-11 19:06 HKT

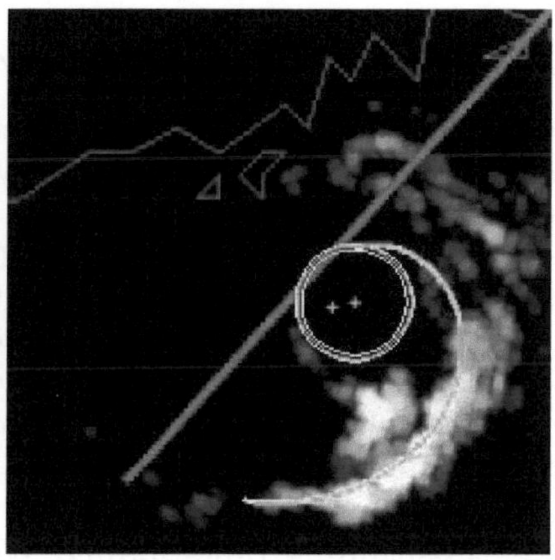

(d) Fig. 9(c): Zoomed in

Fig. 9.9. Results of eye fix: STS Haguput Preprocessed radar reflectivity images overlaid with results of eye fix. The straight line indicates the zero isodop, the polyline indicates the coastline, the spirals are the top 5 GA candidates, and the cross indicates the proposed TC center. The cross in cyan color (the most accurate one) is the final output of the system.

The results demonstrate the efficiency and effectiveness of the GA-based framework in positioning tropical cyclones on remote sensing images. Further information regarding the effect of GA parameters such as $nChild$, $nBest$, p_m and p_x and the use of temporal information for TC eye fix can be found in[56]. The algorithm provides a practical, independent and objective information to assist forecasters to fix TC centers.

9.6 Case Study: Line-Shaped Weather Systems

The generic nature of the proposed framework allows it to be used to position other weather systems. Frontal systems, troughs, ridges, and squall lines are line-shaped by nature. Positioning them on satellite images is usually done by manual identification of a sharp edge of clouds. To locate them on radar images, forecasters look for different features depending on the type of line-shaped weather system being analyzed. A line of squally thunderstorms indicates a squall line, whereas a line of heavy showers often happens in the case of troughs or gust fronts. As remote sensing images are often noisy, and not all the edges correspond to the weather system, automating this process cannot be done by only using simple edge detection or line detection techniques[52]. A fitness function and an arc-shaped model[52, 53] are therefore designed for automated positioning of line-shaped weather system on radar images using the framework introduced in Section 9.4. The model is first introduced in Section 9.6.1, followed by the algorithm in Section 9.6.2. Results are then evaluated in Section 9.6.3.

9.6.1 A Model of Line-Shaped Weather Systems

The line-shaped weather systems are approximated by adjoining segments of arcs, since they can be joined smoothly and each of them can be described by a vector with only five elements $\langle lat, lon, r, \phi, s \rangle$ (Fig. 9.10). The values *lat* and *lon* give the latitude and longitude of the beginning point on the arc. The value r is the radius of the arc, ϕ the angle of the tangent of the arc at (lon, lat), and s the length of the arc drawn clockwise from (lon, lat). The arc at radius r can be extended to an area using two auxiliary arcs with radii $r + d_F$ and $r - d_B$ with the same angular span, where d_F and d_B are both positive constants specified in the experiments.

9.6.2 Positioning Line-Shaped Weather Systems Using Radar Data

With the GA-based positioning framework and the line-shaped weather system model, the steps for positioning line-shaped weather systems on radar data is similar to that for TC eye fix. Unlike TC that has different matching criteria for the spiral rainbands and the TC center, the goal for line-shaped weather system positioning is to locate the line with high reflectivity radar echoes.

The algorithm is as follows. A radar reflectivity image is first preprocessed (Fig. 9.11) by thresholding at 34 dBZ, to retain areas with high reflectivity echoes. Domain-specific information is then used to limit the values of the five model parameters. *lat* and *lon* are limited by the area of coverage of radar, and the values of other template parameters are determined by values of typical weather systems (Table 9.5). Since there is no effective way to determine the ROI, it is set to the area of the radar reflectivity image.

The fitness function for locating the line of high reflectivity echoes on radar reflectivity image is described in Appendix B. Essentially, the weighted spatial average of reflectivity values along the arc is used. The setting of GA parameters

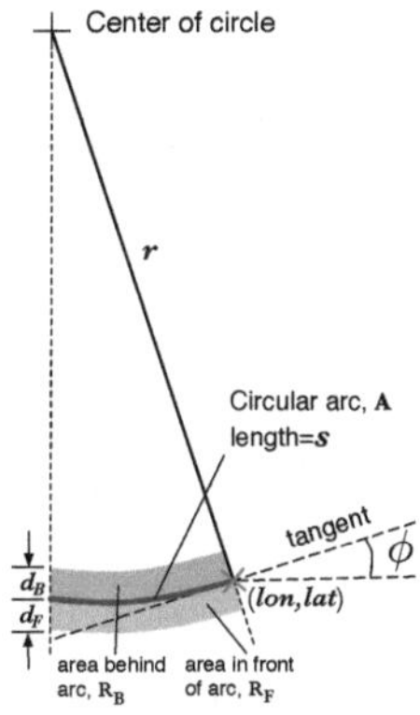

Fig. 9.10. A model of line-shaped weather system

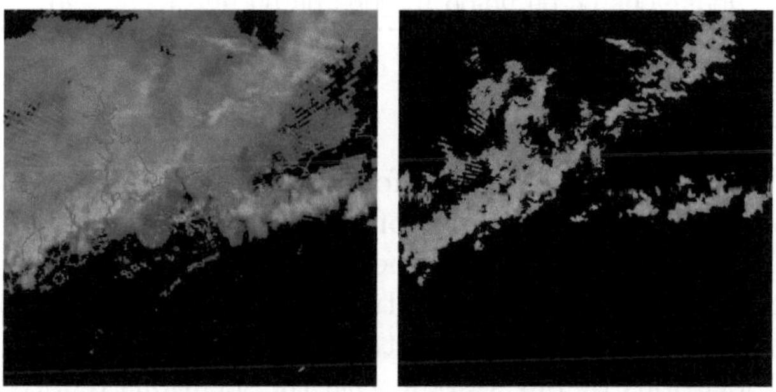

(a) Original (b) Thresholded at 34 dBZ

Fig. 9.11. Original and preprocessed image for line-shaped weather system identification

Table 9.5. Model parameters for line-shaped weather system positioning

Parameter	Min	Max
(lon, lat)	128 km from radar station	
r	80 km	800 km
ϕ	$-\pi$	π
s	105 km	550 km

for locating line-shaped weather systems are listed in Table 9.6. Genetic algorithm mentioned in Section 9.4 is then applied to find the set of parameter values of the arcs that best described the weather system, thus locating the line of active thunderstorms that brings heavy rain to affected areas. The position of the best arc gives the final output of the system. This line of highly active thunderstorms is often located at the front of the whole stormy area, and locating it helps tracking and predicting its movement.

Table 9.6. GA parameters for line-shaped weather system positioning

Parameter	Experiment setting
$nInit$	2000
$nBest$	50
$nChild$	500
$nRetain$	5
$boreCount$	15
$minLoop$	20
p_x	50%
p_m	50%

Table 9.7. Experimental squall line data set

Case	Time	No. of images
May 2005	2005-05-09 08:00 to 2005-05-09 23:54	160
June 2006	2006-06-09 20:00 to 2006-06-09 22:54	30

9.6.3 Evaluation

A Java-based system prototype is used for evaluation. Sequences of 3 km CAPPI radar reflectivity images with a range of 128 km captured every 6 minutes containing squall line cases are used as experimental data sets (Table 9.7).

Since there is no universal ground truth locations for line-shaped weather system positioning, the effectiveness of the algorithm is evaluated by comparing the squall line locations by our system with the locations reported in technical reports[29, 43] of Hong Kong Observatory.

Using typical parameter values (Table 9.6), the positioning results are shown in Fig.s 9.12 and 9.13. It can be seen that the arcs found by the algorithm position at lines of high reflectivity echoes. The positions found are also close to those reported by the Hong Kong Observatory. Moreover, the method discovers arcs that are positionally continuous with respect to time, an important requirement in positioning weather systems.

The efficiency of the algorithms is measured by the system plus user time needed to process the data. Timing tests on a Macintosh PowerMac with Dual 2.5GHz PowerPC G5 processor and 3GB RAM running Mac OS 10.4.8, with only one CPU running the Java program was done. It is found that the automatic positioning process takes less than 278 seconds, including all I/O operations. The average time for parsing the radar file is less than 8 seconds. This is slower than that of the TC cases as there is no effective way of reducing the area of analysis. Yet, the whole process can be done before the next set of radar data is available.

These results demonstrate the ability of the GA-based framework in positioning line-shaped weather systems, such as squall lines, on remote sensing data. Such information allows forecasters to track and predict the movement of weather systems. The arc describes the weather system using position, orientation, length,

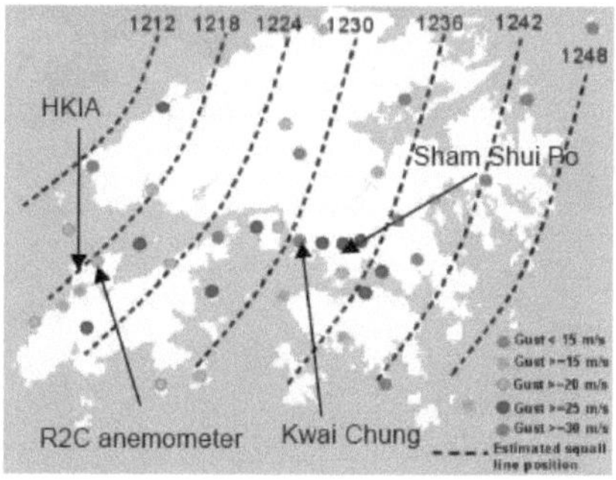

(a) Squall line locations reported in [29]

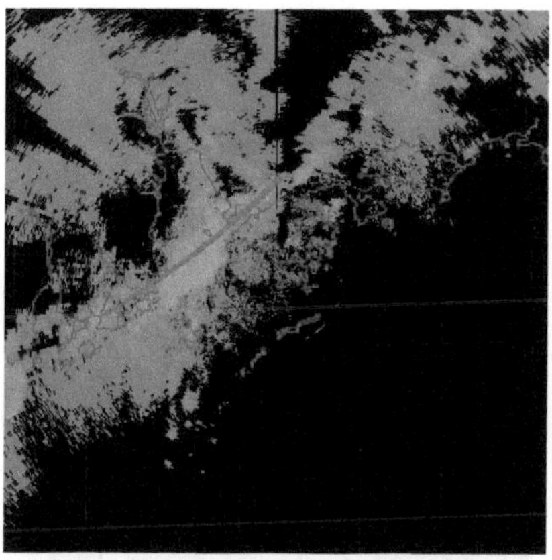

(b) Result, HKT 2005-05-09 12:12

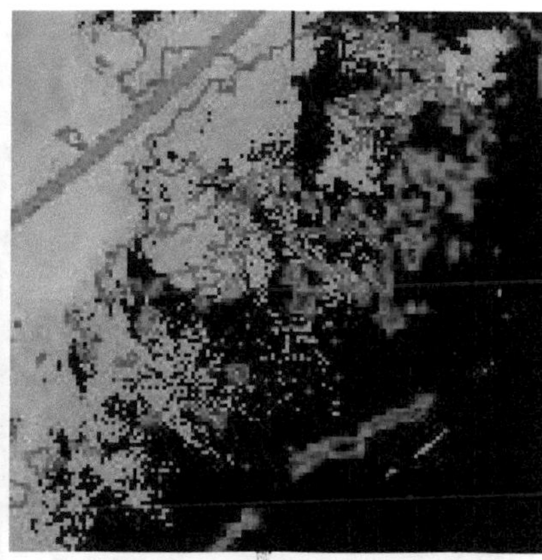

(c) Fig. 12(b): Zoomed in

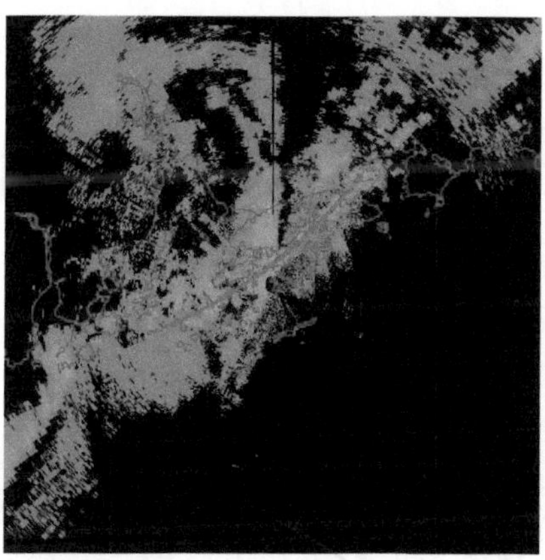

(d) Result, HKT 2005-05-09 12:30

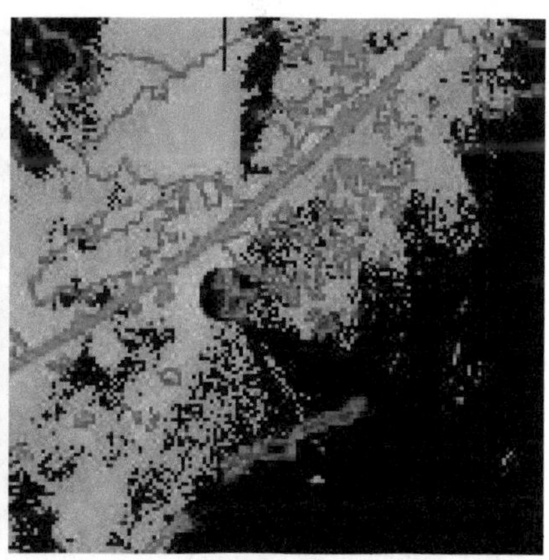

(e) Fig. 12(d): Zoomed in

Fig. 9.12. Case 2005: Preprocessed radar reflectivity images overlaid with results (the arc) of squall line positioning. Fig. (a) is the squall line locations reported by the Hong Kong Observatory.

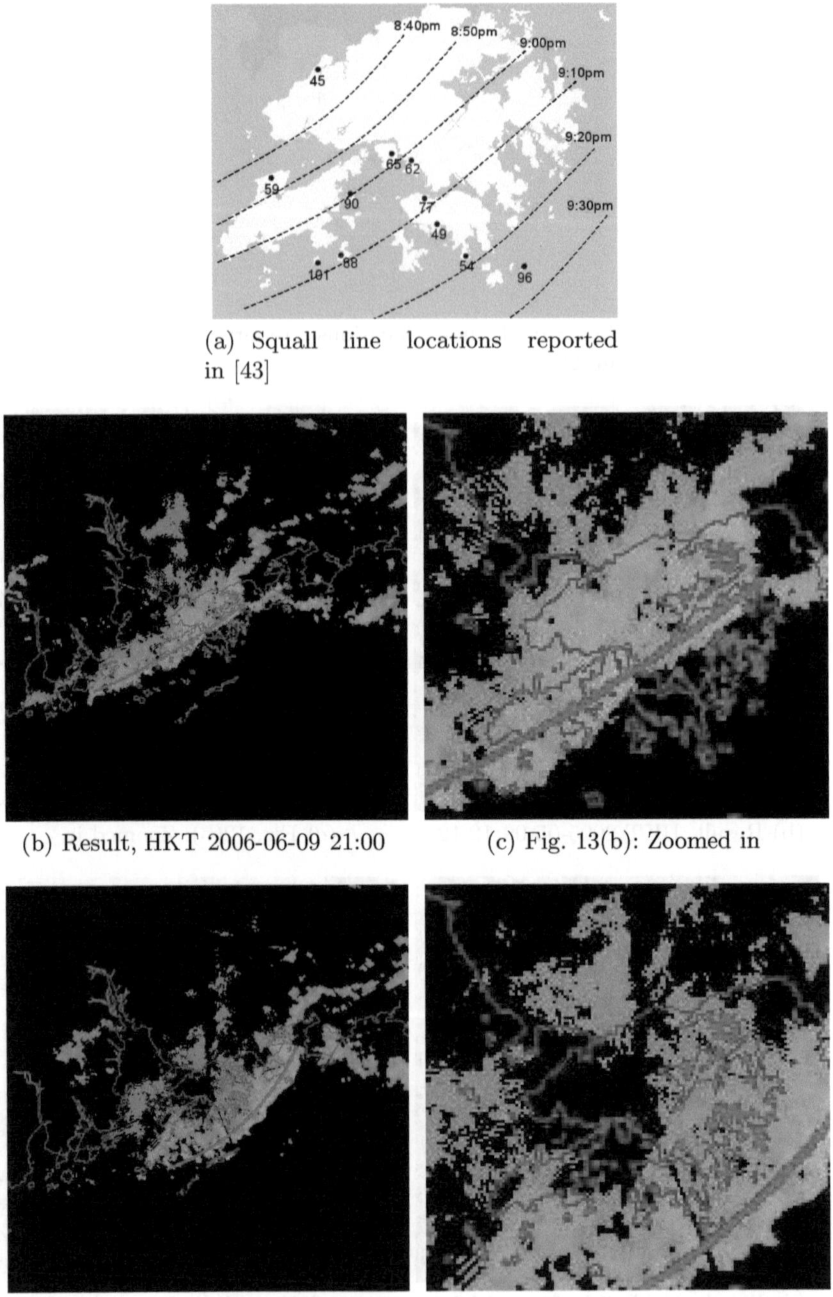

(a) Squall line locations reported in [43]

(b) Result, HKT 2006-06-09 21:00

(c) Fig. 13(b): Zoomed in

(d) Result, HKT 2006-06-09 21:30

(e) Fig. 13(d): Zoomed in

Fig. 9.13. Case 2006: Preprocessed radar reflectivity images overlaid with results (the arc) of squall line positioning. Fig. (a) is the squall line locations reported by the Hong Kong Observatory.

and curvature. Such information is useful for understanding evolution of weather systems. A more detailed study on noise sensitivity and temporal stability of the algorithm can be found in[52]. By adjoining segments of arcs, a more detail description of the weather system can be obtained[53].

9.7 Summary

In this chapter, we have introduced a practical genetic algorithm-based framework for positioning weather systems using remote sensing data. The algorithm adopts the pattern matching approach. With the use of genetic algorithm and a generic weather system model, different fitness functions are formulated for different weather systems according to their properties and features on remote sensing data. The best match between the weather system model and cloud or radar echo pattern on the remote sensing data is then found, giving the position of the weather system.

Examples in positioning tropical cyclones and line-shaped weather systems on radar images demonstrate the practical use of the system framework. The best average error for fixing the center of tropical cyclone using the GA-based system framework is 0.158 degrees, well within the relative error of about 0.3 degrees given by different TC warning centers. The accuracy for locating squall lines on radar images also close to that reported by the Hong Kong Observatory. The system can complete a positioning process before the next set of data available, leaving ample time for further analysis. The system provides a practical, independent and objective information to assist forecasters to locate weather systems. This is a practical application that demonstrates how Computational Intelligence techniques can be used to analyse remote sensing data for weather forecasting.

A Fitness Function: Positioning Tropical Cyclones on Radar Images

A fitness function[51, 56, 59, 60] is defined to evaluate the quality of match for positioning tropical cyclones on radar images. Based on the TC model defined in Fig. 9.3, a modified correlation function is designed, so that high reflectivity areas match the spiral segment BDE and the outer circle of the template, and low reflectivity areas match the inner circle.

To increase the speed of matching, the spiral segment and the two circles of the template are sampled every s_s and s_c km respectively. In our experiments, $s_s = 3$, $s_c = 3$.

Since features of a TC usually change throughout its life, two weighting factors, w_s and w_c, are introduced, to give different weights to the spiral and circles, according to the cloud features and the shape of the TC.

The matching is done on a grayscale-mapped radar image, whose pixel value is in the range of [0, 255]. The match score should be increased when a high reflectivity area (high pixel value) matches the spiral segment BDE and the outer circle of the template, or when a low reflectivity area (low pixel value) matches the inner circle. On the other hand, if a high reflectivity area matches the inner circle, a penalty is given by reducing the score. A constant p_{offset} is defined to be the demarcation pixel value between high and low reflectivities.

The fitness function consists of three parts, evaluating separately how good the match is for the spiral segment, the inner circle, and the outer circle.

For the spiral segment, the score $score_{\text{spiral}}$ is defined as follows:

$$score_{\text{spiral}} = \frac{\sum_{x,y \in \mathbf{S}} p(x,y)}{|\,\mathbf{S}\,|}$$

Here, $p(x,y)$ is the pixel value at point (x,y) on the image, and $\mathbf{S}$ is the set of sample points on the spiral segment. A high score is obtained if the segment matches with high pixel value $p(x,y)$ of the image. It is normalized by the total number of sample points on the spiral segment.

For the inner circle, the score $score_{\text{inner}}$ is defined as follows:

$$score_{\text{inner}} = \frac{\sum_{x,y \in \mathbf{E}} (p_{\text{offset}} - p(x,y))}{|\,\mathbf{E}\,|}$$

Here, $\mathbf{E}$ is the set of sample points on the inner circle circumference. The offset value p_{offset} is used with the intention to deduce the score if the inner circle (eyewall) is matched wrongly with high reflectivity values (stormy area). In our experiments, $p_{\text{offset}} = 64$. This value is considered only for the case of inner circle matching. Otherwise, the score would be dominated by a wrong match of a circle of low reflectivity values at a region outside the tropical cyclone. The score calculated here is normalized by the total number of sample points on the inner circle circumference.

For the outer circle, the score $score_{\text{outer}}$ is defined as follows:

$$score_{\text{outer}} = \frac{\sum_{x,y \in \mathbf{C}} p(x,y)}{|\,\mathbf{C}\,|}$$

Here, $\mathbf{C}$ is the set of sample points on the outer circle circumference.

A high score is obtained if the circle matches high pixel values of the image. It is normalized by the total number of sample points on the outer circumference of the circle.

Finally, the final score of the fitness function is calculated by the weighted sum of these three parts.

$$score = w_s \times score_{\text{spiral}} + w_c \times score_{\text{inner}} + w_c \times score_{\text{outer}}$$

The weightings w_c and w_s are determined empirically using past TC data. In our experiments, $w_c = 1$ and $w_s = 2$. It is found that large values of w_c with respect to w_s favour only the no-rain areas, while the reverse favours rainy areas. To fit the structure of a TC, whose distinct feature is the spiral rainbands, the ratio of w_c to w_s is set to $1:2$.

The higher the score the candidate has, the better it fits the defined TC model.

B Fitness Function: Positioning Line-Shaped Weather Systems on Radar Images

Based on the line-shaped weather system model defined in Fig. 9.10, a fitness function[52, 53] is defined for positioning line-shaped weather systems on radar images. The goal is to locate line-shaped areas of high reflectivity echoes on radar images.

The fitness value calculation is done on a 3 km CAPPI radar reflectivity data, whose values indicate the echo intensity, i.e., reflectivity, measured in dBZ (decibels of Z). Areas in front of, along, and behind the arc are all taken into consideration. To increase the speed of calculating the fitness function, the areas in front of, behind and along the arc are sampled every s_F, s_B, and s_A km respectively. In our experiments, $s_F = s_B = s_A = 1$.

The fitness score is calculated by the spatial average of reflectivity values as follows. The fitness score contributes by the areas behind the arc is defined as

$$score_{\mathrm{R_B}} = \frac{\sum_{x,y \in \mathbf{R_B}} z(x,y)}{\mid \mathbf{R_B} \mid}$$

where $z(x, y)$ is the reflectivity value at location (x, y) of the data, and $\mathbf{R_B}$ is the set of the sample points in the area behind the arc. The score is normalized by the total number of sample points in the area.

For the arc segment, the score $score_{\mathrm{C}}$ is defined as:

$$score_{\mathrm{C}} = \frac{\sum_{x,y \in \mathbf{C}} z(x,y)}{\mid \mathbf{C} \mid}$$

where $\mathbf{C}$ is the set of the sample points along the arc segment.

Similarly, the score for the area in front of the arc is defined as:

$$score_{\mathrm{R_F}} = \frac{\sum_{x,y \in \mathbf{R_F}} z(x,y)}{\mid \mathbf{R_F} \mid}$$

where $\mathbf{R_F}$ is the set of the sample points in the area in front of the arc.

Finally, the final score of the fitness function is calculated by the weighted sum of the three parts.

$$score = w_{\mathrm{R_B}} \times score_{\mathrm{R_B}} + w_{\mathrm{C}} \times score_{\mathrm{C}} + w_{\mathrm{R_F}} \times score_{\mathrm{R_F}}$$

The weights $w_{\mathrm{R_B}}$, w_{C} and $w_{\mathrm{R_F}}$ are determined empirically using past squall line data. In our experiments, $w_{\mathrm{R_B}} = 1$, $w_{\mathrm{C}} = 1$ and $w_{\mathrm{R_F}} = 1$. By changing the values d_F and d_B, the size of the areas in front of and behind the arc (R_F and R_B respectively) can also be adjusted. Here, in our experiments, $d_F = 0$ and $d_B = 10$. Note that when $d_F = 0$ and $d_B = 0$, only reflectivity values along the arc is considered. The higher the score the candidate has, the better it fits the defined line-shaped weather system model.

Acknowledgement. The authors are thankful to Dr. LI Ping Wah for his expert advice on meteorology and the Hong Kong Observatory for their provision of data.

References

1. Images in Weather Forecasting: A Practical Guide for Interpreting Satellite and Radar Imagery. Cambridge University Press, Cambridge (1995), ISBN 0521451116
2. Japan Meteorological Agency. Web page (2008), http://www.jma.go.jp/
3. Ahrens, C.D.: Meteorology Today: An Introduction to Weather, Climate, and the Environment, 7th edn., Brooks Cole (August 2000), ISBN 0534397719
4. Azimi-Sadjadi, M.R., Zekavat, S.A.: Cloud classification using support vector machines. In: Proceedings of International Geoscience and Remote Sensing Symposium, Honolulu, Hawaii, USA, July 2000, vol. 2, pp. 669–671 (2000)
5. Bieringer, P.E., Morgan, J., Hurst, J., Martin, B., Winkler, S., Xie, Y., McGinley, J., Albers, S.: An assessment of automated boundary and front detection to support convective initiation forecasts. In: Proceedings of the 12th Conference on Aviation Range and Aerospace Meteorology, Atlanta, GA, USA, January 2006, American Meteorological Society (2006)
6. Chou, C.Y., Chen, C.H., Chen, C.H.: Economic design of variable sampling intervals T^2 control charts using genetic algorithms. Expert Systems with Appplications 30(2), 233–242 (2006)
7. Corpetti, T., Mémin, É., Pérez, P.: Dense motion analysis in fluid imagery. In: Heyden, A., Sparr, G., Nielsen, M., Johansen, P. (eds.) ECCV 2002. LNCS, vol. 2350, pp. 676–691. Springer, Heidelberg (2002)
8. Corpetti, T., Mémin, É., Pérez, P.: Extraction of singular points from dense motion fields: An analytic approach. Journal of Mathematical Imaging and Vision 19, 175–198 (2003)
9. Darwin, C.: On The Origin of Species by means of Natural Selection. John Murray, London (1859)
10. Dell'Acqua, F., Gamba, P.: A novel technique for hurricane detection in meteosat images bymeans of contour analysis. In: Proceedings of International Geoscience and Remote Sensing Symposium, Sydney, New South Wales, Australia, July 2001, vol. 3, pp. 1512–1514 (2001)
11. El-Zonkoly, A.M.: Power system model validation for power quality assessment applications using genetic algorithm. Expert Systems with Applications 29(4), 941–944 (2005)
12. Fine, S.S., Fraser, A.B.: A simple technique for multiple-parameter pattern recognition with an example of locating fronts in model output. Journal of Atmospheric and Oceanic Technology 7, 896–908 (1990)
13. Hagelgerg, C., Helland, J.: Thin-line detection in meteorological radar images using wavelet transforms. Journal of Atmospheric and Oceanic Technology 12, 633–642 (1995)
14. Hasler, A.F., Palaniappan, K., Kambhammetu, C., Black, P., Uhlhorn, E., Chesters, D.: High resolution wind fields within the inner-core and eye of a mature tropical cyclone from GOES one-minute images. Bulletin of the American Meteorological Society 79(11), 2483–2496 (1998)
15. Hewson, T.D.: Objective fronts. Meteorological Applications 5, 37–65 (1998)
16. Hong Kong Observatory. Hong Kong Observatory — official authority for Hong Kong weather forecast. Web page (2008), http://www.hko.gov.hk/
17. Huang, X., Zhao, F.: Relation-based aggregation: finding objects in large spatial datasets. Intelligent Data Analysis 4(2), 129–147 (2000)
18. Jann, A.: Use of a simple pattern recognition approach for the detection of ridge lines and stripes. Meteorological Applications 9, 357–365 (2002)

19. Kitamoto, A.: The development of typhoon image database with content-based search. In: Proceedings of the 1st International Symposium on Advanced Informatics, Tokyo, Japan, March 2000, pp. 163–170 (2000)
20. Kitamoto, A.: Data mining for typhoon image collection. In: Proceedings of the 2nd International Workshop on Multimedia Data Mining, San Francisco, CA, USA, August 2001, pp. 68–77 (2001)
21. Lai, E.S.T.: TREC application in tropical cyclone observation. In: ESCAP/WMO Typhoon Committee Annual Review 1998, Manila, Typhoon Committee Secretariat, pp. 135–139 (1999)
22. Lai, E.S.T.: Analysing cold fronts for forecasting operations. Technical Report 78, Hong Kong Observatory (2003)
23. Lakshmanan, V.: Using a genetic algorithm to tune a bounded weak echo region detection algorithm. Journal of Applied Meteorology 39, 222–230 (1999)
24. Lam, C.Y.: Operational tropical cyclone forecasting from the perspective of a small weather service. In: Proceedings of ICSU/WMO International Symposium on Tropical Cyclone Disasters, Beijing, October 1992, pp. 530–541 (1992)
25. Li, P.W., Lai, S.T.: Short range quantitative precipitation forecasting in Hong Kong. Journal of Hydrology 288, 189–209 (2004)
26. Li, P.W., Wong, W.K., Chan, K.Y., Lai, E.S.T.: SWIRLS - an evolving nowcasting system. Technical Report 100, Hong Kong Observatory, Hong Kong (1999)
27. Lucas, C., May, P.T., Vincent, R.A.: An algorithm for the detection of fronts in wind profiler data. Weather and Forecasting 16, 234–247 (2001)
28. Maybeck, P.S.: Stochastic models, estimation, and control. Mathematics in Science and Engineering, vol. 1. Academic Press, New York (1979)
29. Hong Kong Observatory. Squall lines and "Shi Hu Feng". Web page (June 17, 2005), `http://www.hko.gov.hk/education/edu01met/wxphe/ele_squalle.htm`
30. Garcia Orellana, C.J., Macias Macias, M., Serrano Perez, A., Gonzalez Velasco, H.M., Gallardo Caballero, R.: A comparison of PCA, ICA and GA selected features for cloud field classification. Journal of Intelligent and Fuzzy Systems 12, 213–219 (2002)
31. Ottenbacher, A., Tomassini, M., Holmlund, K., Schmetz, J.: Low-level cloud motion winds from Meteosat high resolution visible imagery. Weather and Forecasting 12, 175–184 (1997)
32. Palaniappan, K., Faisal, M., Kambhamettu, C., Hasler, A.F.: Implementation of an automatic semi-fluid motion analysis algorithm on a massively parallel computer. In: Proceedings of the 10th International Parallel Processing Symposium, Honolulu, Hawaii, USA, April 1996, pp. 864–872 (1996)
33. Palaniappan, K., Kambhamettu, C., Hasler, A.F., Goldgof, D.B.: Structure and semi-fluid motion analysis of stereoscopic satellite images for cloud tracking. In: Proceedings of the 5th International Conference on Computer Vision (ICCV-1995), Massachusetts, USA, June 1995, pp. 659–665 (1995)
34. Pankiewicz, G.S.: Pattern recognition techniques for the identification of cloud and cloud systems. Meteorological Applications 2, 257–271 (1995)
35. Rinehart, R.E.: Internal storm motions from a single non-Doppler weather radar. Technical Report TN-146+STR, National Center for Atmospheric Research, Springfield, Virginia, USA, pp. 262 (1979)
36. Rodrigues, P.S.S., de Araujo, A.A., Pinotti, M.: Describing patterns in flow-like images. In: Proceedings of the 10th International Conference on Image Analysis and Processing, Venice, Italy, September 1999, pp. 424–429 (1999)

37. Schemetz, J., Holmlund, K., Hoffman, J., Strauss, B., Mason, B., Gaertner, V., Koch, A., Van De Berg, L.: Operational cloud-motion winds from Meteosat infrared images. Journal of Applied Meteorology 32, 1206–1225 (1993)
38. World Meteorological Center Severe Weather Information Center. Terminologies and acronyms used in the region of Western North Pacific and the South China Sea by the Typhoon Committee Members. Web page (2005), http://severe.worldweather.org/tc/wnp/acronyms.html#CIT
39. Shin, K.S., Joo, L.Y.: A genetic algorithm application in bankruptcy prediction modeling. Expert Systems with Applications 23(3), 321–328 (2002)
40. Shu, C.F., Jain, R.C.: Vector field analysis for oriented patterns. IEEE Transactions on Pattern Analysis and Machine Intelligence PAMI-16(9), 946–950 (1994)
41. Sivaramakrishnan, M.V., Selvam, M.: On the use of the spiral overlay technique for estimating the center positions of tropical cyclones from satellite photographs taken over the Indian region. In: Proceedings of the 12th Conference on Radar Meteorology, Norman, Oklahoma, USA, pp. 440–446. American Meteorological Society (1966)
42. Stephanidis, C.N., Cracknell, A.P., Hayes, L.W.B.: The implementation of self organised neural networks for cloud classification in digital satellite images. IEEE Transaction on Neural Networks 4(1), 86–96 (1995)
43. Szeto, K.C., Chan, P.W.: Numerical simulation of a severe squall event in Hong Kong. In: Proceedings of the 23rd Conference on Severe Local Storms, St. Louis, MO, U.S.A., November 2006, American Meteorological Society (2006)
44. Tian, B., Shaikh, M.A., Azimi-Sadjadi, M.R., Von der Haar, T.H., Reinke, D.L.: A study of cloud classification with neural networks using spectral and textural features. IEEE Transactions on Neural Networks 10(1), 138–151 (1999)
45. Torsun, I.S., Kwiatkowska, E.: Neural network system for cloud classification from satellite images. In: International Joint Conference on Neural Networks, Washington, DC, USA, July 1999, vol. 6 (1999)
46. Tuttle, J., Gall, R.: A single-radar technique for estimating the winds in tropical cyclones. Bulletin of the American Meteorological Society 80(4), 653–668 (1999)
47. Wang, Y.Z.: A GA-based methodology to determine an optimal curriculum for schools. Expert Systems with Applications 28(1), 163–174 (2005)
48. Welch, G., Bishop, G.: An introduction to the Kalman filter. In: Proceedings of the ACM Computer Graphics Conference (SIGGRAPH-2001), Los Angeles, California, USA (August 2001)
49. Wimmers, A., Velden, C.: Satellite-based center-fixing of tropical cyclones: New automated approaches. In: Proceedings of the 26th Conference on Hurricanes and Tropical Meteorology, Miami, Florida, USA, May 2004, American Meteorological Society (2004)
50. Wong, K.Y., Yip, C.L.: An intelligent tropical cyclone eye fix system using motion field analysis. In: Proceedings of the 17th IEEE International Conference on Tools with Artificial Intelligence (ICTAI-2005), Hong Kong, November 2005, pp. 652–656 (2005)
51. Wong, K.Y., Yip, C.L.: Tropical cyclone eye fix using genetic algorithm with temporal information. In: Khosla, R., Howlett, R.J., Jain, L.C. (eds.) KES 2005. LNCS (LNAI), vol. 3681, pp. 854–860. Springer, Heidelberg (2005)
52. Wong, K.Y., Yip, C.L.: Comparison of squall line positioning methods using radar data. In: Gabrys, B., Howlett, R.J., Jain, L.C. (eds.) KES 2006. LNCS (LNAI), vol. 4253, pp. 269–276. Springer, Heidelberg (2006)

53. Wong, K.Y., Yip, C.L., Li, P.W.: Identifying weather systems from numerical weather prediction data. In: Proceedings of the 18th International Conference on Pattern Recognition (ICPR-2006), Hong Kong, August 2006, vol. 4, pp. 841–844 (2006)
54. Wong, K.Y., Yip, C.L., Li, P.W.: A novel algorithm for automatic tropical cyclone eye fix using Doppler radar data. Meteorological Applications 14(1), 49–59 (2007)
55. Wong, K.Y., Yip, C.L., Li, P.W.: Automatic identification of weather systems from numerical weather prediction data using genetic algorithm. Expert Systems With Applications (in press, 2008)
56. Wong, K.Y., Yip, C.L., Li, P.W.: Automatic tropical cyclone eye fix using genetic algorithm. Expert Systems With Applications 34(1), 643–656 (2008)
57. Wong, K.Y., Yip, C.L., Li, P.W., Tsang, W.W.: Automatic template matching method for tropical cyclone eye fix. In: Proceedings of the 17th International Conference on Pattern Recognition (ICPR-2004), Cambridge, UK, August 2004, vol. 3, pp. 650–653 (2004)
58. Wood, V.T.: A technique for detecting a tropical cyclone center using a Doppler radar. Journal of Atmospheric and Oceanic Technology 11, 1207–1216 (1994)
59. Yip, C.L., Wong, K.Y.: Efficient and effective tropical cyclone eye fix using genetic algorithms. In: Negoita, M.G., Howlett, R.J., Jain, L.C. (eds.) KES 2004. LNCS (LNAI), vol. 3213, pp. 654–660. Springer, Heidelberg (2004)
60. Yip, C.L., Wong, K.Y.: Tropical cyclone eye fix using genetic algorithm. Technical Report TR-2004-06, Department of Computer Science, The University of Hong Kong, Hong Kong (2004)
61. Yip, C.L., Wong, K.Y., Li, P.W.: Data complexity in tropical cyclone positioning and classification. In: Basu, M., Ho, T.K. (eds.) Data Complexity in Pattern Recognition, December 2006, ch. 13, pp. 249–270. Springer, London (2006)
62. You, J., Bhattacharya, P.: Dynamic shape retrieval by hierarchical curve matching, snakes and data mining. In: Proceedings of the 15th International Conference on Pattern Recognition (ICPR-2000), Barcelona, Spain, September 2000, vol. 1, pp. 1035–1038 (2000)
63. Young, D.: Straight lines and circles in the log-polar image. In: Proceedings of the 11th British Machine Vision Conference (BMVC-2000), Bristol, UK, September 2000, pp. 426–435 (2000)

10

A Computation Reduced Technique to Primitive Feature Extraction for Image Information Mining Via the Use of Wavelets

Vijay P. Shah, Nicolas H. Younan, Surya H. Durbha, and Roger L. King

Mississippi State University, Department of Electrical and Computer Engineering
younan@ece.msstate.edu

10.1 Introduction

NASA's Earth Station receives terabytes of data daily from satellites. However, the usability of the vast amount of data is constrained to users of different domains because data accessibility is limited to queries based on geographical coordinates, time of acquisitions, sensor types, and acquisition mode. Research in knowledge discovery from remote sensing archives has been propelled in the last few years by the adoption of newer technologies. However, processes for archiving, information retrieval, and distribution of different kinds of imagery received from different satellites are still in their infancy. In general, the imagery received from each satellite provides different information that depends on spectral resolution, spatial resolution, temporal resolution, number of bands, and other characteristics that pose greater challenges for image information mining in remote sensing archives then in multimedia image archives. Hence, the adoption of new technologies that allow the accessibility of remote sensing data based on content and semantics is required to overcome this challenge and increase useful exploitation of the data [1, 2, 3].

Over the last few decades, many content-based image retrieval (CBIR) systems have been proposed for multimedia images, yet many problems are left unresolved. Most of these approaches are based on the indexing of imagery in the feature space, typically using "global features" such as color, texture, shape, and region [1, 2, 4, 5]. The IBM QBIC system [6], the MIT Photobook System [7], and the Virage System [8] are few of the available CBIR system, while the Berkeley BlobWorld [9], the UCSB Netra [10], and the Columbia VisualSEEK [11] are few of the region-based image mining systems for multimedia data archives. Inspired by the success of image information mining to other imagery databases, the application of such techniques to remote sensing archives is gaining popularity. Some initial prototype systems include the Knowledge-Driven Information Mining in remote sensing image archives (KIM) system [12], the EO domain-specific Knowledge Enabled Services (KES), and the Intelligent Interactive Image Knowledge Retrieval (I^{3}KR) system [1, 3].

The KIM/KES system differs from traditional feature extraction methods. The features of the image pixel and the areas are stored in a database, providing

M. Graña and R.J. Duro (Eds.): Comput. Intel. for Remote Sensing, SCI 133, pp. 245–266, 2008.
springerlink.com

the user an option to select representative features. The goal is of developing heuristics for an automatic labeling of image regions on region obtained via a hierarchical segmentation approach [12].

10.2 Current State of Knowledge

10.2.1 The I^{3}KR Framework

To specify the scope, objectives, and behavior of a system and its functional components, the Open Geospatial Consortium, Inc. (OGC) has developed an architectural framework for geospatial services on the web [13]. The framework also allows extensibility for specific services and service types.

Figure 10.1 shows the framework for application ontologies driven image mining system (e.g., hydrology, land use/cover, imagery). Built upon the existing OGC Web Coverage Service (WCS), the I^{3}KR will allow a user with a requirement to enable a service that has the capability to extract only the necessary data [4]. The original architecture proposed consists of a three level processing sequences consisting of a Primitive Features Level (PFL), an Intermediate Object Description Level (ODL), and a Higher Conceptual Level (HCL), as shown

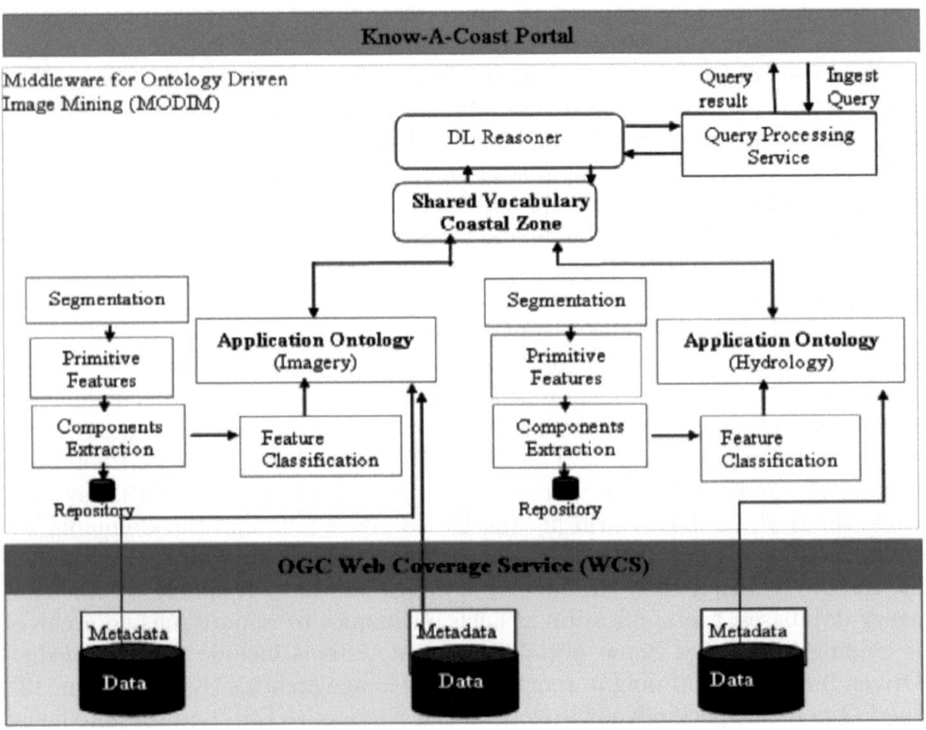

Fig. 10.1. A framework for ontology driven image mining for I^{3}KR

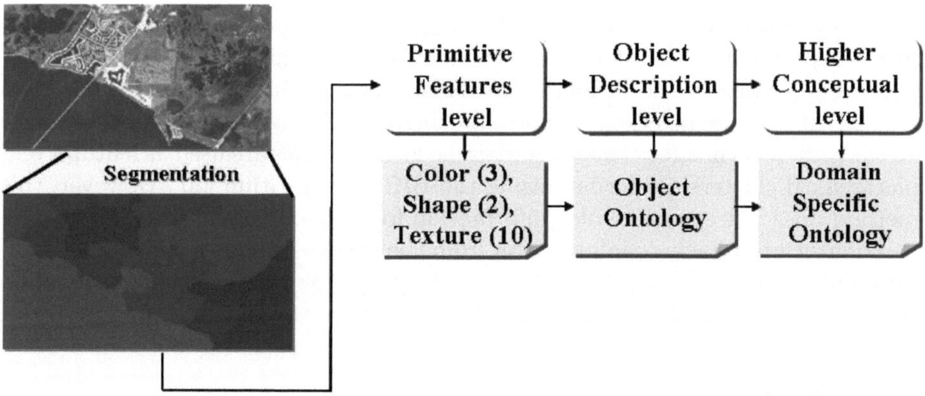

Fig. 10.2. Flow diagram for I^{3}KR - Three levels of processing

in Figure 10.2. Originally, at the primitive level, the regions were indexed based on the color, shape, and texture of each region. Models that quantitatively describe the contents are used to associate machine-centered features (primitives) with a meaningful set of concepts at the intermediate level. The JSEG segmentation algorithm is used to perform the unsupervised segmentation, and each object is stored as a Binary Large Object (BLOB) in the database. Descriptors (primitives) for the color, texture, and shape are calculated for each region and indexed in the database. Further details about the architecture are provided in [2, 3, 4].

10.2.2 Wavelet-Based Retrieval System for Multimedia Archives

The wavelet transform has found successful applications in texture characterization for content-based retrieval of images from multimedia archives. The retrieval systems using the wavelet transform are classified as follows:

- A hierarchical block [14, 15]
- A moving window [16]
- A pixel [17, 18, 19].

Smith et al. [14] proposed a hierarchical block method for feature extraction by generating a quad-tree. Quad trees are used to segment an image into hierarchical blocks, with features being statistics of the wavelet coefficients computed from each sub-band. Although the computation cost of this method is low when features are obtained by applying the wavelet transform only once to an image, this method has low flexibility in the shapes of segmented regions. Remias et al. [15] proposed a nona-tree that provides more flexible image segmentation but comes with a higher computational cost since this method requires performing a wavelet transform on each block. Natsev et al. proposed the WALRUS system for image retrieval [16]. This system uses coefficients from the lowest frequency bands as features after applying the wavelet transform to the region. This is a rather computationally intensive method, since the wavelet transform is performed on each of the many overlapping regions formed by a moving window. This work suffers also from the problem of selecting a proper window size. Suematsu [19] used the wavelet transform for pixel level segmentation. This work computes the logarithmic local energies at all pixel positions after computing n-level wavelet decomposition once on the gray-level images. These logarithmic local energies are used as features and hierarchical clusters are formed using the BIRCH algorithm [20]. However, this method does not consider color information for image segmentation. Extending this algorithm to include information from the spectral bands for image segmentation would lead to a very large number of data points with large dimensionality, thus making this method computationally expensive.

10.3 Motivation

The I^3KR prototype provides a framework that incorporates functionality to carry out data exploratory tasks, leading to knowledge discovery in optical remote sensing data archives. The initial stage consists of segmentation and feature extraction based on color and texture features (co-occurrence matrix, fractal dimension, and shape features). The system also integrates a reasoning engine that can use information contained in the ontologies for the retrieval of queries such as "Find all the eutrophic lakes in year 2000 in Landsat 7 ETM+ imagery for a particular area X" [1, 3]. Considerations of the local spatial properties of the region can help in simplifying such queries. The wavelet transforms take into account the "local" spatial properties, which will help to simplify such queries. Inspired by the successful application of wavelet transform techniques in other imagery archives and their applications in Earth remote sensing imagery classification, this work proposes the use of a wavelet-based method in a semantics-enabled framework for knowledge discovery from geospatial data archives. Geospatial data has different characteristics from multimedia images and poses more challenges; hence, there is a necessity to evaluate the utility of wavelet-based techniques on geospatial data archives. The selection of the appropriate decomposition level for the wavelet transform has also been a major issue. Previous work by Ardizzoni, et. al. that uses wavelet features with multimedia archives makes an experimental assumption that a particular level of decomposition (usually three) will provide better results [17]. Such an assumption may not hold true for remote sensing data archives since images are of different spectral and spatial resolutions. To address this issue, a method for the proper selection of the level of decomposition required for image information mining is developed.

10.4 HSV-Wavelet for Primitive Feature Extraction

The original architecture proposed in [2] consists of three level processing sequences consisting of a Primitive Features Level (PFL), an Intermediate Object Description Level (ODL), and a Higher Conceptual Level (HCL) as described in Section 10.2.1. In this section, a HSV/wavelet-based primitive feature extraction

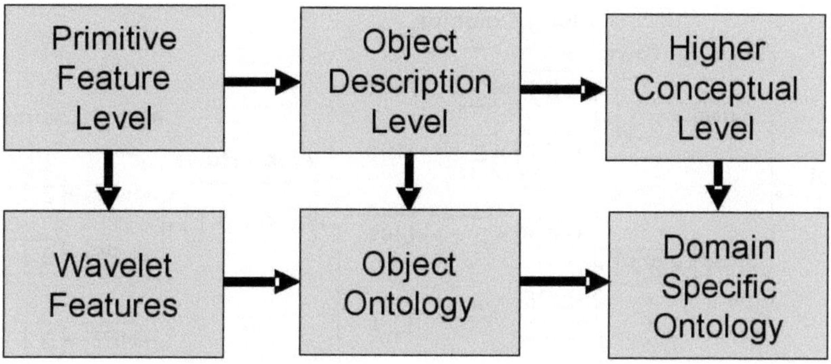

Fig. 10.3. A modified workflow depicting the three levels of processing

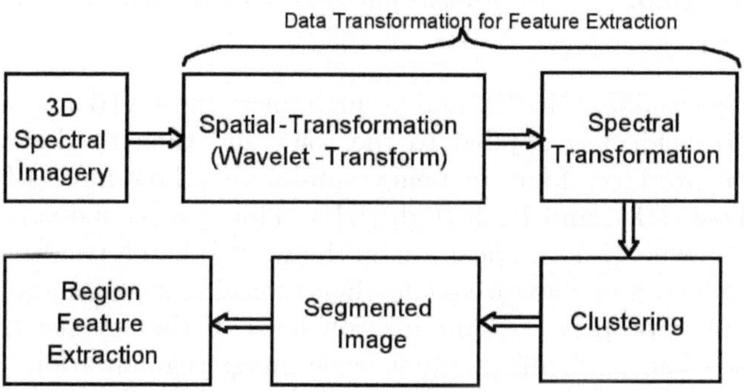

Fig. 10.4. Region feature extraction from an image

process is described to replace the computationally expensive feature extraction process of the I^{3}KR system as shown in Figure 10.3.

Primitive feature extraction is the first step in any image mining process for knowledge discovery. For efficient and quick recognition, the extracted features must be low in dimension and optimal to differentiate various objects. The second step consists of clustering these extracted features, as the feature extraction process produces a huge amount of data. The k-means clustering algorithm is an unsupervised learning process used to form the object. The final stage consists of assigning labels to those regions using support vector machines (SVM). The performance of the presented system is evaluated in terms of recall, precision, and F-measures. Figure 10.4 shows the flow diagram for extracting features from a desired region.

10.4.1 Spatial 2D-Wavelet Transform

Mallat first proposed an efficient implementation of the two-dimensional wavelet transform (2D-DWT) [21]. Figure 10.5 shows the implementation of a 2D-DWT

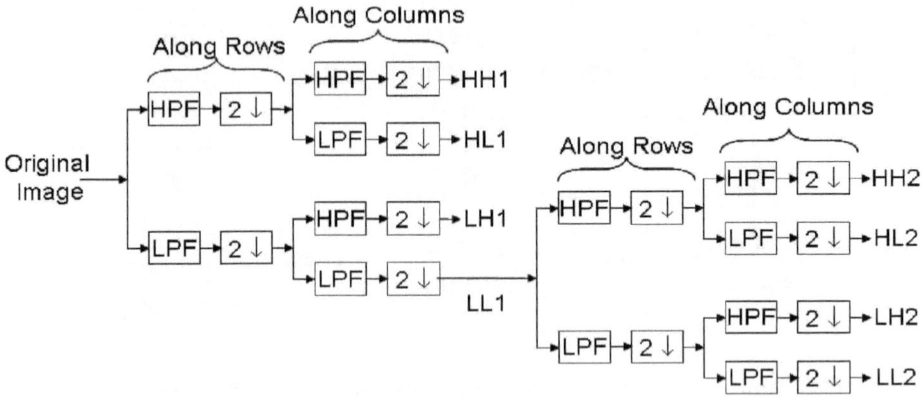

Fig. 10.5. A 2-dimensional wavelet decomposition by 2 levels

using a low pass filter (LPF) and a high pass filter (HPF), where a one-dimensional transform is applied to the rows and then to the columns. The decomposition produces four frequency sub-bands - Low-Low (LL), Low-High (LH), High-Low (HL), and High-High (HH). This is a recursive process that is repeated for N-levels of decomposition on the LL sub-band, resulting in 3N high-frequency bands and one low-frequency band forming a pyramidal structure.

Selecting an appropriate decomposition level of the wavelet-transform has been an issue when performing coarse scale image segmentation. The wavelet-based retrieval system now popularly used for multimedia archives makes an experimental assumption that a particular level of decomposition (usually three) would provide better results [17]. This work develops an automated approach for the selection of the appropriate wavelet decomposition level required for coarse scale image segmentation.

10.4.2 HSV Spectral Transformation

The RGB space is non-linearly converted to the HSV (Hue, Saturation, Value) space to make color components perceptually independent and uniform. Here, Hue represents the color type (such as red, blue, or green), Saturation represents the purity of the color, and Value represents the intensity of the color [22].

10.4.3 Feature Extraction

The wavelet transform has been popularly used for texture analysis [21, 23, 24]. The most common features extracted from the coefficients are the cross-correlation energies and the channel energies to capture, respectively, the color and texture information from the approximate coefficients of the 2D-DWT transform. For the one-level DWT, a 3-D vector is generated for the j-th wavelet coefficient of each sub-band and is given by

$$w_j^{l;B} = \left(w_{0j}^{l;B}, w_{1j}^{l;B}, \cdots, w_{nj}^{l;B}\right), \tag{10.1}$$

where each component refers to a spectral component c $[c \in \{0, 1, 2\}]$. The energy of $w_j^{l;B}$ on c and d is then defined as:

$$e_{cdj}^{l;B} = w_{cj}^{l;B} \cdot w_{dj}^{l;B}. \tag{10.2}$$

When $c = d$, the energy is known as the channel energy of channel c, whereas the energy for the $c \neq d$ case corresponds to the cross-correlation energy between channels c and d. Thus, the final energy vector that captures both color and texture properties is given by [17, 25, 26]

$$e_j^{l;B} = \left(e_{00j}^{l;B}, e_{01j}^{l;B}, e_{02j}^{l;B}, e_{11j}^{l;B}, e_{12j}^{l;B}, e_{22j}^{l;B}\right). \tag{10.3}$$

10.4.4 Clustering Via the k-Means Algorithm

This work uses the k-means algorithm for unsupervised segmentation. This algorithm provides relative scalability and very efficient processing for very large datasets compared to the hierarchical clustering algorithm. The algorithm is based on the squared-error function criterion and is defined as [27]

$$E = \sum_{i=1}^{k} \sum_{x \in C_i} |x - m_i|^2, \tag{10.4}$$

where x is the feature value and mi is the mean of the cluster C_i.. The k-means algorithm is an iterative algorithm where a starting condition is set to K clusters and the center of these clusters are chosen arbitrarily. The repetitive steps of the algorithm are:

- Compute the association of the new feature according to the present result.
- Update clusters according to the membership of the feature.

These steps are repeated until the criterion function E remains constant. For in depth details regarding this algorithm refer to [27].

10.4.5 Support Vector Machines (SVM) for Region Identification

This work uses a new-generation learning system, Support Vector Machines (SVMs), for the identification of the region in an image. The SVMs have been successfully used for solving non-linear classification and regression problems [28]. A few of the applications using the SVMs for classification are handwritten digit recognition [29], object detection [30], and face recognition [31]. A complete mathematical detail of the SVM model formulation can be obtained from

[28, 29, 32]. Different kernel functions available to convert the input space into the feature space, thus requiring the selection of a kernel and its parameters for good generalization. For example, a RBF kernel has a parameter (standard deviation), which must be selected a priori. A recent study by Chang and Lin shows that the RBF is a useful kernel function, since the kernel matrix using a sigmoid may not be positive and definite, and in general its accuracy is not better than RBF [33].

10.5 Overall Performance of Image Mining System

The efficiency of the wavelet-based system is evaluated in terms of precision and recall measures [34], which are defined as follows:

$$\text{Precision} = \frac{|\text{Relevant and retrieved images}|}{|\text{Retrieved images}|} \tag{10.5}$$

and

$$\text{Recall} = \frac{|\text{Relevant and retrieved images}|}{|\text{Relevant images}|} \tag{10.6}$$

Precision is the system's ability to find images that are most relevant, while recall is the ability of the search to find all relevant images in the database. For the ideal system, one strives to increase both precision and recall to 1. For image mining systems, it is useful to report both precision and recall in order to evaluate the system's efficiency. However, there is a trade-off between those quantities. For example, if a system retrieves all images in the database, it would have 100% recall, but may have less precision, and vice-versa (i.e., a system can have 100% precision if only a few documents are retrieved). However, if recall is low, many relevant documents may not be retrieved. Hence, the system efficiency is also evaluated in terms of the F-measure, which takes into account both precision and recall. The higher F-measure indicates better system. The F-measure is given by [34]

$$\text{F-measure} = \frac{|2{*}\text{Precision}{*}\text{Recall}|}{|\text{Precision}+\text{Recall}|} \tag{10.7}$$

10.6 A Systematic Approach to Wavelet Decomposition Level Selection

10.6.1 Insight on the Wavelet Decomposition Level Selection Process

The selection of the appropriate decomposition level for the wavelet-transform has been an issue for image mining. Experimental assumption of commonly using three level of wavelet decomposition has been used for the wavelet-based image mining system for multimedia archives [17]. This section provides an in-depth

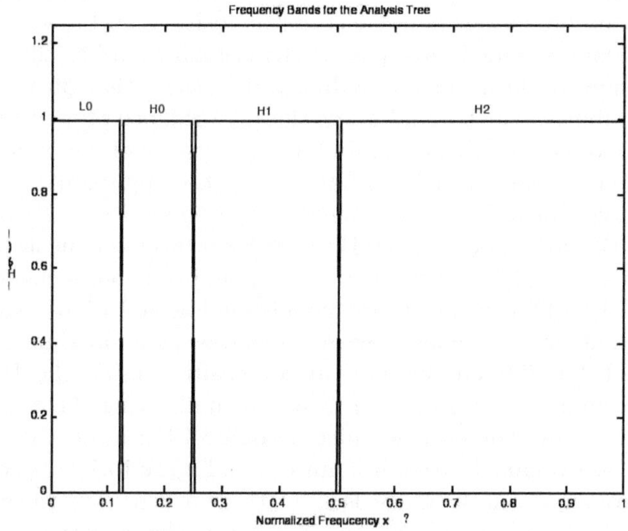

Fig. 10.6. Frequency bands for the analysis tree

understanding required for a good selection of the decomposition level for the image mining in geospatial data archives, a problem that hasn't been addressed in previous research.

As shown in section 10.4.1, the wavelet transform consists of applying low-pass filtering and high-pass filtering to the original image, which results in the approximate and detail coefficients at a particular level, i.e., the spectrum is divided into two equal parts at each level of decomposition. Furthermore, the frequency bands for the analysis tree are shown in Figure 10.6 for the three levels of decomposition. For example, for a three level decomposition, the approximate coefficients are contained in the first 1/8-th frequency spectrum, i.e., $\pi/8 = \pi\ /2^3$ radian, where $n = 3$ is the number of the decomposition level. As the number of decomposition level increases, the frequency resolution increases, while the spatial resolutions decreases. Good frequency resolution is necessary for discriminating the two images. Hence, one would see the advantage of using a higher level of decomposition. However, the DWT approximate coefficients are obtained by low-pass filtering and decimation of the original image. This is similar to obtaining samples by reducing the sampling frequency of the image to half at each level. This results in aliasing of the high-frequency components into the lower frequency spectrum. For the reconstruction problem, this would not matter because these aliasing effects are cancelled by details from the higher-frequency components [35]. For image classification in general, where only the approximate coefficients are used, aliasing can create a problem and there is no assurance about the content of the images, thus reducing the chance of correct classification. On the other hand, reducing the wavelet decomposition level does not help because of low frequency resolution. The inclusion of detail components might be useful only if the wavelet decomposition level is set high and the wavelet coefficients obtained have passed the critical sampling interval. Otherwise, its inclusion would mean the addition of features that do not offer good discrimination ability, which could lower the performance of a system and also would require more computation time. Artificial images of 256 x 256 pixels are generated to demonstrate the effect of aliasing in the approximate components.

The first image consists of two cosines with frequencies of 50 Hz & 78 Hz respectively in the x-direction and 60 Hz & 68 Hz respectively in the y-direction. Figure 10.7(a) shows this image. The corresponding two-dimensional Fourier transforms (2D-FFT) for the image are included in Figure 10.7(b) to mainly show the frequency content of the image texture. The wavelet approximate coefficients and its 2D-FFT for different levels of decomposition using the Haar mother wavelet are shown in Figure 10.7 (a-j) to see the underlying effect in the spatial and spectral domains. For each wavelet decomposition level, the spectrum is divided in half, as explained earlier in Figure 10.6. Figure 10.7(f) shows that for a two level of decomposition, the image is correctly represented by the approximate coefficients as the frequency peaks are observed at the correct location. When the level of decomposition is increased to three, only one peak is observed in the 2D-FFT. At this level, the frequency spectrum of the approximate coefficients is within $\pi/8$ radian (normalized), i.e., $512/8 = 64$ Hz, since the sampling frequency is 1024 Hz. That means if the frequency is 78 Hz, the replication of the frequency peak would be observed at $64 * 2 - 78 = 50$ Hz. Similarly, the 68 Hz frequency is aliased to 60 Hz. However, the wavelet approximate coefficients do not represent the frequency content of the image. Further decomposition to level 4 shows more aliasing effects in the wavelet approximate coefficients. Accordingly, an image consisting of two cosines (50 & 78 Hz in the x-direction and 60 & 68 Hz in the y-direction) is represented by the wavelet coefficients showing a frequency of 14 Hz in the x-direction and 4 Hz in the y-direction.

The second artificial texture generated consists of cosines with frequencies of 14 Hz in the x-direction and 4 Hz in the y-direction and is generated with the same sampling rate. Figure 10.8(a-j) shows the texture image, its 2D-FFT, the approximate wavelet coefficients at different levels and its 2D-FFT. The original images (Figures 10.7 (a) and 10.8 (a)) show that both of these images have completely different textures. However, the approximate wavelet coefficients of both images obtained using a decomposition level of four look very interesting. Figure 10.7(e) and Figure 10.8(e) show that the approximate wavelet coefficients of both these images at the wavelet decomposition level 4 are similar, thus indicating that both textures are similar. This can be verified from the 2D-FFT of the wavelet coefficients at this level. Hence, both texture images are likely to classify as the same texture. In order to discriminate between these two textures using the approximate coefficients, a decomposition level of 2 seems to be an appropriate trade-off between the frequency resolution and spatial resolution. This selection is based on the fact that the frequency spectrum of the image is contained in $\pi/4 = \pi/2^2$ where $n = 2$ is the number of decomposition levels.

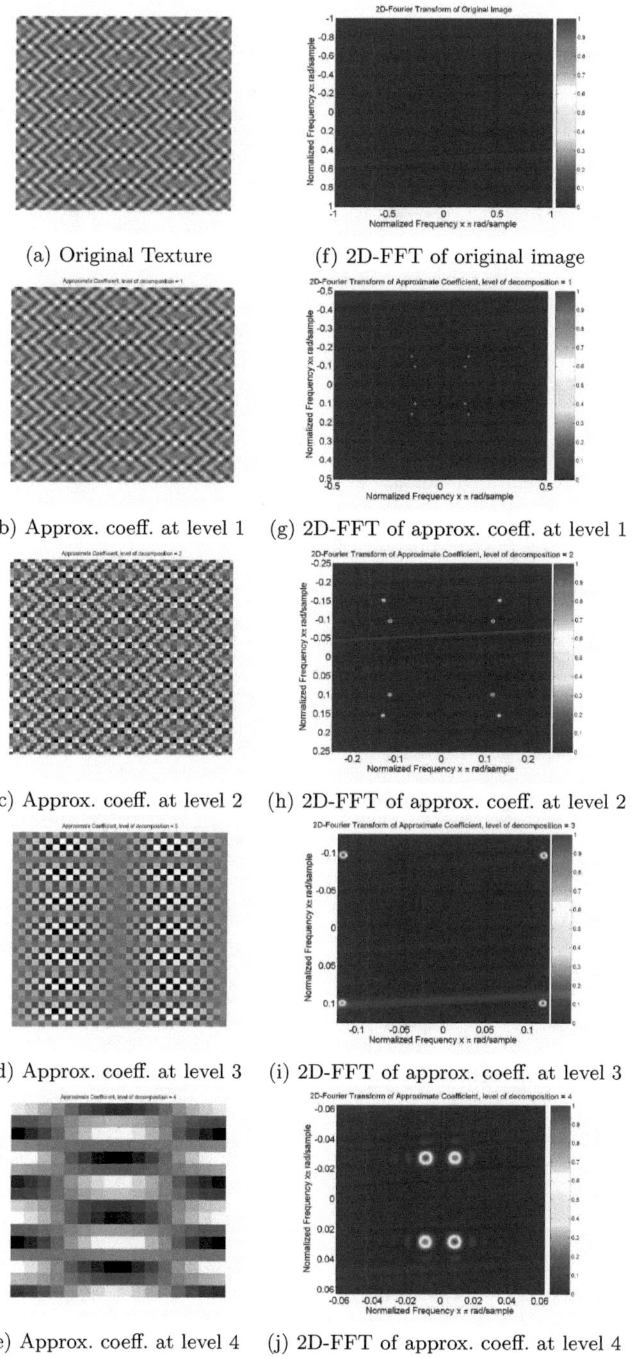

(a) Original Texture (f) 2D-FFT of original image

(b) Approx. coeff. at level 1 (g) 2D-FFT of approx. coeff. at level 1

(c) Approx. coeff. at level 2 (h) 2D-FFT of approx. coeff. at level 2

(d) Approx. coeff. at level 3 (i) 2D-FFT of approx. coeff. at level 3

(e) Approx. coeff. at level 4 (j) 2D-FFT of approx. coeff. at level 4

Fig. 10.7. Texture image with cosines of Fx = 50 & 78 Hz, Fy = 60 & 68 Hz. See text for explanation.

(a) Original Texture

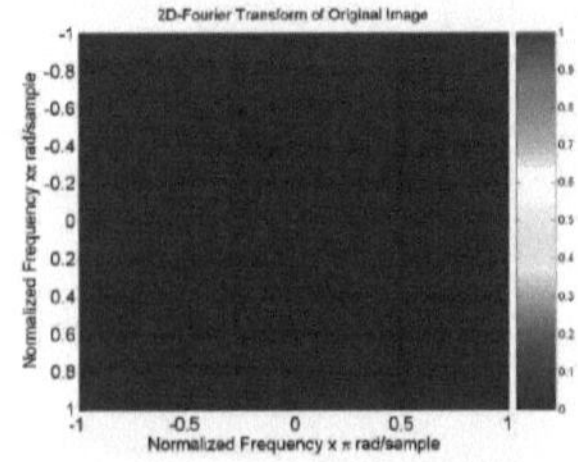

(f) 2D-FFT of original image

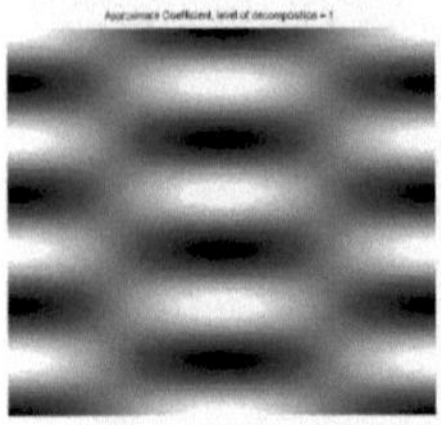

(b) Approx. coeff. at level 1

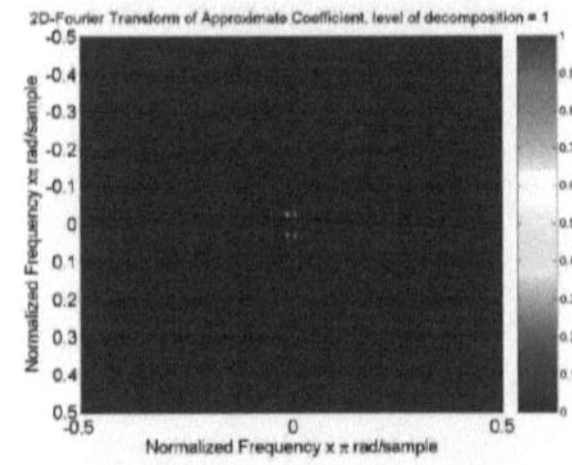

(g) 2D-FFT of approx. coeff. at level 1

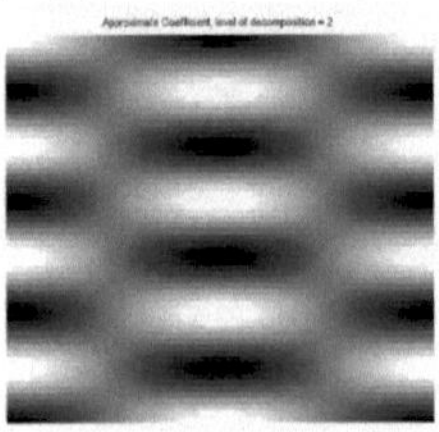

(c) Approx. coeff. at level 2

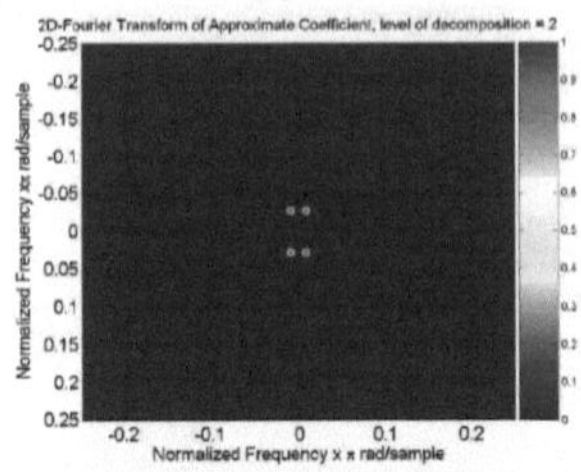

(h) 2D-FFT of approx. coeff. at level 2

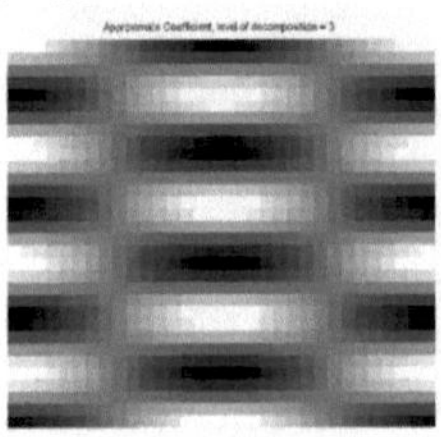

(d) Approx. coeff. at level 3

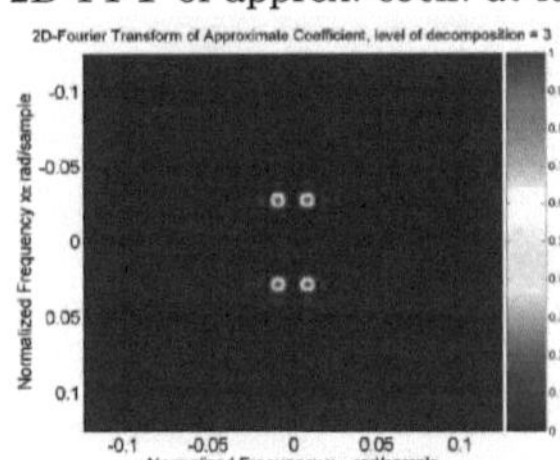

(i) 2D-FFT of approx. coeff. at level 3

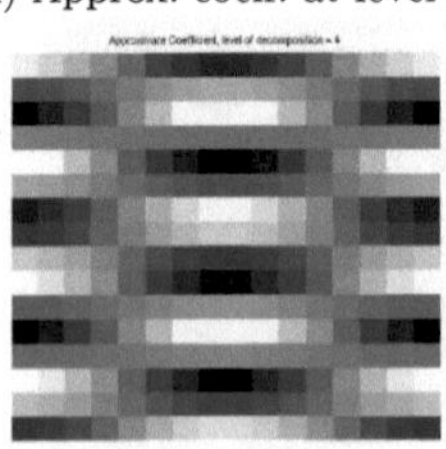

(e) Approx. coeff. at level 4

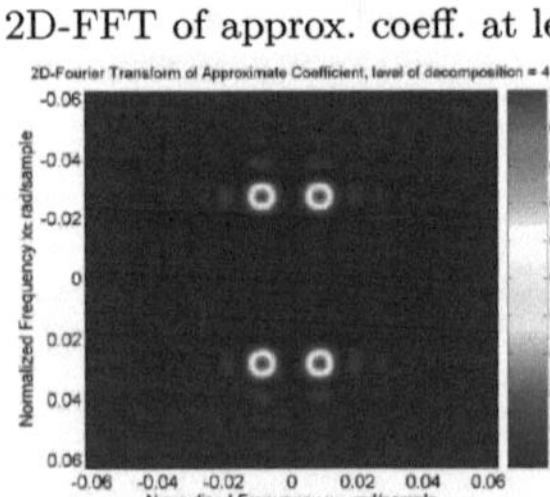

(j) 2D-FFT of approx. coeff. at level 4

Fig. 10.8. Texture image with cosine of Fx = 14 Hz, Fy = 4 Hz. See text for explanation.

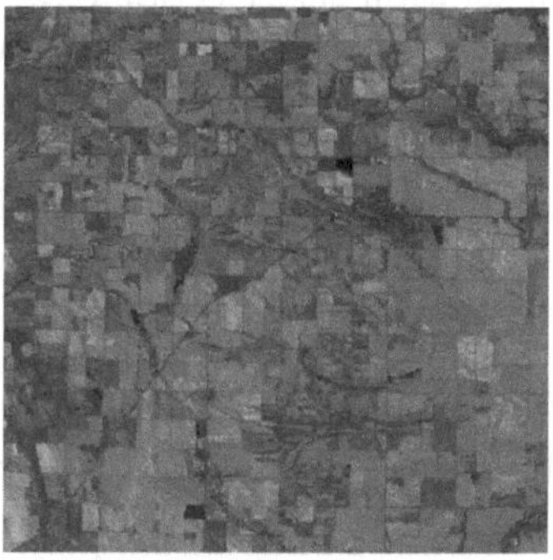

(a) Original Image

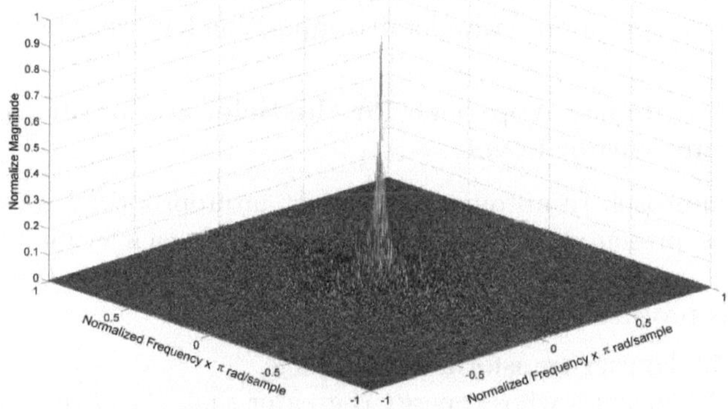

(b) 2D-FFT of the Original Image

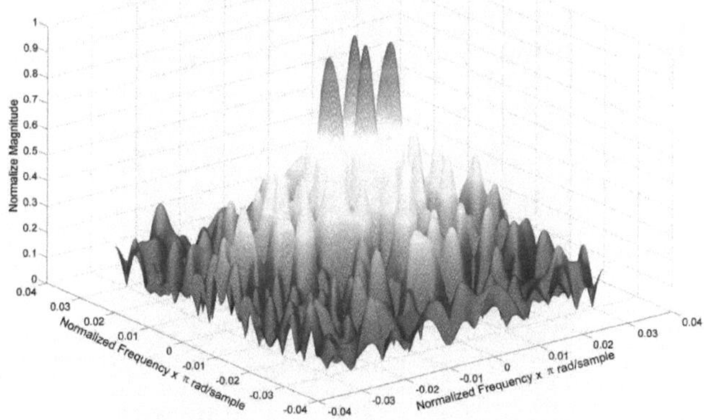

(c) 2D-FFT of the approximate wavelet coefficient at the 5th level of decomposition

Fig. 10.9. Landsat 7 ETM+ image consisting of agricultural land, fallow land, and forest

This study provides an insight on the selection of a good decomposition level for the wavelet transform especially for image mining and suggests that the determination is based on the frequency content of the image. The selected wavelet decomposition level might have the least aliasing effects in the approximate coefficients from the high frequency components. Figure 10.9 shows a sample Landsat 7 ETM+ scene (the value component from the HSV image), its 2D-FFT, and 2D-FFT of the approximate coefficients at fifth level of decomposition.

Note that the major frequency contents of the image, shown in Figure 10.9(b), are within the 0 to $\pi/32 = \pi/2^5$ radian range. Hence, a wavelet decomposition level 5 might provide good results compared with other levels of decomposition, as the aliasing of the high frequency components to the lower frequency components is least, and would have a good frequency resolution compared to the wavelet decomposition level of 1 through 4. However, such manual selection for each individual image in the database is practically impossible and is not quantitative. Hence, based on this study, the next section develops an automatic approach for the selection of an appropriate wavelet decomposition level selection.

10.6.2 An Automatic Approach for the Selection of an Appropriate Decomposition Level

A systematic approach to automatically select an appropriate wavelet decomposition level is presented in this section. This approach is mainly based on quantifying the energy in each frequency sub-band of Figure 10.10. The steps involved in this process are described as below:

- Calculate the Fourier transform of the image.
- Retain the frequencies whose energy is greater than 1% of the main peak, as the frequency components of the image.

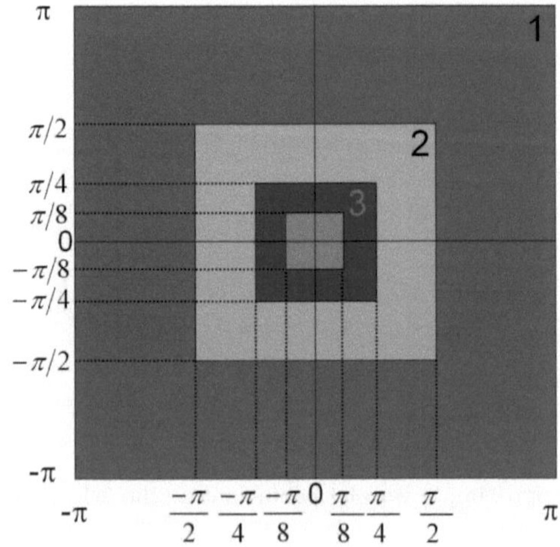

Fig. 10.10. Frequency spectrum for different wavelet decomposition levels

- Calculate the total energy E of the spectrum.
- Calculate the total energy E_i in sub-band (i), over the region of $\pm(\pi \text{ to } \pi/2^i)$ with $i = 1, 2, , N$.
- If $E_i/E < threshold$, increase i and repeat the previous step, else return $(i-1)$ as the appropriate level of decomposition for the image. The threshold value is typically chosen to be close to 0 to guarantee that there is no loss in the frequency components of an image, i.e. higher threshold values will disregard low magnitude frequency components as being significant to the image content.

10.7 Experiments and Results

10.7.1 Experiment 1 - Estimating Wavelet Decomposition Level for an Artificial Texture

To test the efficacy of the presented method in approximating the decomposition level based on image texture content, a 512 x 512 artificial image is generated. This artificial image consists mainly of two different textures, each 256 x 256 pixels, which are concatenated together to form a single image. The first texture is made up of two cosines with frequencies of 50 Hz & 78 Hz in the x-direction and 60 Hz & 68 Hz in the y-direction using a sampling frequency of 1024 Hz. The second texture consists of cosines with frequencies of 14 Hz in the x-direction and 4 Hz in the y-direction and is generated with the same sampling rate. Figure 10.11 illustrates the frequency content of this imagery for decomposition levels of 2 and 3 respectively using the Haar and Biorthogonal9.7 mother wavelets. A visual inspection of Figure 10.11 reveals that an appropriate decomposition level would be 2 since aliasing occurs at a higher decomposition level. When the presented method is applied to the synthetic imagery to systematically perform the selection, an optimum wavelet decomposition level of 2 is obtained for a threshold value of 0.01, as illustrated in Figure 10.12. Note that since the decomposition level is selected based on the frequency content of the image (see Figure 10.11), the selection process is independent of the choice of mother wavelet (kernel filter) used for feature extraction.

10.7.2 Experiment 2 - Varying Wavelet Decomposition Level for Images with Different Spatial Resolution

This experiment illustrates the concept of systematically selecting an appropriate decomposition level based on geospatial imagery with different resolutions. A 512 x 512 pixels Landsat7 ETM+ image with a spatial resolution of 28.5 m is considered and synthetic images of different spatial resolutions (42 m, 57m, 85m, 114m and 228 m) are then generated from the sample image. Figure 10.13 illustrates that for images with resolutions of 28.5 m, 42 m, 57 m, 114 m, and 228 m, an appropriate choice, based on the presented method, for a wavelet decomposition level is 3, 2, 2, 1, and 0 respectively. The zero decomposition

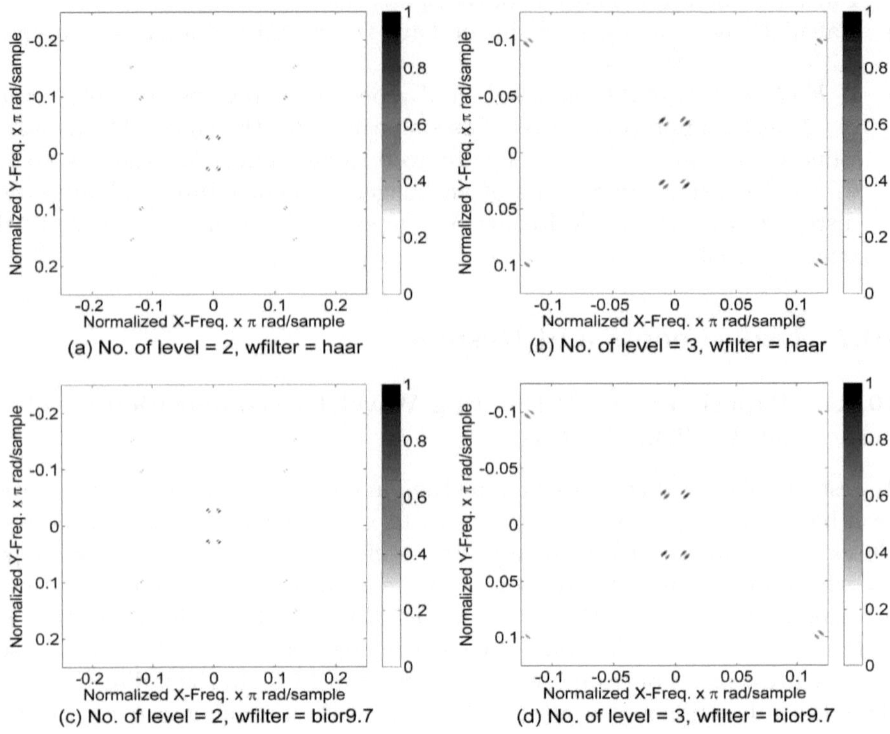

Fig. 10.11. Frequency representation of the approximate wavelet coefficients

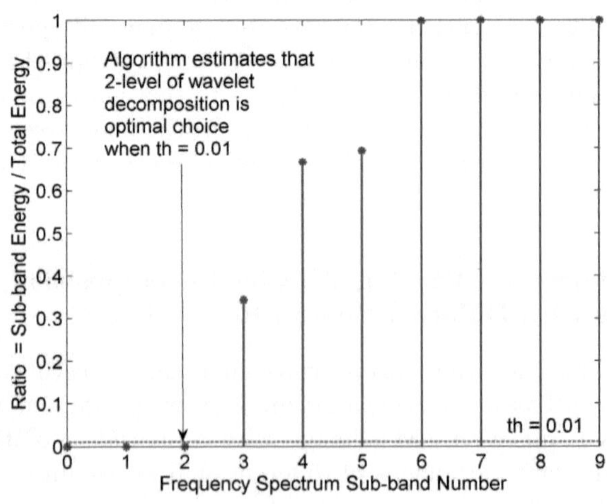

Fig. 10.12. Optimal wavelet decomposition level selection for an artificial sample image

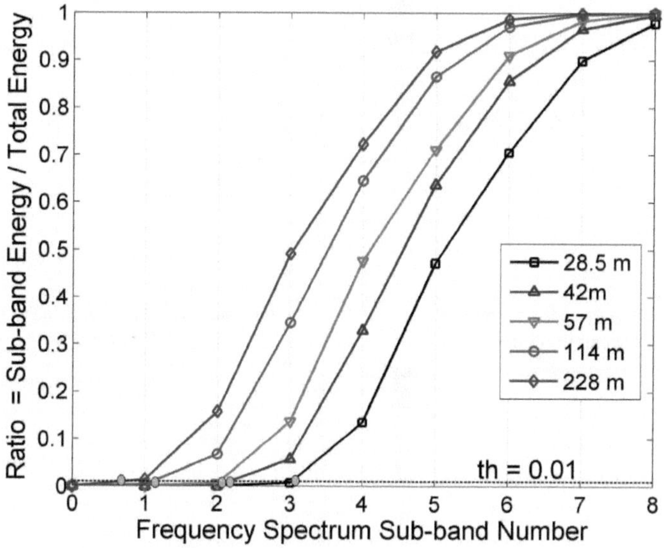

Fig. 10.13. Wavelet decomposition level selection for a Landsat7 ETM+ imagery with different resolutions

level of the 228 m resolution imagery indicates that the wavelet transform would not be a good choice for primitive feature extraction to perform fast coarse segmentation.

10.7.3 Experiment 3 - Overall System Performance Using the Adaptive Approach to Wavelet Decomposition Level Selection

The image archive used in this study consists of 200 false color Landsat 7 ETM+ scenes. Each image is 512 x 512 pixels. The water-body is present in 118 images, agricultural land in 170 images, forest in 183 images, and fallow land in all the images. In false color images, water-body is generally dark color objects with a smooth texture, agricultural land with healthy vegetation is dark pinkish-red with a smooth texture, forest is dark red with a coarse texture, and fallow land is yellowish in color with a coarse texture. Figure 10.14 illustrates the frequency content of the four different classes. It can be seen that the forest has two major low frequency components, while other classes have one low frequency component. Furthermore, since forest has a coarse texture, there are frequency component activities around the main peaks. For the image archive considered in this study, the selection of a wavelet decomposition level is performed based on the method of section 10.6.2. Figure 10.15 illustrates the distribution of the image archive for different decomposition levels. This figure shows that the appropriate selection of the wavelet decomposition level depends on the threshold selection.

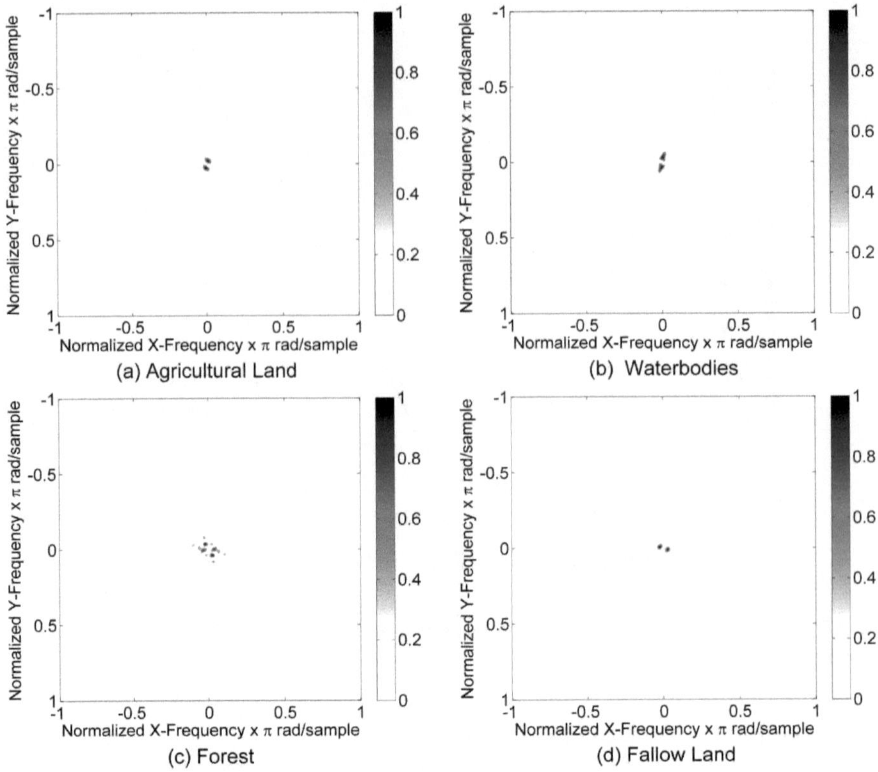

Fig. 10.14. Frequency components for different type of classes

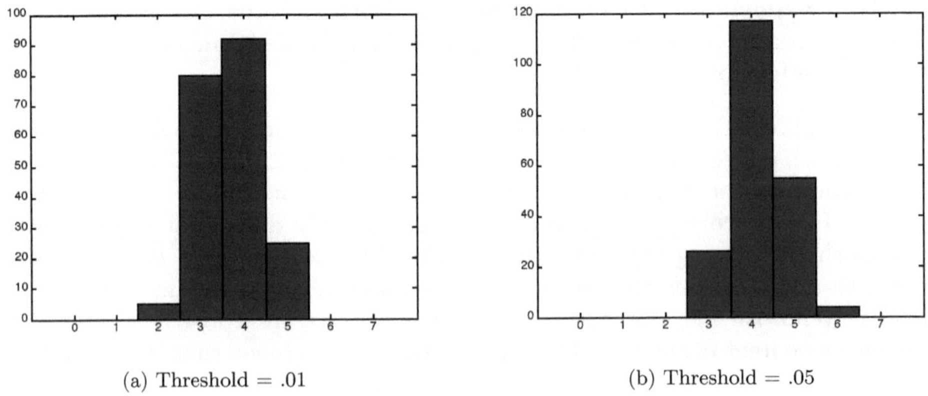

Fig. 10.15. Image archive distribution

For instance, for a threshold of 0.01 (Figure 10.15(a)), a wavelet decomposition level range from 2 to 5 is observed, with most of the images using a wavelet decomposition of level 3 or 4. Increasing the threshold to 0.05 (Figure 10.15(b)),

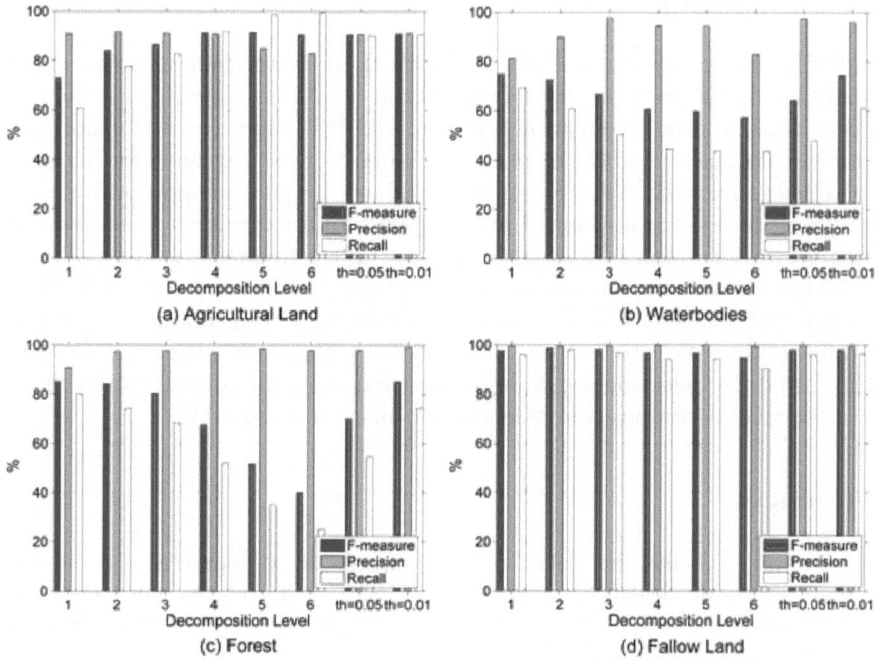

Fig. 10.16. Performance retrieval using different levels of decomposition

a decomposition level range from 3 to 5 is observed, with most of the images using decomposition level of 4.

The performance of the system is examined in terms of the F-measure by adaptively selecting the appropriate decomposition level. Figure 10.16 shows the accuracy results obtained for different classes using different levels of decomposition with the Haar filter. The results obtained with the adaptive selection correspond to threshold values of 0.01 and 0.05 respectively, as indicated in figure 10.16. Note that the F-measure results are, in general, better with the adaptive selection process compared to a fixed level of decomposition for all four classes. Furthermore, these results show that, for the forest class, as the number of decomposition level increases, the recall accuracy decreases, indicating that many of the forest texture details are lost. Also, for higher levels of decomposition, agriculture and forest areas are segmented into one region, which results in lower recall accuracy for the forest class and lower precision accuracy for agriculture. The selection of higher wavelet decomposition levels affected the recall accuracy of water bodies as well. Images having water-body of a very small size is not correctly recalled as the decomposition level increases. Finally, fallow land, which also has different spectral characteristics and a sufficiently larger area, is easily recalled at any decomposition level.

10.8 Summary

The I^3KR system using a semantic frame-work for image-mining uses computationally expensive methods for feature extraction. This chapter introduces the concept of using primitive feature obtained via HSV and wavelet transformation for image information mining from Earth observation data archives for the semantic enabled framework. A systematic approach to selecting an appropriate wavelet decomposition level is developed for image information mining from geospatial data archive. The selection procedure is based on the frequency content of the image. Three experiments (Synthetic imagery, Landsat7 imagery with different resolutions, and Landsat7 ETM+ imagery archive) have been conducted to validate the applicability of the technique. In all cases, it is shown that the results obtained with an adaptive selection are more robust than using the general approach of selecting wavelet decomposition of level 3 for image segmentation.

References

1. Datcu, M., Daschiel, H., Pelizzari, A., Quartulli, M., Galoppo, A., Colapicchioni, A., Pastori, M., Seidel, K., Machetti, P.G., D'Elia, S.: Information mining in remote sensing image archives: system concepts. IEEE Trans. On Geoscience and Remote 41(12), 2922–2936 (2003)
2. Durbha, S.S., King, R.L.: Knowledge Mining in Earth Observation Data Archives: A Domain Ontology Perspective. In: IEEE Int'l Conf. Geoscience and Remote Sensing Symposium, September 2004, vol. 1, pp. 172–173 (2004)
3. Shah, V.P., Younan, N.H., Durbha, S.S., King, R.L.: Wavelet Features for Information Mining in Remote Sensing Imagery. In: IEEE Int'l Conf. Geoscience and Remote Sensing Symposium, July 2005, vol. 8, pp. 5630–5633 (2005)
4. Durbha, S.S., King, R.L.: Semantics-enabled framework for knowledge discovery from Earth observation data archives. IEEE Transactions on Geosciences and Remote Sensing 43(11), 2563–2572 (2005)
5. Pass, G., Zabih, R.: Histogram refinement for content-based image. In: Proc. of the IEEE workshop on Applications of Computer Vision, Sarasota, Florida (December 1996)
6. Flickner, M., et al.: Query by image and Video content: The QBIC system. IEEE Computer Society 28(9), 23–32 (1995)
7. Pentland, A., Picard, R.W., Sclaroff, S.: Photobook: Content-Based Manipulation for Image Databases. Int'l J. Comp. Vision 18(3), 233–254 (1996)
8. Gupta, A., Jain, R.: Visual Information Reterival. Comm. Journal of ACM 40(5), 70–79 (1997)
9. Carson, C., Belongie, S., Greenspan, H., Malik, J.: Blobworld: Image Segmentation Using Expectation - Maximization and its Application to Image Querying. IEEE Trans. Pattern Anal. Machine Intell. 24(8), 1026–1038 (2002)
10. Ma, W.Y., Manjunath, B.: NeTra: A Toolbox for Navigating Large Image Databases. In: Proc. IEEE Int'l Conf. Image Processing, pp. 568–571 (1997)
11. Smith, J.R., Chang, S.F.: VISUALSEEK: A Fully Automated Content Based Query System. In: Proc. 4th ACM Int'l Conf. On Multimedia, pp. 87–98 (1996)

12. Tilton, J.C., Marchisio, G., Datcu, M.: Knowledge Discovery and Data Mining Based on Hierarchical Segmentation of Image Data, a research proposal, in response to NRA2-37143 from NASA's Information System Program (October 23, 2000) (submitted)
13. Open Geospatial Consortium (OGC) (2004), http://www.opengeospatial.org/ (last visited May 6th, 2006)
14. Smith, M., Chang, S.F.: Quad-tree segmentation for texture-based image query. In: ACM multimedia, pp. 279–286 (1994)
15. Remias, E., Sheikholeslami, G., Zhang, A., Syeda-Mahmood, T.F.: Supporting content-based retrieval in large image database systems. Multimedia Tools and Applications 4(2), 153–170 (1997)
16. Natsev, A., Rastogi, R., Shim, K.: WALRUS: a similarity retrieval algorithm for image databases. In: Proceedings of the 1999 ACM SIGMOD International Conference on Management of Data, pp. 395–406. ACM Press, New York (1999)
17. Ardizzoni, S., Bartolini, I., Patella, M.: Wind-surf: Region-based image retrieval using wavelets. In: Proceedings of the 10th Int'l Workshop on Database & Expert Systems Applications (1999)
18. Sun, Y., Ozawa, S.: Efficient Wavelet-Based Image Retrieval Using Coarse Segmentation and Fine Region Feature Extraction. IEICE Trans. Inf. & Syst. E88-D(5) (May 2005)
19. Suematsu, N., Ishida, Y., Hayashi, A., Kanbara, T.: Region-Based Image Retrieval using Wavelet Transform. In: 15th Int. Conf. on Vision Interface, Calgary (Canada), pp. 9–16 (2002)
20. Zhang, T., Ramakrishnan, R., Livny, M.: BIRCH: A new data clustering algorithm and its applications. Data Mining and Knowledge Discovery 1(2), 141–182 (1997)
21. Mallat, S.G.: A Theory of multiresolution signal decomposition: The Wavelet representation. IEEE Trans. Patt. Anal Machine Intell. 11(7), 674–693 (1989)
22. HSV color space, http://en.wikipedia.org/wiki/HSV_color_space
23. Fukuda, S., Hirosawa, H.: A Wavelet-Based Texture Feature Set Applied to Classification of Multifrequency Polarimetric SAR Images. IEEE Trans. on Geoscience and Remote sensing 37(5), 2282–2286 (1999)
24. Smith, J.R.: Integrated Spatial and Feature Image Systems: Retrieval, Analysis, and Compression. Phd Thesis, Columbia University (1997)
25. Chang, T., Kuo, C.-C.J.: Texture analysis and classification with tree-structured wavelet transform. IEEE Trans. On Image Proc. 2(4), 429–440 (1993)
26. Van de Wouver, G., Scheunders, P., Livens, S., Van Dyck, D.: Color Texture Classification by Wavelet Energy-Correlation Signatures. In: Proc. ICIAP Intern. Conf. On Computer Analysis and Image Processing, Florence, Italy, September 17-19, 1997, pp. 327–334 (1997)
27. Fukunaga, K.: Introduction to Statistical Pattern Recognition, 2nd edn. Elsevier Publisher, Amsterdam (1990)
28. Vapnik, V.: The Nature of Statistical Learning Theory. Springer, New York (1995)
29. Burges, C.J.C., Schšlkopf, B.: Improving the accuracy and speed of support vector learning machine. Advances in neural information processing systems, pp. 375–381. MIT Press, Cambridge (1997)
30. Blanz, V., Schšlkopf, B., Bulthoff, H., Burges, C., Vapnik, V., Vetter, T.: Comparison of view-based object recognition algorithms using realistic 3d models. In: Vorbrüggen, J.C., von Seelen, W., Sendhoff, B. (eds.) ICANN 1996. LNCS, vol. 1112, pp. 251–256. Springer, Heidelberg (1996)

31. Osuna, E., Freund, R., Girosi, F.: ÔTraining support vector machines: an application to face detection. In: Proc. Computer vision and pattern recognition, pp. 130–136 (1997)
32. Shawe-Taylor, J., Cristianini, N.: Kernel Methods for Pattern Analysis, 1st edn. Cambridge University Press, Cambridge (2004)
33. Chang, C.-C., Lin, C.-J.: LIBSVM: a library for support vector machines, `http://www.csie.ntu.edu.tw/~cjlin/libsvm/`
34. Baeza-Yates, R., Ribeiro-Neto, B.: Modern Information Retrieval. Addison-Wesley-Longman, Harlow (1999)
35. Burrus, C.S., Gopinath, R.A., Guo, H.: Introduction to Wavelets and Wavelet Transforms - A primer. Prentice Hall, Englewood Cliffs (1998)

11

Neural Networks for Land Cover Applications

Fabio Pacifici[1], Fabio Del Frate[1], Chiara Solimini[1], and William J. Emery[2]

[1] Department of Information, Systems and Productions - Tor Vergata University
f.pacifici@disp.uniroma2.it
[2] Aerospace Department - University of Colorado at Boulder

11.1 Neural Networks in Remote Sensing

Following upon their resurgence as a research topic in the second half of the eighties [1][2], neural networks burst out in the remote sensing community by the end of the same decade when a bunch of investigators started looking at them as an effective alternative to more traditional methods for extracting information from data collected by airborne and spaceborne platforms [3] [4]. Indeed, for researchers used to handle multidimensional data, the possibility of building classification algorithms avoiding the assumptions required by the Bayesian methods on the data probability distributions seemed rather attractive. For the case of parameter retrieval, neural networks owned the advantage of determining the input-output relationship directly from the training data with no need to seek for an explicit modelling of the physical mechanisms, often nonlinear and with poorly understood factors involved [5]. Moreover, it was shown that multilayer feedforward networks formed a class of universal approximators, capable of approximating any real-valued continuous function provided a sufficient number of units in the hidden layer was considered [6]. There were enough reasons to explore the potentialities of such neurocomputational models, also known as associative memory based models, in a wide range of remote sensing applications. This was actually what happened throughout all nineties and, even if with less emphasis, is continuing up to date. Image classification, from SAR imagery to latest hyperspectral data, has probably been one of the most investigated fields in this context. However, the use of neural network for the retrieval of bio-geophysical parameters from remotely sensed data has also played a significant role. In particular, the synergy between these algorithms and RT (Radiative Transfer) electromagnetic models represented a new and effective way to replace the empirical approaches, often based on limited seasonally or regionally data, hence with little generalization capabilities. The use of the RT models is twofold. They can help in the sensitivity analysis of the electromagnetic quantities to the bio-geophysical parameters. They can be exploited for the synthetic generation of training data in case of a lack of experimental data and ground-truth. The combined use of electromagnetic models and neural networks is a topic treated in many published studies taking into account the most diverse applicative scenarios: from the inversion of radiance spectra to infer atmospheric ozone concentration profiles [7], to the retrieval of snow parameters from passive microwave

M. Graña and R.J. Duro (Eds.): Comput. Intel. for Remote Sensing, SCI 133, pp. 267–293, 2008.
springerlink.com

measurements [8] or to the estimation of sea water optically active parameters from hyperspectral data [9], just to mention one case for the three main different domains: atmosphere, earth surface and ocean. As far as the network topology is concerned, MLP probably are on the top of the authors preference ranking, but the choice is widely spread over the numerous alternatives which include RBF (Radial Basis Function), recurrent architectures, as well as the Kohonen SOM [10] if we focus on the unsupervised image classification field. Such an extensive research activity confirmed that neural networks, besides representing a new and easy way of machine learning, possess particularly interesting properties, such as the capability of capturing subtle dependencies among the data, an inherent fault tolerance due to their parallel and distributed structure, a capability of positively merging pieces of information stemming from different sources. Only advantages? Obviously not. The conventional backpropagation learning algorithm can be stuck in a local minimum. Moreover, the choice of network architecture (i.e. number of hidden layers and nodes in each layer), weight initialization and number of iterations required for training may significantly affect the learning performance. Addressing these issues is one of the dominant themes of the current research. Some authors [11] suggest that SVM (Support Vector Machines) may reach in terms of accuracy performances similar to those obtainable by means of neural networks without suffering from the problem of local minima and with limited effort required for architecture design. Other authors [12] remain focused on neural models, aiming at automating the selection of the parameters characterizing the algorithm or searching for improvements in the learning procedures. A different ongoing line of research is more sensitive to the opportunities that should be provided by the data of the last generation EO (Earth Observation) missions. Hence it is more dedicated to exploit the aforementioned neural networks potentialities for more complex applications [13][14]. A couple of interesting examples belonging to this area of research will be illustrated in section 11.3 of this chapter, while section 11.2 recalls some theoretical aspects about neural models.

11.2 Neural Networks Models

An artificial Neural Network (NN) may be viewed as a mathematical model composed of non-linear computational elements, named neurons, operating in parallel and connected by links characterized by different weights. A single neuron computes the sum of its inputs, adds a bias term and drives the result through a general non-linear activation function to produce a single output. NNs models are mainly specified by:

- neuron structure
- net topology
- training or learning rules

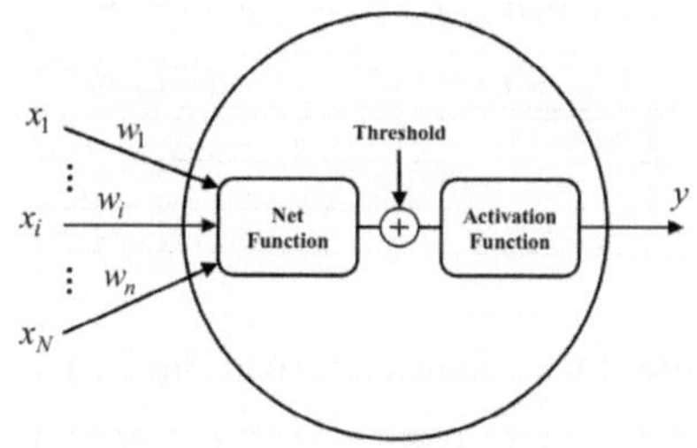

Fig. 11.1. Neuron model

11.2.1 Neuron Structure

The basic building block of a NN is the neuron, which is inspired by principles and characteristics observed in the human brain. As described by many models over the years [15]-[19], a single neuron is an information processing unit generally characterized by several inputs and one output as shown in Figure 11.1. Each neuron consists of three main parts:

- the weight vector
- the net function
- the activation function

The input vector is combined with the weight vector through the net function. Then, the activation function is applied to the output of the net function modified with an additional bias, producing the neuron output.

An example of a commonly used net function is the weighted linear combination of inputs such as:

$$y = \sum_{i=1}^{N} x_i w_i + \theta \tag{11.1}$$

where y denotes the net function output, $x_1, x_2, ..., x_N$ and $w_1, w_2, ..., w_N$ are the components of the neuron input and synaptic weight vectors, respectively, while θ is called bias or threshold. The biological interpretation of this threshold is to inhibit the firing of the neuron when the cumulative effect of the inputs does not exceed the value of θ. Many other net functions have been proposed in the literature, which vary with the combination of inputs. A brief list of net functions is shown in Table 11.1.

The neuron output is produced by the activation function related to the net function output through a linear or non-linear transformation. In general, these activation functions are characterized by saturation at a minimum and a maximum value and by being non-decreasing functions. The fundamental types of activation functions are shown in Table 11.2.

Table 11.1. Net functions

Net Functions	Formula
Linear	$y = \sum_{i=1}^{N} x_i w_i + \theta$
Higher order (2nd order)	$y = \sum_{i=1}^{N} \sum_{j=1}^{N} x_i x_j w_{ij} + \theta$
Delta ($\sum - \prod$)	$y = \prod_{i=1}^{N} x_i w_i$

Table 11.2. Neuron activations functions

Activation Functions	Neuron Output
Sigmoid	$f(x) = \frac{1}{1+e^{-x}}$
Hyperbolic tangent	$f(x) = tanh \frac{x}{T}$
Inverse tangent	$f(x) = \frac{2}{\pi} tan^{-1} \frac{x}{T}$
Threshold	$f(x) = \begin{cases} 1 & x > 0 \\ -1 & x < 0 \end{cases}$
Gaussian radial basis	$f(x) = exp[-\|\|x - m\|\|^2/\sigma]$
Linear	$f(x) = ax + b$

11.2.2 Network Topology

Although a single neuron can resolve simple information processing functions, the advantage of the neural computation approach comes from connecting neurons in networks, where several units are interconnected to form a distributed architecture. In general, the way in which the units are linked together is related to the learning algorithm used to train the network: neurons can be fully connected, which means that every neuron is connected to every other one, or partially connected, where only particular connections between units in different layers are allowed.

The feedforward and the feedback architectures are the two types of approaches that can be distinguished, depending on the absence or presence of feedback connections in the network. In a feedforward architecture there are no connections back from the output. In particular, this network does not remember its previous output values: the Multi Layer Perceptrons (MLP) is one of the most popular feedforward models among all the existing paradigms. In a feedback architecture there are connections from output to input neurons. Distinct from the feedforward system, such a network remembers its previous states and the actual state does not depend only on the input signals: the Elman network [20] is of this type. The MLP and the Elman networks are briefly recalled in the next two subsections.

Multilayer Perceptron

Due to its high generalization ability, the MLP is the most widely used neural network for solving decision-making problems for many different applications. An MLP is a more complex version of the original perceptron model proposed in the early 50's which consists of a single neuron that utilizes a linear net function and a threshold activation neuron function, respectively (see Table 11.1 and 11.2). To overcome the linear separability limitation of the perceptron model, the MLP neural networks were designed to have continuous value inputs and outputs, and nonlinear activation functions. In fact, in many practical applications, such as pattern recognition, the major problem of the simple perceptron model is related to the learning algorithm, which converges only for linearly separable samples. An MLP approximates an unknown input-output relationship providing a nonlinear mapping between its input and output. The MLP architecture is a feedforward type, therefore there are no connections back from the output to the input neurons and each neuron of a layer is connected to all neurons of the successive layer without feedback to neurons in the previous layer. The architecture consists of different layers of neurons. The first layer merely distributes the inputs to the internal stages of the network: there is no processing at this level. The last layer is the output which provides the data processed. The layers between the input and the output are called hidden layers. The number of neurons which compose the input and output layers are directly related to the dimension of the input space and to the dimension of desired output space, respectively. Even if the MLP is one of the most used neural network models, there are several problems in designing the whole topology of the network since it is necessary to decide the number of units in the hidden layers, which in general may have any arbitrary number of neurons. Therefore, the network topology providing the optimal performance should be selected to achieve the desired behavior. If the number of neurons is too small, the input-output associative capabilities of the net are too weak. On the other hand, this number should not be too large; otherwise, these capabilities can show a lack of generality being tailored too much to the training set, and the computational complexity of the algorithm would be increased in vain. In Figure 11.2 is illustrated the case of a two hidden layers network.

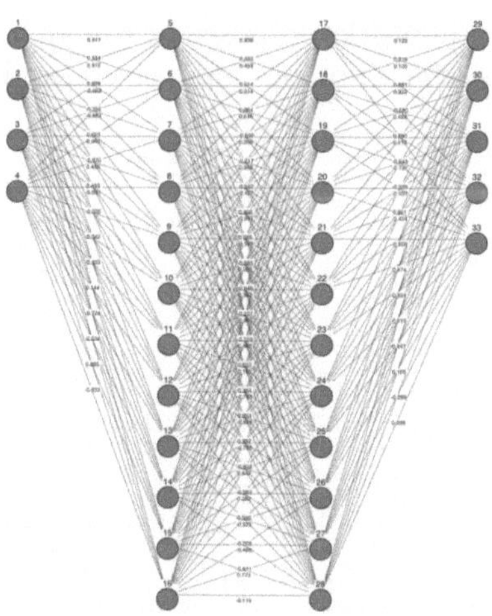

Fig. 11.2. Two hidden layers network topology. Each circle represents a single neuron.

Elman Neural Network

The MLP model discussed in the previous section provides an output that strictly depends on the actual vector input to the network. In this sense, the MLP model has no memory since it does not take into account the dependence of an individual input on the others processed previously. In fact, the capability of exploiting the information embedded in multitemporal measurements may be remarkable for a classification algorithm, since time series data provide significant contributions to understanding processes in engineering applications. This is especially true in many remote sensing cases, where it is often necessary to deal with objects that continuously evolve their state following precise dynamical conditions. For example, the backscattering of an agricultural field changes during the year due to the phenological cycle of the plants which covers sprouting, growth and the decay or harvest of the vegetation. As a consequence, little difference can be observed between samples belonging to the same scene and taken at short time distance, while inter-annual variations may be large. Recurrent networks can be used when the temporal variability of the inputs contains useful information since they have feedback connections from the neurons in one layer to neurons in a previous layer. The successive state of the network depends not only on the actual input vectors and connection weights, but also on the previous states of the network. Depending on the architecture of the feedback connections, there are two general models [18]:

- fully recurrent neural networks, in which any node is connected to any other node
- partially recurrent neural networks, in which only specific feedback connections to the input layer are established

The partially recurrent networks proposed by Elman [21] at the beginning of the 90's for string recognition consist of four main layers, as illustrated in Figure 11.3:

- input (I)
- hidden (H)
- output (O)
- context (C)

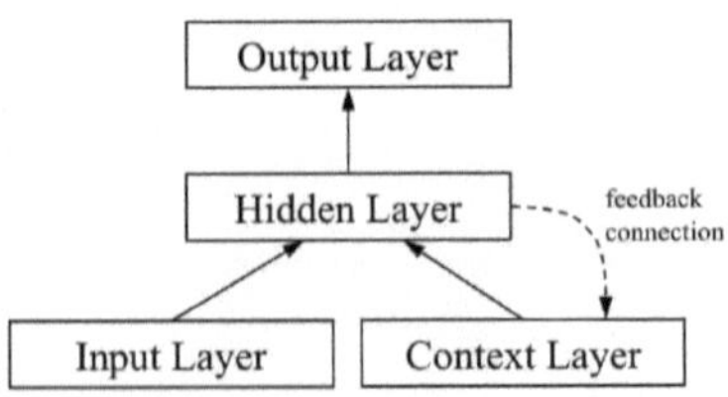

Fig. 11.3. Architecture of the Elman NN. Context units can be considered as additional input units.

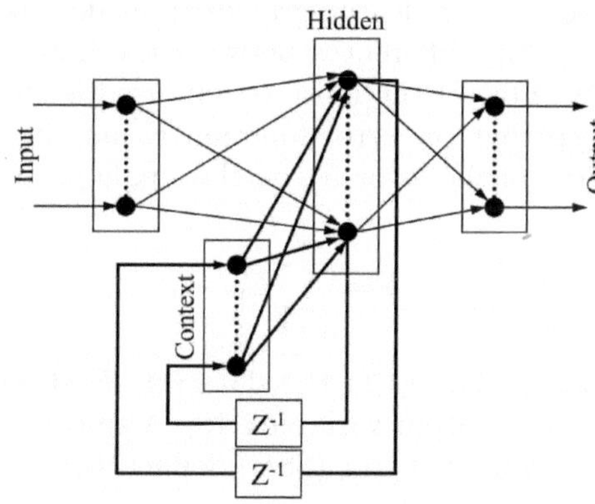

Fig. 11.4. Feedback connections between layers in the Elman architecture (Z^{-1} is the unit delay)

The weights assigned to the feedback connections are not all adjustable, in fact:

- hidden neurons are feedback connected to the context units with fixed weights (dashed line in Figure 11.3)
- feedforward connections (solid line in Figure 11.3) require training

The recurrent structure of the Elman NNs is better detailed in Figure 11.4. As discussed previously, there are recurrent connections from the hidden to the context layers by means of the unit delay. Furthermore, the context neurons are fully forward connected with all hidden units. Since the context layer consists of supplementary input units whose values are feedbacks from the hidden layer, this architecture is considered as a more general feedforward neural network with additional memory units and local feedback.

11.2.3 Learning Algorithms

In the most common network architectures, the error minimization is pursued by an iterative non-linear optimization algorithm. The performance of the learning algorithms is related to their abilities to converge in a relatively small number of iterations to a minimum of the global error function which depends on the network weights. Many of these optimization methods are based on the gradient descent algorithm which requires that the activation function is continuously differentiable.

Back Propagation

During the training phase, the network learns how to approximate an unknown input-output relation by adjusting the weight connections. This is done by receiving both the inputs and the desired outputs which are used to minimize an error function. There are several learning algorithms designed to minimize this error function related to the differences between the inputs $x_1, ..., x_N$ and the target outputs $t_1, ..., t_N$. One of the most widely used is the Back Propagation algorithm. It is assumed that the error function E may be written as the sum over all training patterns of the error E_p on the single pattern:

$$E = \sum_{p=1}^{N} E_p \tag{11.2}$$

where N is the number of patters in the data set. E should be differentiable with respect to the network output variables [19]. A suitable function with these behaviors is the Sum-of-Squares Error (SSE) defined by:

$$E_p(w) = \frac{1}{2}\sum_{i=1}^{N}[t_i(p) - y_i(p)]^2 \tag{11.3}$$

There are numerous nonlinear optimization algorithms available to minimize the above error function. Basically, they adopt a similar iterative formulation:

$$w(t+1) = w(t) + \Delta w(t) \tag{11.4}$$

where $\Delta w(t)$ is the correction made to the current weights $w(t)$. Many algorithms differ in the form of $\Delta w(t)$, among them, the well-known *gradient descent procedure* changes the weight vector by a quantity proportional to η and in the direction of the negative gradient as in:

$$\Delta w = -\eta \nabla E_p(w) \tag{11.5}$$

where η governs the overall speed of weight adaptation. In general, η should be sufficiently large to avoid oscillations within a local minimum of E_p. At the same time, η should to be sufficiently small so that the minimum does not jump over the correct value.

The backpropagation algorithm uses a step by step procedure [22]. During the first phase the weights are initialized to random, small values and the training patterns are propagated forward through the network. Then, the obtained outputs are compared with the desired values, computing the preliminary sum-of-square cost function for the data applied. During the second phase, after an appropriate value of η has been chosen, a backward pass through the network permits the computation of appropriate weight changes.

With the years, many different algorithms have been proposed in the literature to optimize the performance of the standard backpropagation. Among them, Conjugate Gradient Methods (CGM) play a significant role. The CGM belong to the second order techniques that minimize the error function. Second order indicates that these methods exploit the second derivative of the error function instead of the first derivative. While the standard backpropagation always proceeds down the gradient of the error function, CGM proceed to a direction which is conjugate to the directions of the previous steps. This approximation gives a more complete information about the search direction and step size. In particular, the SCG algorithm [23] adds to CGM a scaling approach to the step size in order to speed up the training. Basically in most applications problems, the error function decreases monotonically towards zero for a increasing number of learning iterations.

Back Propagation Through Time

The behavior of a recurrent network can be generally achieved in a feedforward network at the cost of duplicating its structure an appropriate number of times [1]. For the Elman case, context units can be considered as input units and, therefore, the total network input vector consists of two parts [22]: the first is the actual input vector, which is the only input of the partial recurrent network. The second is the context state vector, which is given through the next-state function in every step.

Many practical methodologies based on the back propagation algorithm have been proposed in the literature for recurrent networks. One popular technique, known as Back Propagation Through Time (BPTT), has been developed to train recurrent networks that are updated in discrete time steps. In fact, the original recurrent network is transformed to a virtual network in discrete time by replication of every processing unit in time as shown in Figure 11.5. The connections are established between a unit at time t to another at time $t+1$. The value of each unit at $t+1$ is function of the state of the units at the previous times. Basically, this is done by running the recurrent network for different time steps and successively unrolling the network in time. Then, the back propagation algorithm is applied to the virtual network and the result is used to update the weights of the original network.

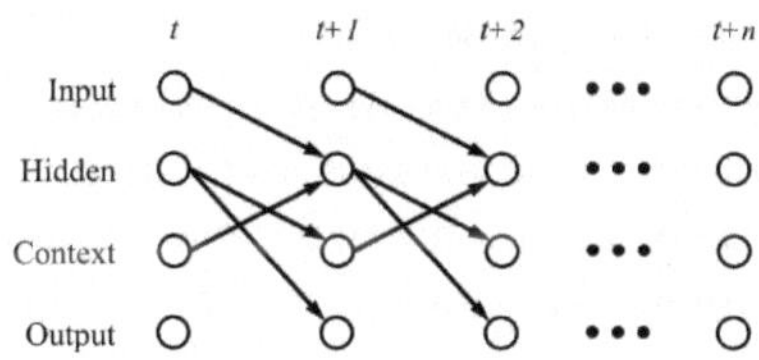

Fig. 11.5. Equivalent discrete time model of the original recurrent network

The standard back propagation algorithm needs to train only the input/output pairs, which is different from the back propagation through time method that needs to memorize the input and output vectors of the whole training data set before the learning phase. The buffer used to memorize the n input/output pairs before the back propagation of the error through the network increases the computation time, which makes the algorithm harder to use for long training sequences. To avoid this problem, it is possible to memorize only a shorter sequence $n' < n$ of the input/output pairs. The appropriate dimension of n' depends mostly on the particular application.

11.3 Application to Land Cover

11.3.1 Change Detection in Very High Resolution Optical Imagery

Introduction

Detection of land cover changes is one of the most interesting aspects of the analysis of multi-temporal remote sensing images. In particular, change detection is very useful in applications such as land use changes analysis, assessment of burned areas, studies of cultivation shifting or assessment of deforestation. Further, the availability of commercial very high resolution optical satellite imagery has enlarged the number of applications especially in urban monitoring where the spatial resolution plays a key role related to the detection of fine-scale objects such as a single houses or small structures. The urban environment represents one of the most challenging areas for the remote sensing community due to its high spatial and spectral diversity: a typical urban surface is mainly composed by a wide range of roofs and roads characterized by different age, quality and composition. Moreover, bare soil and vegetation, which may include trees, plants and parks, make the landscape more complicated. An advantage of the very high spatial resolution imagery, is that different categories of change can be identified when comparing two or more acquisitions of the same scene. They may include, for example, newly built houses, roof variations and widened roads which are relevant for public housing authorities. On the other hand, streets with or without traffic jams, trees with more or less leaves, or parking areas where cars occupy different places can result in a simple modification of an existing object and not a conversion from one object to another. Therefore, all these changes may appear as irrelevant for public institutions. Moreover, technical aspects should be further taken into account. For example, shadowing effects and different off-nadir acquisitions may cause false signals which increase the difficulty of interpretating of changes [24].

In the remote sensing literature, two main approaches to the change detection problem have been proposed: unsupervised and supervised. The former performs change detection by transforming the two separate multispectral images into a single or multi-band image in which the areas of land cover change can be successively detected. The latter is based on supervised classification methods, which require the availability of suitable training sets for the classifier learning process.

Many unsupervised techniques perform change detection using simple procedures to extract the final change map, e.g., by subtracting, on a pixel basis, the two images acquired at different times. More sophisticated techniques, analyzing the difference image using a Markov Random Field (MRF) approach, exploit the interpixel class dependency in the spatial domain by considering the spatial contextual information included in the neighborhood of each pixel [25]. It may also combine the use of a MRF and a Maximum a Posteriori Probability (MAP) decision criterion in order to search for an optimal image of changes [26].

In the category of supervised techniques, the simplest is Post Classification Comparison (PCC) [27]. It performs change detection by comparing the classification maps obtained by independently classifying two remote-sensing images of the same area acquired at different times. In this way, the separate classification of multi-temporal images avoids the need to normalize for atmospheric conditions, sensor differences etc. However, the performance of the PCC technique critically depends on the accuracies of the classification maps. In particular, the final change detection map exhibits an accuracy close to the product of the classification accuracies given at the two times. This is principally due to the fact that PCC does not take into account the dependence existing between two images of the same area acquired at two different times. The supervised Direct Multidata Classification (DMC) is able to partially overcome this problem [28]. In this technique, pixels are characterized by a vector obtained by stacking the single feature related to the images acquired at two times. Then, change detection is performed by considering each change as a single class and by training the classifier to recognize these transitions. Appropriate training sets are required for the success of these supervised methods: the training pixels at the two times should be related to the same points on the ground and should accurately represent the proportions of all the transitions in the entire images. Usually, in real applications, it is difficult to have training sets with such characteristics, but in general, this approach is more flexible than that based on unsupervised classification. In addition, it reduces the effects of different acquisition conditions and allow change detection using different sensors at different times. In this context, neural network algorithms can be successfully applied to change detection problems.

In the following section, we apply a neural network scheme to generate change detection maps of very high spatial resolution optical imagery.

The NAHIRI Change Detection Scheme

The NAHIRI (Neural Architecture for HIgh-Resolution Imagery) technique is a change detection scheme designed to process very high resolution satellite imagery and produce change detection maps [29][30]. The distinctive feature of NAHIRI with respect to the methodologies already proposed in literature is its ability to simultaneously exploit both the multi-spectral and the multi-temporal information contained in the input images. Moreover, it not only detects the different kinds of changes that have occurred, but also explicitly distinguishes, on a pixel basis, the various typologies of class transitions. The change detection is carried out using a neural architecture with different stages. As shown in Figure 11.6, the NAHIRI scheme uses a parallel approach that includes three different neural networks: NN 1 and NN 2 generate the so called *Change Map* by using the multispectral information, while NN 3 produces the *Change Mask* exploiting the multi temporality. Then, the resulting outputs are merged together generating the final change map. This method was designed to use the imagery provided by DigitalGlobe QuickBird (QB) satellite which collects both multi-spectral and panchromatic imagery concurrently. The panchromatic resolution is 0.61 m at nadir to 0.72 m at 25° off-nadir and the multi-spectral is from 2.44 m at nadir to 2.88 m at 25° off-nadir. The panchromatic bandwidth spans

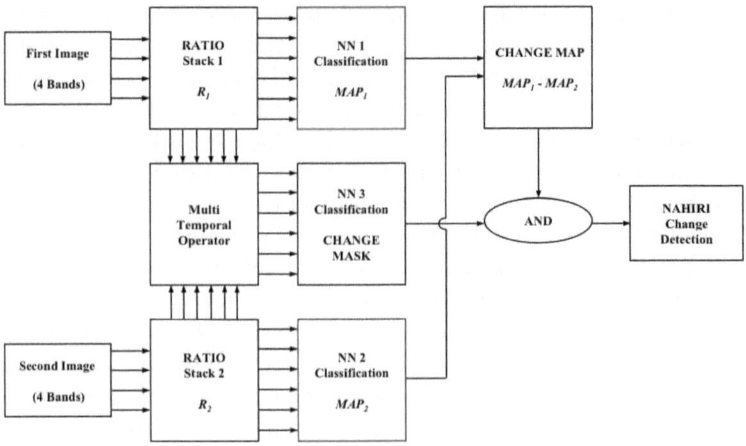

Fig. 11.6. NAHIRI flow chart: each arrow of the input stage represents one of the four multi-spectral bands of the image; in the inner stage, each arrow represents one of the six components of the vector R_i and of the output vector of the multi-temporal operator

from 0.45 μm to 0.90 μm, while the multi-spectral images are acquired in four channels: Blue (0.45 – 0.52 μm), Green (0.52 – 0.60 μm), Red (0.63 – 0.69 μm) and Near-Infrared (NIR) (0.76 – 0.90 μm).

The input vectors of the algorithm contain the reflectance information provided by the different channels of the multi-spectral images. The next step generates of the ratio images R_i which gives two distinct advantages. From one side it reduces the effects of different scene illuminations, atmospheric and ground moisture conditions. In addition, the information provided to the neural classifier is not the simple absolute value of the multi spectral channels, but, for each pixel of the multi spectral QB images, the following input vector is computed:

$$R_i = |B_{iB}/B_{iG}, B_{iB}/B_{iR}, B_{iB}/B_{iNIR}, B_{iG}/B_{iNIR}, B_{iR}/B_{iNIR}|^T \qquad (11.6)$$

whose six components are the non-redundant ratios of the spectral bands. In fact, the same spectral information is carried out by the ratio vector R_i regardless of the choice of the band order. For this reason, the number of elements of the vector R_i is $N(N-1)/2$, where N is the number of bands in the imagery used (e.g., $N = 4$ for QB), i represents the acquisition date and R, G, B, NIR denote the QB bands. Consequently, the ratio images represent the new inputs used to train the three neural networks. The *Change Map* is obtained by direct comparison of MAP1 and MAP2 which correspond to the classification maps produced by independently training NN 1 and NN 2, respectively. For this particular classification exercise, the various types of surfaces have been grouped into the following three classes of interest:

- man-made surfaces, including buildings, concrete, asphalt, gravel and sites under construction

- green vegetation
- bare soil including low density and dry vegetation, and unproductive surfaces

The number of classes was kept small to simplify the image classification and enable an easy interpretation of the quantitative change detection. Moreover, water surfaces were not present in the scene. Once the input and the output layers of the networks NN 1 and NN 2 are established, the number of neurons to be considered in the hidden layers is the remaining critical choice. In practical cases, these networks can be characterized by the same topology, but, in general, the optimization of the number of neurons is always mandatory for a precise classification map output. To avoid saturation within the networks, the values of the input ratios vector are normalized to the range between -1 and 1. This step also helps to mitigate single-image effects if pixels belonging to different images are included in the training set. During the learning phase, the cost function may be minimized using a scaled conjugate gradient (SCG) algorithm, while the component of the output vector corresponding to the true class can be set to 1 and the others to 0. A competitive approach (*winner take all*) can be used to decide on the final classification. The information provided by the *Change Map* is composed by the $N_{cl}^2 - N_{cl} + 1$ possible transitions from one class to another between the two classification maps, where N_{cl} is the number of chosen classes ($N_{cl} = 3$, in this case). It should be noted that the *Change Map* is a PCC output performed by comparing MAP1 with MAP2.

While NN 1 and NN 2 produce the classification maps, NN 3 is designed to classify only the changed and unchanged features, exploiting the multi temporality. Therefore, this third output, which directly corresponds to the so called *Change Mask*, is the map representing only the changed pixels: the output values vary according to whether the individual pixel has changed or not. The multitemporal operator combines the R_1 and R_2 ratios in a unique vector used to feed NN 3. The k_{th} component of this new multi temporal vector is given by:

$$\left| Log \frac{R_1(k)}{R_2(k)} \right| \tag{11.7}$$

The choice of the Log operator allows an extended dynamic range of the network input values, while the absolute value assures invariance with respect to the direction of the transition. The NN 3 topology is then composed of six inputs and two outputs and similar considerations are made for NN 1 and NN 2 regarding the choice of the number of neurons in the hidden layers and the learning strategy.

To discriminate between real changes and false alarms, for each pixel in both *Change Mask* and *Change Map* it is possible to associate the logic value 1 with the changes and the logic value 0 with no-changes. These values are the input of the AND gate shown in Figure 11.6. Only when both inputs of the AND gate have the value 1, the variation in the Change Map considered to be a valid change. The resulting final NAHIRI change detection output is an image having different colors associated with different class transitions and gray scale background in areas where no changes have taken place. Therefore, this method merges the Post

Classification Comparison product, corresponding to the *Change Map*, with a change analysis performed by the *Change Mask*.

Although NAHIRI have been designed to solve change detection problems in very high spatial resolution optical imagery, it can be also used with different data types and its extension to moderate resolution images is straightforward and accurate. The NAHIRI architecture has been successfully applied to both QB and Landsat images over two test areas, corresponding to different landscapes. One is located in Rock Creek-Superior, Colorado, U.S.A., which includes single and multi family houses, shopping centers and low-density commercial buildings and corresponds to a typical American urban area. The other one is located next to the campus of the Tor Vergata University, in Rome, Italy. The region is a typical European suburban landscape including: dense residential, commercial and industrial buildings. The NAHIRI performance has been evaluated against a PCC technique based on neural network architecture for the one-step multispectral image classification. The relative output maps were subsequently compared in terms of classification accuracies as shown in Table 11.3.

Table 11.3. Comaparison of the change detection accuracies between NAHIRI and PCC over the different study areas

Site Information		**k-Coefficient**		**Overall Error (%)**	
Location	*Spatial Res. (m)*	*NAHIRI*	*PCC*	*NAHIRI*	*PCC*
Rock Creek-Superior, Colorado, U.S.A.	0.6	0.722	0.568	11.9	23.3
Rock Creek-Superior, Colorado, U.S.A.	30	0.811	0.619	5	18.3
Tor Vergata Campus, Rome, Italy	2.8	0.783	0.444	5.5	22.2
	Mean	0.795	0.544	7.5	21.3

Experimental results, obtained from both very high and high spatial resolution images, confirm that NAHIRI, unlike other techniques presented in the literature, is a general approach that can be applied to a wide range of spatial resolutions and land cover types. The mean of the k-coefficient for the PCC method is 0.544, while it is 0.795 for NAHIRI, which correspond to an accuracy increase of about 50%.

As already explained, change detection in very high spatial resolution imagery over urban areas presents a difficult challenge. One factor is the presence of different materials within the same class, the effects of temporary objects such as cars, daily markets or short-lived vegetation, imperfect image co-registration, and effects due to different image acquisition angles and solar conditions. All these problems may occur frequently in the scene and mainly yield false positives in change detection as shown in Figures 11.7, 11.8 and 11.9, which represent examples of the problems discussed.

The scenes in Figures 11.7(a) and 11.7(b) contain twelve-floor buildings and no relevant changes have occurred between the two acquisitions. Due to the different solar conditions and height of the buildings, the area covered by shadow presents different conditions. Moreover, further displacements of the objects are caused by the different acquisition angles and imperfect image registration. In Figure 11.7(c) it is shown the PCC output of the scene where, as expected, many false alarms

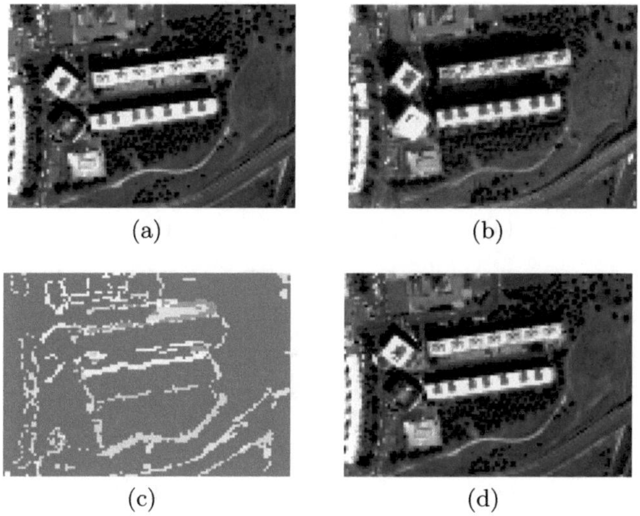

Fig. 11.7. Effects due to different acquisition angles and solar conditions: (a) and (b) are the RGB representation of the area, while (c) and (d) are the PCC and the NAHIRI products, respectively

Table 11.4. Color table of the PCC and NAHIRI change maps

	AFTER		
BEFORE	*Vegetation*	*Man-made*	*Soil*
Vegetation	Gray	Cyan	Orange
Man-made	Green	Gray	White
Soil	Red	Yellow	Gray

are present, especially around the buildings, which is a very typical change detection problem in urban areas. The color table for the change maps is reported in Table 11.4. The NAHIRI product is shown in Figure 11.8(d) in which the gray scale satellite image background denotes areas where no changes have taken place and colors represent the various class transitions. In this case, the noise present in the PCC output has been totally filtered out.

Figure 11.8 shows a change which is the construction of a new building in the middle of the scene. The height of buildings does not exceed four floors and the noise resulting from shadows is less evident compared with Figure 11.7. The PCC output recognizes well the changed area shown in cyan in Figure 11.8(c), but also in this case, several false alarms appear along the edges of the buildings. The NAHIRI map shown in Figure 11.8(d) still exhibits the construction of the new building, but at the same time the false alarms have been largely filtered out. This exercise demonstrates the effectiveness of the NAHIRI algorithm to distinguish between real changes and false alarms.

No changes have occurred between the two acquisitions of the area shown in Figure 11.9. This contains several buildings and a soccer field in the lower corner and represents well a typical dense urban scene mainly affected by imperfect image

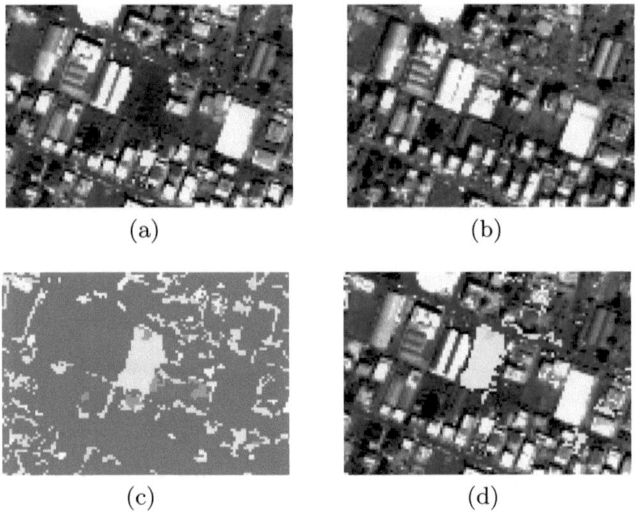

Fig. 11.8. Effects due to real changes: (a) and (b) are the RGB representation of the area, while (c) and (d) are the PCC and the NAHIRI products, respectively

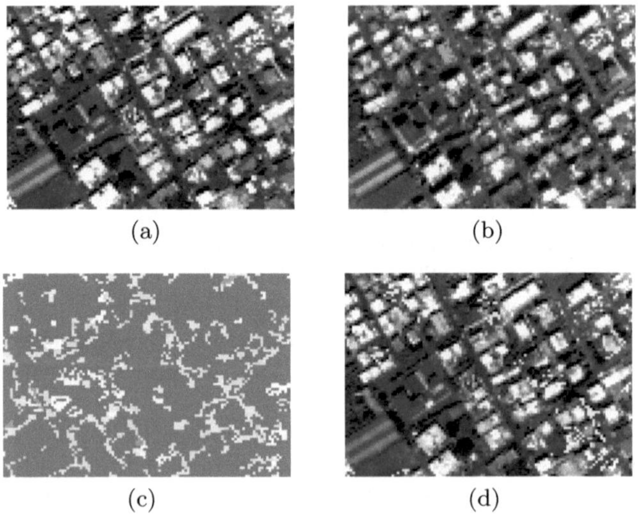

Fig. 11.9. Effects due to imperfect registration of the images: (a) and (b) are the RGB representation of the area, while (c) and (d) are the PCC and the NAHIRI products, respectively

registration, which is commonly considered in the literature as an additional error. Also in this case, the NAHIRI output shown in Figure 11.9(d) greatly reduces the occurrence of false alarms present in the PCC map of Figure 11.9(c).

11.3.2 Real-Time Classification of Multitemporal SAR Data

A short overview concerning the design and the optimization of the Elman NNs for simple time series cases is addressed in the following paragraph. Then, the real-time classification of multitemporal SAR data will be introduced.

Simulation with Elman Neural Networks

This section introduces the Elman model in two practical time series cases. The first exercise compares the performances of MLP and Elman neural networks for the classification of sequences. The second concerns an analysis of the number of hidden units for the Elman model.

Let "ROMA" and "RAMO" be two strings to be classified on the base of a single character provided as input. Before starting the training of the neural network, each letter should be conveniently codified. For this exercise, the input layer is composed of four units, which means that each character is codified by four bits: the first bit of each letter is set to 1 for consonants and 0 for vocals, as reported in Table 11.5(a). The output layer of the network is composed by two units (the relative code is in Table 11.5(b)).

Table 11.5. (a) Code for the input letters, (b) code for the output words

(a)

Input Letter	Code
A	0101
M	1001
O	0110
R	1010

(b)

Code	Output Word
10	ROMA
01	RAMO

The architecture of the recurrent network is shown in Figure 11.10 in which the units of the hidden and context layers are four, resulting in a 4-4-2-4 (I-H-O-C) topology.

The minimization of the error functions for 10 MLP networks implementing the same classification problem and initialized with different random values is shown in Figure 11.11(a). We note that the minimum of the error function is

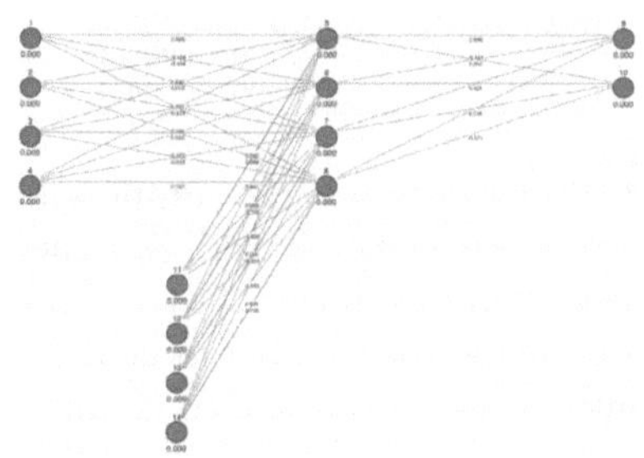

Fig. 11.10. Architecture of the recurrent network for string recognition used in this exercise

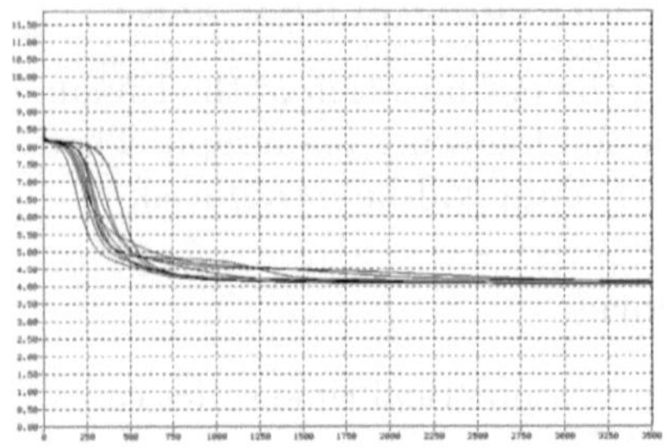

(a)

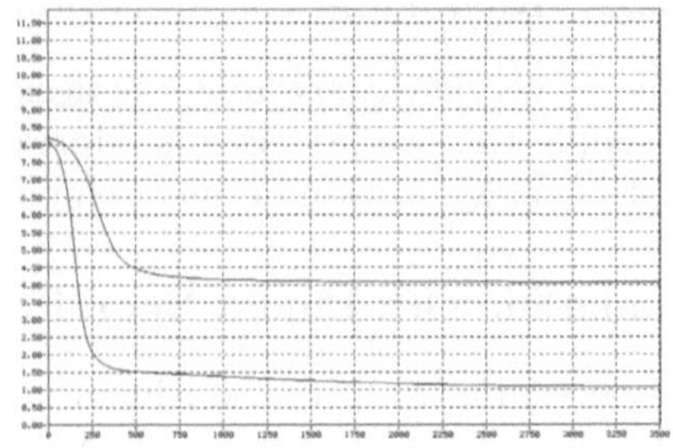

(b)

Fig. 11.11. (a) SSE minimization functions for 10 MLP networks whose initial conditions were different and (b) their comparison (upper function) with respect of the Elman model (lower function)

reached after only a few iterations of the algorithm and it is never less than 4, corresponding to patterns. This is caused by the presence of the same four letters in both words: for the MLP architecture, it is not possible to understand the context of the sequence. In this case, 50% of the characters are classified incorrectly. The upper error function of Figure 11.11(b) corresponds to the error provided by the best (in terms of accuracy) MLP network trained previously, while the lower is the error function for the Elman model. In this case, there is a drastic improvement in terms of accuracy with respect to MLP: the classification error is significant only when the first character, R, is provided to the network since it is the initial letter in both words. The error function of the Elman architecture asymptotically reaches SSE=1, which represents the ambiguity of the first character.

At this point, it is interesting to investigate the effects of varying the number of units in the hidden layer, and, the number of units in the context layer. Lets consider a new exercise where the time series in Table 11.6 are the input sequences of an Elman network.

Table 11.6. Code for the input letters

Input Time Series	Code
S_1	0000000000
S_2	0000000001

For this exercise, the training data set was composed of 5000 patterns derived by feeding the network randomly with S_1 and S_2. The validation data set was built in the same way, with 2500 patterns. The network was composed of one input and two output units, while the number of elements in the hidden layer, as well as for the context layer, varied from one to twenty. In Figure 11.12 (a), the minimization of the error functions for the training (black lines) and validation (red lines) data sets are shown for the case of only one hidden unit: the Elman architecture, as well as MLP, is not able to properly classify the input series,

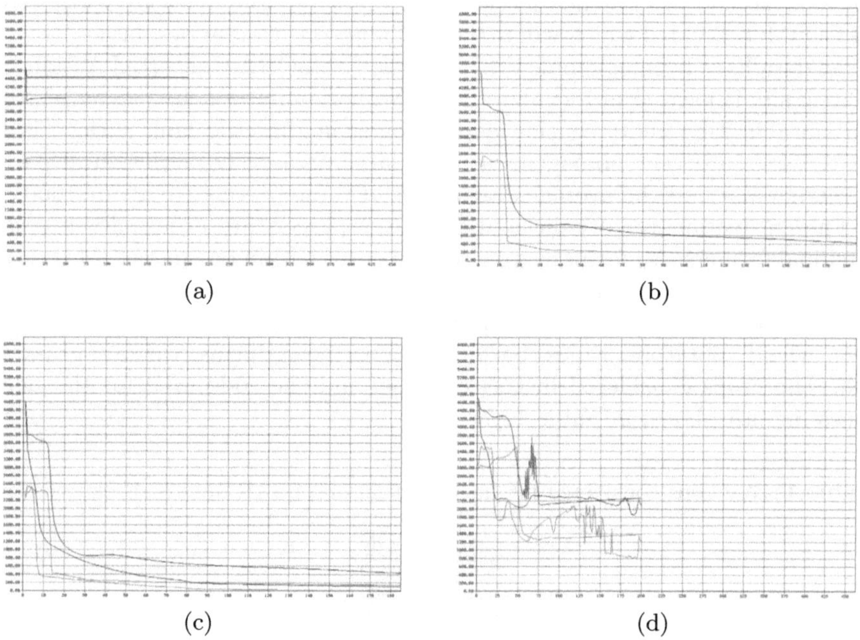

Fig. 11.12. (a) One hidden unit error minimization functions for training and validation sets in the case of MLP and Elman neural network; (b) two, (c) three/four and (d) twenty hidden units error minimization functions for training and validation sets in the case of Elman neural network. Back and red lines indicate training and validation sets, respectively.

since it is not possible to distinguish if one zero occurs in the first or in the second sequence, in fact, the only value that allows the separation of the sequences is 1 in S_2. The classification accuracy changes drastically when two hidden units are used. In Figure 11.12(b), only the Elman case is considered: black and red lines represent the error functions for the training and the validation sets, respectively. In a very short time, the Elman network is able to reach the desirable accuracy, classifying all sequences correctly. It is possible to obtain a faster training of the network by increasing the number of hidden units, as shown in Figure 11.12(c) for three and four hidden units. The network training becomes more unstable, as shown in Figure 11.12 (d) for the case of twenty hidden units.

In both cases a trade off among the requirements is mandatory.

The Elman Real-Time Classification Scheme

The previous section reported on the effectiveness of the Elman NNs for time series classification. This section aims at evaluating the performance of the Elman model in a real classification of synthetic aperture radar (SAR) data acquired by the European Remote Sensing Satellite 1 (ERS 1) in the years 1993, 1994 and 1995.

This analysis can be extended in view of the new satellite missions TerraSAR-X and COSMO-SkyMed which will make available long time-series data.

The data have been assembled through the ERA-ORA Concerted Action [31]. The test site is located in Flevoland (The Netherlands), which is an agricultural area suitable for remote sensing experiments, in that the land surface is flat and the soils are very homogeneous over large areas. The agricultural parcels are rectangular and the average dimensions of these parcels are about 1600x500 m. The cultivation is periodic and the large parcels are eventually leased to individual farmers. All farmers grow several crops within one parcel. This results in the parcels being subdivided into fields characterized by a single crop type. The crops present in the scenes in the mentioned years, include maize, sugar beets, wheat, potatoes, rapes, barleys, onions, peas, beans, carrots, grass and bush. The crops considered in this exercise, and the code used for the neural processing, are shown in Table 11.7.

Table 11.7. Crops considered for the classification of SAR data and relative code

Crops	Code
barley	10000
grass	01000
potato	00100
sugar beet	00010
wheat	00001

The ERS 1 data cover the whole year with 12 scenes, therefore covering the early stages and the senescence of the various plants. The analysis of SAR time series, generally, indicates that both the average and the standard deviation of the backscattering coefficient change from year to year [32]. However, the observed dispersion is not wide enough to confound the patterns of the growth cycle of each type of plant, as shown in Figure 11.13.

The number of patterns, which in this context can be associated with the pixels of the image to be classified, is 485 for years 1993 and 1994, and 128 for year 1995. A structure containing 12 temporal samples was built resulting in a training set composed of 5820 (485x12) patterns, and a validation set of 1536 (128x12) patterns. This structure is reported in Figure 11.14, where σ_i^0 is the backscattering coefficient for the measurement i_{th} of the year and o_j^0 can assume values 0 or 1 according to Table 11.7. Each pattern is composed by 10 input elements which means that the Elman model is characterized by a buffer memory of $n^{'} = 10$ units.

For a real time classification model, it is desirable to provide an inventory of the crop fields as soon as the discrimination is accomplished. The structure illustrated previously simulates a real time analysis, in fact, in the first temporal pattern only the first backscattering coefficient is used to feed the Elman network, while the rest of the 12 acquisitions are sequentially introduced in successive patterns. Therefore, this procedure reproduces a continuous updating of the classification maps

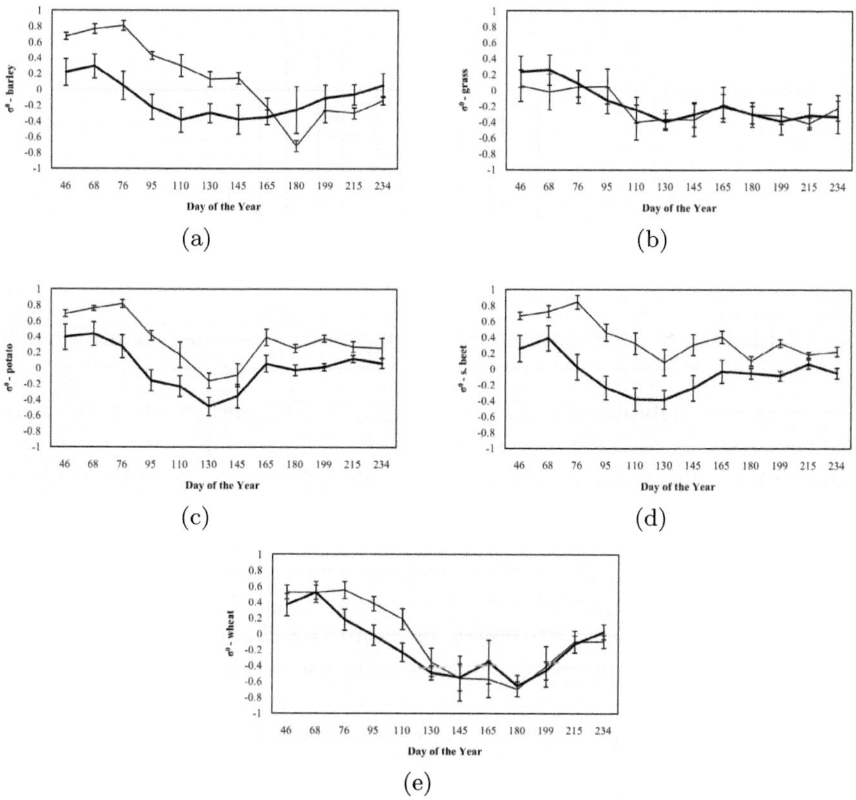

Fig. 11.13. Trends of the normalized backscattering coefficients measured by ERS 1 for the corps considered: (a) barley, (b) grass, (c) potato, (d) sugar beet and (e) wheat. Bold lines indicate the training data set, while light lines indicate the test data set. The training data set is composed of the measurements of years 1993 and 1994, while the test data set by year 1995. The normalization is performed within the interval $[-1; 1]$, which corresponds to the minimum and maximum value of the data set considered, respectively.

pattern 1
$\sigma_1^0 \quad \sigma_1^0 \quad \sigma_1^0 \quad \sigma_1^0 \quad \sigma_1^0 \quad \sigma_1^0 \quad \sigma_1^0 \quad \sigma_1^0 \quad \sigma_1^0 \quad \sigma_1^0$
$o_1 \quad o_2 \quad o_3 \quad o_4 \quad o_5$
pattern 2
$\sigma_1^0 \quad \sigma_1^0 \quad \sigma_1^0 \quad \sigma_1^0 \quad \sigma_1^0 \quad \sigma_1^0 \quad \sigma_1^0 \quad \sigma_1^0 \quad \sigma_1^0 \quad \sigma_2^0$
$o_1 \quad o_2 \quad o_3 \quad o_4 \quad o_5$
pattern 3
$\sigma_1^0 \quad \sigma_1^0 \quad \sigma_1^0 \quad \sigma_1^0 \quad \sigma_1^0 \quad \sigma_1^0 \quad \sigma_1^0 \quad \sigma_1^0 \quad \sigma_2^0 \quad \sigma_3^0$
$o_1 \quad o_2 \quad o_3 \quad o_4 \quad o_5$
⋮
pattern 12
$\sigma_3^0 \quad \sigma_4^0 \quad \sigma_5^0 \quad \sigma_6^0 \quad \sigma_7^0 \quad \sigma_8^0 \quad \sigma_9^0 \quad \sigma_{10}^0 \quad \sigma_{11}^0 \quad \sigma_{12}^0$
$o_1 \quad o_2 \quad o_3 \quad o_4 \quad o_5$

Fig. 11.14. Structure of the 12 patterns

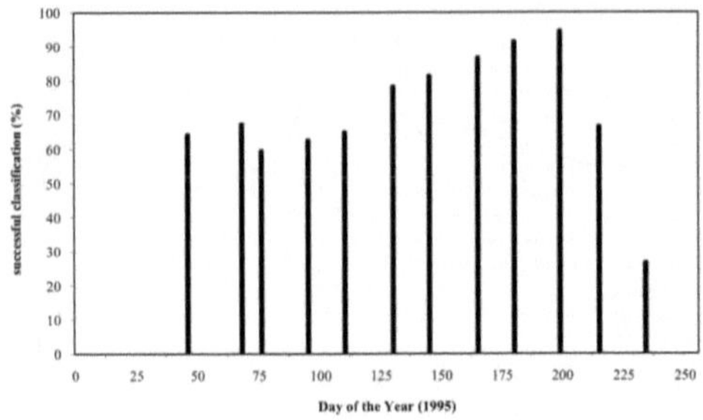

Fig. 11.15. Percentage of successfully classified fields versus measurement number

following the availability of a new measurement. The performances in terms of overall accuracy at each date are summarized in Figure 11.15.

The results obtained show that the overall accuracy in classifying all fields in 1995 is above 50% up to Day of Year (DoY) 215: the fraction of successfully classified fields increases gradually to the maximum value of 94.5%, reached on DoY 199, then it starts to decrease with the senescence or the harvest of the crops, which makes further classification meaningless. The following confusion matrices (Tables 11.8 - 11.10) report in detail the classification accuracies for each crop using the 12 measurements one after the other.

Table 11.8. Confusion matrices for dates **1**, **2**, **3** and **4**

(a) Confusion matrix for date 1.

	Classified as				
True class	*barley*	*grass*	*potato*	*s. beet*	*wheat*
barley	0	0	0	6	2
grass	0	30	0	0	0
potato	0	0	13	16	1
s. beet	0	0	11	15	4
wheat	0	3	1	2	24
			Accuracy (%)		64.1

(b) Confusion matrix for date 2.

	Classified as				
True class	*barley*	*grass*	*potato*	*s. beet*	*wheat*
barley	0	0	3	5	0
grass	0	29	0	0	1
potato	0	0	11	19	0
s. beet	0	0	7	21	2
wheat	0	1	0	4	25
			Accuracy (%)		67.2

(c) Confusion matrix for date 3.

	Classified as				
True class	*barley*	*grass*	*potato*	*s. beet*	*wheat*
barley	0	0	5	3	0
grass	0	28	0	0	2
potato	0	0	17	13	0
s. beet	0	0	21	8	1
wheat	0	1	0	6	23
			Accuracy (%)		59.4

(d) Confusion matrix for date 4.

	Classified as				
True class	*barley*	*grass*	*potato*	*s. beet*	*wheat*
emph barley	0	0	1	7	0
emph grass	0	28	0	0	2
emph potato	0	0	3	27	0
emph s. beet	0	0	2	27	1
emph wheat	0	1	0	7	22
			Accuracy (%)		62.5

The time dependence of the fraction of individual crops successfully classified depends on the type of vegetation as illustrated in Figure 11.16. Spontaneous grass exhibits a peculiar behavior, since more than 96% of these fields can be detected in the first two images. Indeed, the spontaneous vegetation present in uncultivated fields yields values consistently lower than those of the other types of rough surfaces, which initially are mainly bare and moist. In subsequent measurements, the backscattering from grass merges into that of the other sprouting crops, and only after DoY 110 can the uncultivated parcels be discriminated with

Table 11.9. Confusion matrices for dates **5**, **6**, **7** and **8**

(a) Confusion matrix for date 5.

	Classified as				
True class	*barley*	*grass*	*potato*	*s. beet*	*wheat*
barley	0	0	2	6	0
grass	0	28	0	0	2
potato	0	0	18	12	0
s. beet	0	0	12	17	1
wheat	0	1	0	9	20
				Accuracy (%)	64.8

(b) Confusion matrix for date 6.

	Classified as				
True class	*barley*	*grass*	*potato*	*s. beet*	*wheat*
barley	0	0	0	8	0
grass	0	29	0	0	1
potato	0	0	28	2	0
s. beet	0	0	9	21	0
wheat	0	2	4	2	22
				Accuracy (%)	78.1

(c) Confusion matrix for date 7.

	Classified as				
True class	*barley*	*grass*	*potato*	*s. beet*	*wheat*
barley	0	0	2	6	0
grass	0	29	0	0	1
potato	0	0	29	1	0
s. beet	0	0	9	21	0
wheat	0	2	2	1	25
				Accuracy (%)	81.2

(d) Confusion matrix for date 8.

	Classified as				
True class	*barley*	*grass*	*potato*	*s. beet*	*wheat*
barley	0	0	7	1	0
grass	0	29	0	0	1
potato	0	0	28	2	0
s. beet	0	0	5	25	0
wheat	0	0	0	1	29
				Accuracy (%)	86.7

Table 11.10. Confusion matrices for dates **9**, **10**, **11** and **12**

(a) Confusion matrix for date 9.

	Classified as				
True class	*barley*	*grass*	*potato*	*s. beet*	*wheat*
barley	6	0	0	0	2
grass	0	30	0	0	0
potato	0	0	26	4	0
s. beet	0	1	3	26	0
wheat	0	0	1	0	29
				Accuracy (%)	91.4

(b) Confusion matrix for date 10.

	Classified as				
True class	*barley*	*grass*	*potato*	*s. beet*	*wheat*
barley	8	0	0	0	0
grass	0	30	0	0	0
potato	0	0	29	1	0
s. beet	1	2	2	25	0
wheat	0	0	1	0	29
				Accuracy (%)	94.5

(c) Confusion matrix for date 11.

	Classified as				
True class	*barley*	*grass*	*potato*	*s. beet*	*wheat*
barley	7	0	0	6	1
grass	0	30	0	0	0
potato	10	18	0	2	0
s. beet	1	9	0	20	0
wheat	0	2	0	0	28
				Accuracy (%)	66.4

(d) Confusion matrix for date 12.

	Classified as				
True class	*barley*	*grass*	*potato*	*s. beet*	*wheat*
barley	0	7	0	0	1
grass	0	21	0	0	9
potato	0	29	1	0	0
s. beet	0	26	0	2	2
wheat	0	19	0	1	10
				Accuracy (%)	26.56

high accuracy (up to 100%). Potatoes fields can be discriminated with an accuracy greater than 90% after DoY 110, followed by wheat fields, with an accuracy above 80% starting from DoY 145. More than 80% of barley fields can be classified after DoY 164.

Similar results in terms of overall accuracy could be found using the MLP approach [33][34]. This method is applied to an increasing sequence of measurements taken in years 1993 and 1994 and tested on the correspondingly increasing sequence of measurements in the subsequent year. The initial procedure consisted of training the MLP neural network with the first measurements taken in 1993 and 1994 and applying it to classify the scene by using the first measurement acquired in 1995: in this case, the network is composed of one input unit, two

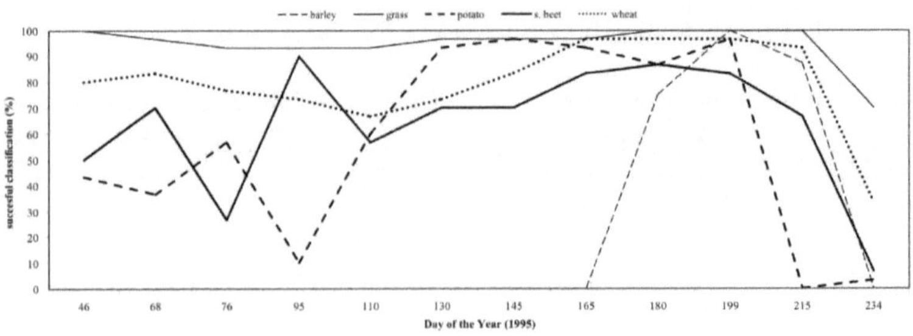

Fig. 11.16. Detailed percentage of successfully classified crops versus measurement number

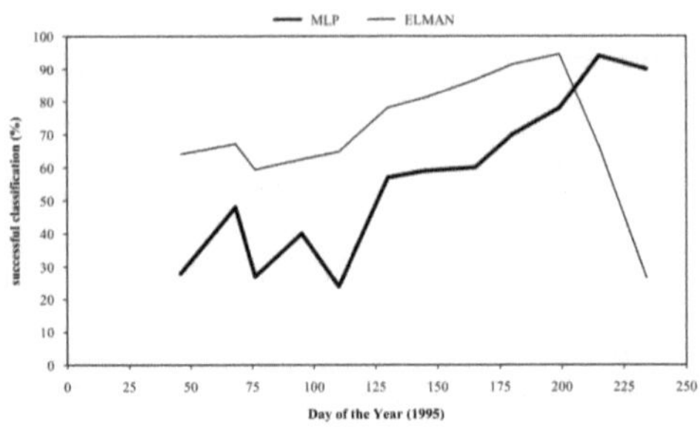

Fig. 11.17. MLP and Elman comparison of the percentage of successfully classified fields versus measurements

hidden layers and five output units, and the training strategy follows a scaled conjugate gradient (SCG) algorithm. Then, the number of training measurements is progressively increased by adding the subsequent measurements and consequently, also the topology of the neural network changes introducing new inputs until the total number of measurements for the year is reached: in this final case, the network is composed of twelve input units. Also in this case, the MLP procedure was implemented with the aim of updating the classification maps as new measurements were available, simulating real time processing. The fraction of correctly classified fields obtained by applying this approach reaches 94% accuracy on DoY=215; the comparison accuracy between the MLP and Elman is shown in Figure 11.17. For DoY=234, the accuracy of the MLP network is close to 90% even if very modest crops were still present at that date due to decay or harvest of vegetation. This can be explained considering that the last measurement introduces only 1/12 of new information in this architecture, resulting in a minor impact on the final classification response.

From the time series point of view, the advantage of the Elman implementation is that this is a *unique* neural network independent of the number of measurements used in the input stage. In fact, to insure the same flexibility, a series of 12 different MLP networks have to be trained, one with 1 input in the case of 1 measurement, another with 2 inputs in the case of 2 measurements and so on until all measurements are used in a network with 12 inputs. Supposing to use, as in this exercise, twelve networks with ten units in each of two hidden layers and five outputs, it is necessary in the learning phase, to optimize 2680 different connection weights, while the same results are obtained by updating only 250 connection weights in the unique Elman network.

References

1. Rumelhart, D.E., Hinton, G.E., Williams, R.J.: Learning internal representations by error propagation. In: Rumelhart, D.E., McClelland, J.L. (eds.) Parallel Distributed Processing. The MIT press, Cambridge (1986)
2. Lippmann, R.P.: An introduction to computing with neural nets. IEEE Acoust. Speech Signal Process. Mag. 4(2), 4–22 (1987)
3. Benediktsson, J.A., Swain, P.H., Ersoy, O.K.: Neural network approaches versus statistical methods in classification of multisource remote sensing data. IEEE Trans. Geosci. Remote Sens. 28(4), 540–552 (1990)
4. Bishof, H., Schneider, W., Pinz, A.J.: Multispectral classification of Landsat-images using neural networks. IEEE Trans. Geosci. Remote Sens. 30(3), 482–490 (1992)
5. Dawson, M.S.: Applications of electromagnetic scattering models to parameter retrieval and classification. In: Fung, A.K. (ed.) Microwave Scattering and Emission Models and Their Applications, pp. 527–557. Artech House, Norwood, Mass. (1994)
6. Hornik, K., Stinchcombe, M., White, H.: Multilayer feedforward networks are universal approximators. Neural Netw. 2(5), 359–366 (1989)
7. Del Frate, F., Ortenzi, A., Casadio, S., Zehner, C.: Application of neural algorithms for a real-time estimation of ozone profiles from GOME measurements. Geoscience and Remote Sensing, IEEE Transactions 40(10), 2263–2270 (2002)
8. Tsang, L., Chen, Z., Oh, S., Marks, R.J., Chang, A.T.C.: Inversion of snow parameters from passive microwave remote sensing measurements by a neural network trained with a multiple scattering model. Geoscience and Remote Sensing, IEEE Transactions 30(5), 1015–1024 (1992)
9. Cipollini, P., Corsini, G., Diani, M., Grasso, R.: Retrieval of sea water optically active parameters from hyperspectral data by means of generalized radial basis function neural networks. Geoscience and Remote Sensing, IEEE Transactions 39(7), 1508–1524 (2001)
10. Kohonen, T.: Self-Organizing Maps. Springer Series in Information Sciences, vol. 30. Springer, Heidelberg (2001)
11. Bruzzone, L., Melgani, F.: Robust multiple estimator systems for the analysis of biophysical parameters from remotely sensed data. Geoscience and Remote Sensing, IEEE Transactions 43(1), 159–174 (2005)
12. Huang, G.B., Zhu, Q.Y., Siew, C.K.: Extreme learning machine: Theory and applications. Neurocomputing (70), 489–501 (2006)
13. Del Frate, F., Petrocchi, A., Lichteneggger, J., Calabresi, G.: Neural Networks for oil spill detection using ERS-SAR data. IEEE Trans. Geosci. Remote Sens. 38(5), 2282–2287 (2000)

14. Del Frate, F., Lichteneggger, J., Solimini, D.: Monitoring urban areas by using ERS-SAR data and neural networks algorithms. In: Proc. IEEE Geosci. Remote Sens. Symposium, Hamburg, Germany, vol. 5, pp. 2696–2698 (1999)
15. Hu, Y.H., Hwang, J.N.: Handbook of Neural Network Signal Processing. CRC press LLC, Boca Raton (2002)
16. Kasabov, N.K.: Foundations of Neural Networks, Fuzzy Systems and Knowledge Engineering. The MIT press, Cambridge (1996)
17. Kecman, V.: Learning and Soft Computing: Support Vector Machines. In: Neural Networks and Fuzzy Logic Models, The MIT press, Cambridge (2001)
18. Arbib, M.A.: The handbook of brain theory and neural networks. The MIT press, Cambridge (2003)
19. Bishop, C.: Neural Networks for pattern recognition. Oxford University Press, New York (1995)
20. Cheng, Y.C., Qi, W.M., Cai, W.Y.: Dynamic properties of Elman and modified Elman neural network. In: Proceedings of IEEE Machine Learning and Cybernetics, November 4-5, 2002, vol. 2, pp. 637–640 (2002)
21. Elman, J.L.: Distributed representations, simple recurrent networks, and grammatical structure. Machine Learning 7, 195–225 (1991)
22. Zell, A., et al.: SNNS, Stuttgart Neural Network Simulator, user manual v. 4.2, University of Stuttgart, Stuttgart, Germany
23. Møller, M.F.: A scaled conjugate gradient algorithm for fast supervised learning. Neural Networks 6(4), 525–533 (1993)
24. Addink, E.A., Thonon, I., De Jong, S.M., Gorte, B.G.H., Beck, R., Van Persie, M., Kramer, H.: The Mutatis Mutandis Project: single detection, multiple users - change detection using high spatial-resolution imagery in Dutch urban areas. In: Proceedings of EARSeL Workshop of the SIG Urban Remote Sensing, Humboldt-Universität, Berlin, Germany, March 2-3 (2006)
25. Bruzzone, L., Prieto, D.F.: An adaptive semiparametric and contextbased approach to unsupervised change detection in multitemporal remote sensing images. IEEE Trans. Image Process. 11(4), 452–466 (2002)
26. Kasetkasem, T., Varshney, P.K.: Image change detection algorithm based on Markov random field models. IEEE Trans. Geosci. Remote Sens. 40(8), 1815–1823 (2002)
27. Del Frate, F., Pacifici, F., Schiavon, G., Solimini, C.: Use of Neural Networks for Automatic Classification From High-Resolution Images. IEEE Trans. Geosci. Remote Sens. 45(4), 800–809 (2007)
28. Singh, A.: Digital change detection techniques using remotely-sensed data. Int. J. Remote Sens. 10(6), 989–1003 (1989)
29. Pacifici, F., Del Frate, F., Solimini, C., Emery, W.J.: An Innovative Neural-Net Method to Detect Temporal Changes in High-Resolution Optical Satellite Imagery. IEEE Trans. Geosci. Rem. Sens. 45(9), 2940–2952 (2007)
30. Chini, M., Pacifici, F., Emery, W.J., Pierdicca, N., Del Frate, F.: Comparing statistical and neural network methods applied to very high resolution satellite images showing changes in man-made structures at Rocky Flats. IEEE Trans. Geosci. Rem. Sens.(in press)
31. Ferrazzoli, P., Guerriero, L., Schiavon, G., Solimini, D.: European RAdar-Optical Research Assemblage: Final Report (March 2002), `http://eraora.disp.uniroma2.it/`
32. Saich, P., Borgeaud, M.: Interpreting ERS SAR signatures of agricultural crops in Flevoland, 1993-1996. IEEE Trans. Geosci. Remote Sens. 38(2), 651–657 (2000)

33. Borgeaud, S., Del Frate, F., Hoekman, H., Schiavon, G., Solimini, D., Vissers, M.: Crop classification using multi-configuration C band SAR data. IEEE Trans. Geosci. Remote Sens. 41(7), 1611–1619 (2003)
34. Del Frate, F., Schiavon, G., Solimini, D., Borgeaud, S., Hoekman, H., Vissers, M.: The potential of SAR in crop classification using multi-configuration data. In: Proceedings of the Third International Symposium on Retrieval of Bio- and Geophysical Parameters from SAR Data for Land Applications, Sheffield, UK, September 11-14, 2001, pp. 93–98 (2001)

12

Information Extraction for Forest Fires Management

Andrea Pelizzari[1], Ricardo Armas Goncalves[1], and Mario Caetano[2]

[1] Critical Software SA
www.criticalsoftware.com
[2] Portuguese Geographic Institute (IGP),
Instituto Superior de Estatística e Gestão de Informação
(ISEGi), Universidade Nova de Lisboa (UNL)
www.igeo.pt

12.1 Introduction

Forest fires are a recurrent environmental and economic emergency worldwide. In Southern Europe the year 2003 is considered one of the worst ever with a total burnt area of 740,000 hectares [22]. Even if in some countries according to the national burnt area statistics the problem seems to be under control and of steady or decreasing magnitude, in regions like the western Iberian Peninsula or large parts of North America the trend is worryingly opposite [23, 29]. The causes of such catastrophic events depend on a number of factors, ranging from meteorology to societal changes to vegetation type and emergency response readiness and efficiency.

The main characteristic of the forest fires emergency is its extreme dimension in terms of affected area and number of events. The only technology capable of responding to the requirements of cost effective monitoring and impact estimation is remote sensing. Since the 50s aerial photography has been used to collect information on forest type and health [35]. With the advent of Earth Observation satellites, the forest environment has become a favourite target of scientific observations and image analysis. Thanks to the global coverage, increasing spatial and temporal resolution, and near real time availability satellite imagery is nowadays considered an essential source of information for the fire emergency management activities.

Beside increasingly efficient sensors, high performance data processing and computer aided image analysis have been the key technological development for information extraction from satellite images. The large volume of remote sensing data and the intrinsic variability and complexity of the observed phenomena can only be efficiently analysed by intelligent agents. The automatic computational methods give a relevant support to the human expert for solving the specific requirements related to forest fires [36]. Several image processing techniques like neighbourhood filtering and statistical thresholding can be adapted and applied to the automatic interpretation of images of forest regions both for evaluating

M. Graña and R.J. Duro (Eds.): Comput. Intel. for Remote Sensing, SCI 133, pp. 295–312, 2008.
springerlink.com

the vegetation status before the fire as well as for monitoring and mapping the fire severity and burn scars during and after the event [36].

This chapter describes two case studies of operational and autonomous processing chains in place for supporting forest fires management in Europe. The authors focused on the prevention and damage assessment phases of the wildfire emergency cycle, showing how computational intelligence can be effectively used for:

- Fire risk estimation, based on structural and dynamic risk information fusion from in-situ monitoring and remote sensing;
- Burn scars mapping, based on automatic change detection with medium resolution multispectral satellite data.

Conclusions will be drawn on the current state of the art computational tools for fire emergency management and future development derived by remote sensing technological progress.

12.2 Forest Fire Risk Mapping

12.2.1 Motivation for Service Development

Forest fire risk mapping is not a recent practice; it has been used for several decades by diverse institutes and agencies with competences in the prevention and combat of forest fires. Forest fire management actors recognize that the forest fire risk measure is a very powerful tool for supporting the prevention and combat activities.

Although forest fire risk is intimately related to meteorological conditions, the production of an operational forest fire risk map must take into consideration other elements such as land cover, topography, demography, infra-structures or forest-urban interface. The importance of these factors is related to the influence in the occurrence, extension and intensity of forest fires, being the vegetation type and condition the most important elements in the evaluation of the forest fire risk, since it characterizes the forest fuel [15].

During the Year of 2002, in the scope of the PREMFIRE (PREvention and Mitigation of FIRE hazards) project, sponsored by the European Space Agency (ESA Ref. AO/1-3880/01/I-SB), the Remote Sensing Group of the Portuguese Geographic Institute and Critical Software developed and implemented a methodology for the production of a daily forest fire risk mapping for Mainland Portugal, based on satellite imagery and ancillary data [11].

After the 2003 forest fire season, where more then 400 000 hectares burn in Portugal Mainland [22], COTEC Portugal - Associação Empresarial para a Inovação, an enterprise association aiming the improvement of the competitiveness of the enterprises located in Portugal, promoted an initiative to encourage the best practices on the management of the Portuguese forest against fires. Within this initiative [2004-2005], the forest fire risk methodology previously developed in the scope of the PREMFIRE project was improved, operated and tested on a semi-operational scenario.

12.2.2 Data Requirements

The fire risk methodology developed under this scope is an adaptation of the methodology suggested by the European Union's (UN) Joint Research Centre [21] for the entire European region, allowing it to be applied to national level scales.

Regarding the temporal scale, the fire risk methodologies can be classified into [21]:

- Structural or long-term indices, derived from factors that do not change in short lapse of time, i.e. topography or land-cover.
- Dynamic or short-term indices, relying on parameters that change fairly continuously over time, i.e. vegetation condition or weather.
- Integrated or Advanced indices, including both structural and dynamic variables. The integrated approach is based on the assumption that the start and progression of a forest fire is affected by different factors, therefore requiring an integrated analysis [14].

The forest fire risk methodology herein presented, the Integrated Forest Fire Risk (IFFR), is an integrated fire risk index based on the combination of a structural fire index and a dynamic index [12, 19, 10].

The Structural Fire Index (SFI), produced on an annual basis, before each forest fire season, is based on five variables considered as the most influent in forest fire occurrence in the Mediterranean basin [14]:

- Land cover map extracted from Landsat TM (Thematic Mapper) imagery and ancillary information. This land cover map provides two base maps for the system: vegetation classes and distance to urban areas;
- Digital elevation model (DEM). This DEM provides three different base maps: slope, elevation and aspect.
- Road network, providing the accessibility base map.

The dynamic index applied in the integrated methodology is the Fire Potential Index (FPI) [9, 24], incorporating both satellite imagery and weather forecast in a danger index updated on a daily basis. The FPI initially implemented under the scope of the PREMFIRE project used AVHRR (Advanced Very High Resolution Radiometer) satellite imagery from NOAA (National Oceanic & Atmospheric Administration). However, due to the existence of medium spatial resolution MODIS (Moderate Resolution Imaging Spectroradiometer) sensors, acquiring data in the spectral channels applied in this index, an improvement was made during the COTEC initiative, in order to adapt the system allowing the use of MODIS imagery.

Nowadays, the FPI applies daily imagery acquired by MODIS sensors, on board of the TERRA and AQUA satellites from NASA (National Aeronautics and Space Administration). MODIS products used in this index are the surface reflectance (250 meters of spatial resolution), and the acquisition angles. Acquisition angles product is applied to eliminate parts of images acquired with high solar and sensor zenith angles.

The inputs to the FPI model are:

- Dead fuel moisture of extinction, providing, for each fuel model, the values above which fire will no longer spread due to the amount of moisture content [31].
- A maximum live ratio map and a relative greenness map [8] providing, respectively, the percentage of live fuel load for a given pixel when the vegetation is at maximum greenness, and how green each pixel currently is in relation to the range of historical maximum and minimum greenness observations for it. The maximum live ratio and the relative greenness maps are assessed using the Normalized Difference Vegetation Index (NDVI) [32].
- A Ten-hour timelag dead fuel moisture [18] map produced with weather forecast data (temperature, relative humidity and nebulosity). Ten-hour timelag fuels are defined as dead woody vegetation in the size range of 0.6 to 2.5 cm in diameter. Fine fuels, namely grass, leaves, bark, and twigs are those materials which ignite and burn easily. Heavy fuels, generally more than 6mm in diameter, will not burn unless fine fuels are present to support fire. While large fuels can take a whole season to dry up, forest fires usually originate in the light fuels. These fine fuels react almost immediately to changes in relative humidity and temperature, making the estimation of the equilibrium moisture feasible [5].

12.2.3 Methods

Structural Fire Index

The SFI is based on the combination of five variables:

- Vegetation classes;
- Slope;
- Aspect;
- Distance to roads and urban areas;
- Elevation.

To rank each variable's original values into different levels of danger, taking into account the relative importance of each class as a factor for increasing the fire danger, these variables are reclassified, and a coefficient of 0, 1 or 2 is assigned to each level according to their fire danger (respectively from high to low fire danger) [14]. After the reclassification, the following formula is applied, in order to integrate all the variables:

$$\mathrm{SFI} = 100v + 30s + 10a + 5u + 2e \tag{12.1}$$

Where v, s, a, u, and e represent vegetation, slope, aspect, distance to roads and urban areas, and elevation variables, respectively. SFI provides a map with values ranging from 0 to 255, which needs to be further integrated with the dynamic risk (Fire Potential Index) in order to provide the Integrated Forest Fire Risk index.

Dynamic Fire Index

The dynamic fire index (FPI) applied in this integrated approach is based on several intermediate information layers.

- Dead Fuel Moisture of Extinction (MX_d) is based on the assignment of dead fuels extinction moisture values to each fuel class on the yearly land cover map. The dead fuels extinction moisture values were identified according to the specific characteristics of each fuel class and indicate moisture values above which fire will no longer spread[31].
- Maximum Live Ratio (LR_{max}) is calculated as a function of the live and dead loads assigned to each fuel class in order to determine the live ratio for a given pixel when vegetation is at its maximum greenness [9, 26].

  This process is assessed, for each pixel, recurring to the maximum value of NDVI occurring on a certain fuel class (ND_{absmax}) and to the NDVI value occurring on the pixel. The NDVI values applied are based on 7 day maximum value composite and are reclassified to values between 0 and 200. To calculate the Maximum Live Ratio, the following formula is then applied:

$$LR_{max} = \left(0.25 + \frac{ND_{max}}{2ND_{absmax}}\right) \cdot 100 \qquad (12.2)$$

  Where ND_{max} is the maximum historical NDVI value for a certain pixel.
- Relative Greenness (RG) is computed by comparing, for each pixel, the current 7 day NDVI composite value to the available historical range of NDVI data [8]. This process allows comparing the actual value of greenness of a certain pixel to its historical values. To calculate the Relative Greenness, the following formula is applied:

$$RG = \frac{(ND_0 - ND_{min})}{(ND_{max} - ND_{min})} \cdot 100 \qquad (12.3)$$

  Where ND_{min} is the minimum historical NDVI value for a certain pixel (based on MODIS images acquired over that area since the launch of AQUA and TERRA satellites) and ND_0 is the NDVI 7 day MVC (Maximum Value Composite) composite value for a certain pixel.
- By scaling the Maximum Live Ratio map (LR_{max}) with Relative Greenness map (RG), the Live Fuel Ratio map (LR), which expresses the current live fuel ratio for each pixel, can be computed. In order to calculate LR, RG has to be converted to fractional value ($RG_f = RG/100$), after which the following formula is applied:

$$LR = \frac{RG_f \cdot LR_{max}}{100} \qquad (12.4)$$

- Ten-hour Timelag Dead Fuel Moisture (FM_{10}) is based on weather forecast (air temperature, relative humidity, precipitation and nebulosity) for the following two days, allowing calculating the fire risk for the current day, as well as a forecast for the following day. Its calculus is based on a system presented by Fosberg and Deeming, [18] that allows crossing all these forecast variables.

- After calculating FM_{10}, the Fractional Ten-hour Fuel Moisture map (TN_f) is calculated for each pixel by normalizing FM_{10} values on Dead Fuel Moisture of Extinction (MX_d), applying the following formula:

$$TN_f = \frac{FM_{10} - 2}{MX_d - 2} \tag{12.5}$$

After calculating the Fractional Ten-hour Fuel Moisture map (TNf) and the Live Fuel Ratio map (LR), the Fire Potential Index (FPI) is calculated with the following formula:

$$FPI = (1 - TN_f)\,(1 - LR) \cdot 100 \tag{12.6}$$

Fire Potential Index provides a map with values ranging from 0 to 100 to be further integrated with the Structural Fire Index in order to provide the Integrated Forest Fire Risk.

Integrated Forest Fire Index

After calculating the Structural Fire Index and the Fire Potential Index, these two indices have to be integrated in order to provide an efficient integrated forest fire risk map. Figure 12.1 presents the main schema of the entire process allowing providing the Integrated Forest Fire Risk service.

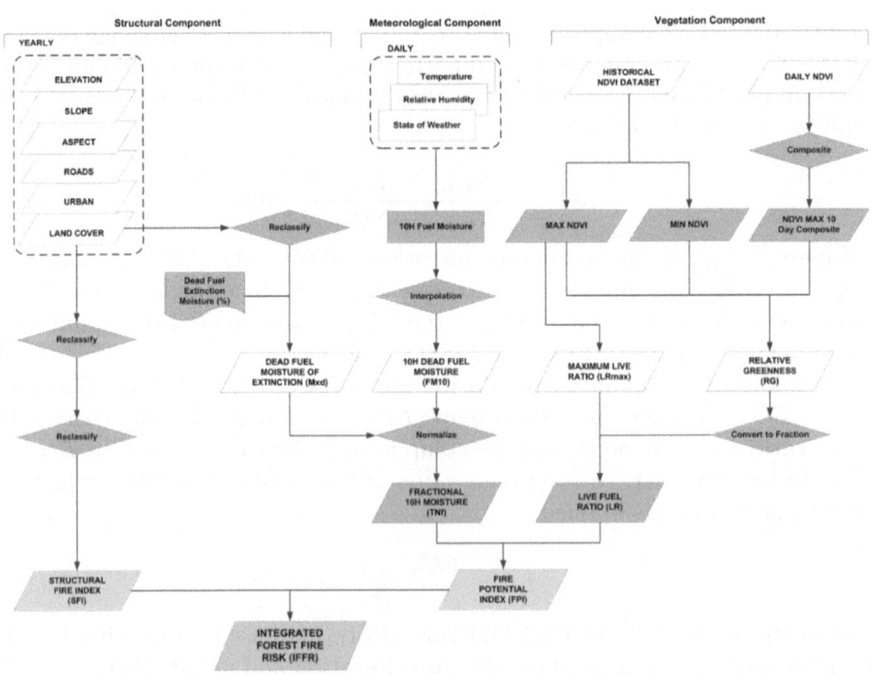

Fig. 12.1. Integrated Forest Fire Risk Index main schema [10]

Table 12.1. Integration of SFI and FPI providing the Integrated Forest Fire Risk

	FPI									
SFI	]0_10]	]10_20]	]20_30]	]30_40]	]40_50]	]50_60]	]60_70]	]70_80]	]80_90]	]90_100]
]0_25]										
]25_50]										
]50_75]										
]75_100]										
]100_125]										
]125_150]										
]150_175]										
]175_200]										
]200_225]										
]225_250]										
]250_275]										

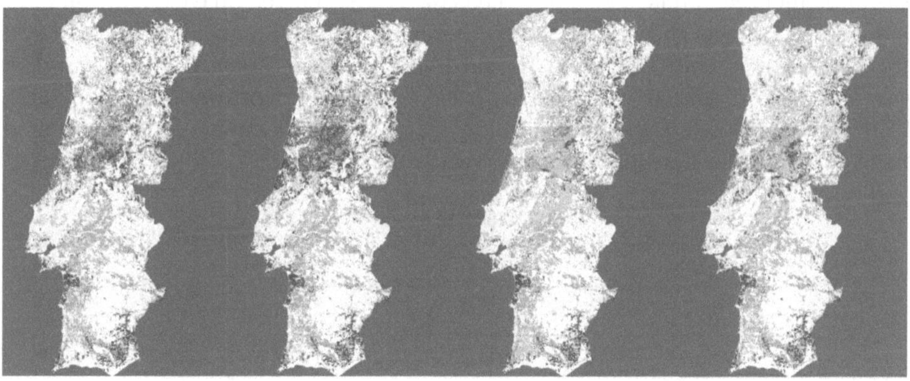

Fig. 12.2. Examples of Integrated Forest Fire Risk index in Portugal Mainland for four different days of 2005

During the PREMFIRE project, the integration was made in order to provide four fire risk classes. However, due to the new Portuguese legislation, the system had to be improved in order to provide five risk classes (low, medium, high, very high and extreme). This improvement was made during the COTEC initiative and the fire risk classes' distribution was made by evaluating several possible combinations of FPI and SFI, trying to achieve the best possible compromise between fires and the extent of each fire risk class, with data concerning forest fires occurred during the years of 2004 and 2005 [27]. Table 12.1 presents the chosen combination after the year 2005, where each colors represent fire risk classes and the arrow direction indicates the increasing of fire risk.

The result provided to final users is a map in the raster data model (Figure 12.2) with information regarding the five risk classes of the Integrated Forest Fire Risk.

12.2.4 Validation

During the Year of 2005, the Integrated Forest Fire Risk methodology has been operated during the Portuguese forest fire season (15th May to 15th of October) and validated against two types of data (ignition points and burnt areas). The validation process used the following data [27]:

- 137 ignition points resulting in fires larger then 6 ha, occurred during the period of 15th of July to 15th of September;
- All burnt areas larger then 48ha, occurred during the period of 21st of July to 13th September.

Since the ignition points provided by the Portuguese Forest Department of the Ministry of Agriculture usually referred to the closest urban area, and since the IFFR does not provide information for urban or agriculture areas, a system to relocate these ignition points to places where they could actually have occurred, was implemented. Thus, a grid of 16 cells (4*4) was centred on each ignition point and each point was then relocated to the pixel with the higher integrated forest fire risk value within that grid. Regarding the burnt areas, the data applied was manually digitized by ISA (Instituto Superior de Agronomia) from daily TERRA/AQUA MODIS images, allowing the identification of the amount of burnt area for the days presented on the Table 12.2.

Table 12.2. Days in which burnt areas were collected

July	21, 23
August	2, 4, 7, 8, 11, 15, 16, 17, 19, 22, 23, 24, 25, 26, 28, 31
September	2, 4, 8, 13

In order to evaluate the Integrated Forest Fire Risk Maps against the burnt areas, the following procedure was applied [27]:

1. For each period occurred between two days, identify all new burnt areas;
2. For each pixel, identify the value of Structural Fire Index (SFI);
3. For each pixel, identify, for the period occurred between two burnt areas collection, the average value of Fire Potential Index (FPI);
4. Based on the values identified in the two previous lines, assign to each pixel the class of Integrated Forest Fire Risk (IFFR).

After completing the previous procedures, the percentage of ignition points and burnt areas affected to each IFFR class was identified (Table 12.3).

Table 12.3. Results obtained for the Integrated Forest Fire Risk (IFFR) map

Risk Class	Ignition Points (%)	Burnt Areas (%)
Extreme	78	33
Very High	17	27
High	6	25
Medium	0	14
Low	0	1

As it can be observed on Table 12.3, the great majority of the ignition points (78%) were found in the Extreme IFFR class and all the ignition points occurred in the three higher risk IFFR classes. Although, the ignitions points' analysis should be done with some reserve, since its inaccuracy was countered with a generalization process. Regarding the evaluation against the burnt areas, it can be observed that during the period under analysis, 60% of the total burnt areas occurred on the two most dangerous IFFR classes.

12.3 Automatic Burned Area Mapping

12.3.1 Motivation for Service Development

Several studies and operational programs based on satellite imagery have been carried out with the aim of producing cartography and statistics of burnt areas [1, 2, 3, 6, 7, 17]. In these projects, several methodologies have been applied to satellite imagery acquired by sensors with different characteristics.

These different sensors characteristics, such as the spatial resolution, have obvious consequences in the technical specifications of the cartography and statistics of burned area (e.g., minimum mapping unit). Even though these effects shouldn't be judged as erroneous, once they are a consequence of the characteristics inherent to each image, it is important to define a balance point between final users needs and the information capable of being provided [1, 2, 3].

After evaluating Portuguese user's requirements and Earth Observation (EO) imagery availability, a system balancing these two factors was developed in order to provide valuable information, on useful time, to final users. The system developed is completely automatic and provides final users with burnt areas cartography, on a daily basis, concerning information acquired four days before. This delay is due to the service chain of the United States Geological Survey (USGS) that provides the surface reflectance product to users worldwide. [4].

12.3.2 Data Requirements

In different situations, there are different needs and possibilities to produce information about burnt areas. During forest fire season, in a first temporal phase, the priority is based on obtaining the information in the shortest period of time as possible, to carry out a fast assessment of damages allowing the provision of public general information and for an eventual redefinition of strategies on fire fighting. This first information is directly influenced by the temporal resolution of the sensors, which often implies the use of lower spatial resolution imagery due to the fact that these two resolutions (spatial and temporal) are indirectly related.

After crisis season, there is the need to identify, with the maximum accuracy as possible, the affected areas, aiming the accurately quantification of the damages and allowing reparation efforts and financial resources allocation. During this phase, there is a greater concern with the spatial resolution, instead of the temporal one. This greater spatial resolution acquired by satellites such as Landsat, SPOT or QUICKBIRD and IKONOS allows the production of cartography with smaller minimum mapping units, i.e. identification of burned areas with smaller extensions [17, 25, 28, 30, 34]. Though this kind of data is commonly used in many studies, it is almost impossible to use during the forest fires occurrence because of the low temporal resolution. In fact, there is a need of balance between the spatial resolution, accuracy, acquisition cost, and images availability [13].

After testing several methodologies and evaluating the spatial resolution influence on the information to be extracted from the satellite imagery [1, 2, 3], an automatic system based on medium spatial resolution imagery, acquired by Moderate Resolution Imaging Spectroradiometer (MODIS) hyperspectral sensors, installed in TERRA (1999) and AQUA (2002) satellites from the National Aeronautics and Space Administration (NASA), was developed and implemented in order to provide, on a daily basis, burn scars maps during the forest fire season. Besides land cover data collected by Landsat TM imagery, this automatic system applies the MODIS daily surface reflectance (250 meters of spatial resolution) and acquisition angles products to detect the burn areas.

12.3.3 Methods

Figure 12.3 depicts the high level methodology architecture of the burn scar maps service, as well as the different types of data used for their production.

The automatic burn scars detection methodology entails two main steps [2]:

- Forest change detection, where changes on forest cover are identified by multi-temporal approach based on vegetation index differences;
- Forest change discrimination, where forest changes detected in the first step are now discriminated by type (i.e., burn areas are separated from other changes such as clear cuts and new plantations).

The selection of this methodology was due to the fact that it allows to implement the system in a systematic way, without human intervention, and also because it has been successfully implemented in several other circumstances (e.g. [16, 20, 33]).

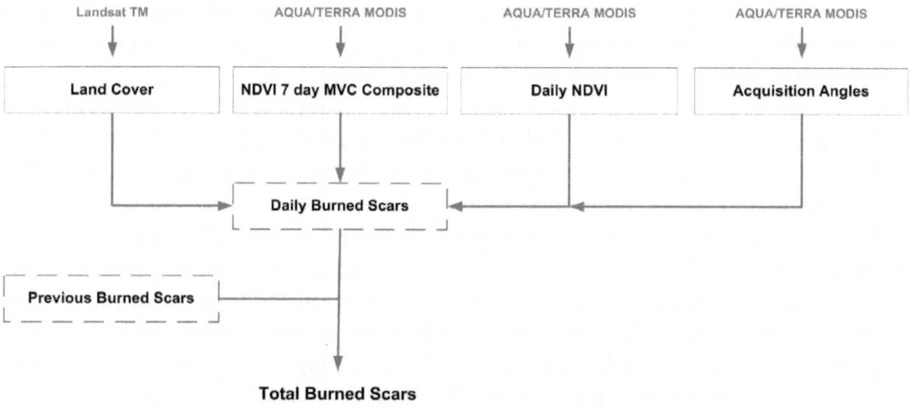

Fig. 12.3. Main Scheme of the Burn Scar Map Production

Forest Change Detection

In the change detection methodology, Normalized Difference Vegetation Index (NDVI) from two different dates are subtracted, resulting in a new image with values around zero for places where no changes occurred and positive or negative values for areas with change (depending if there was a decrease or increase in

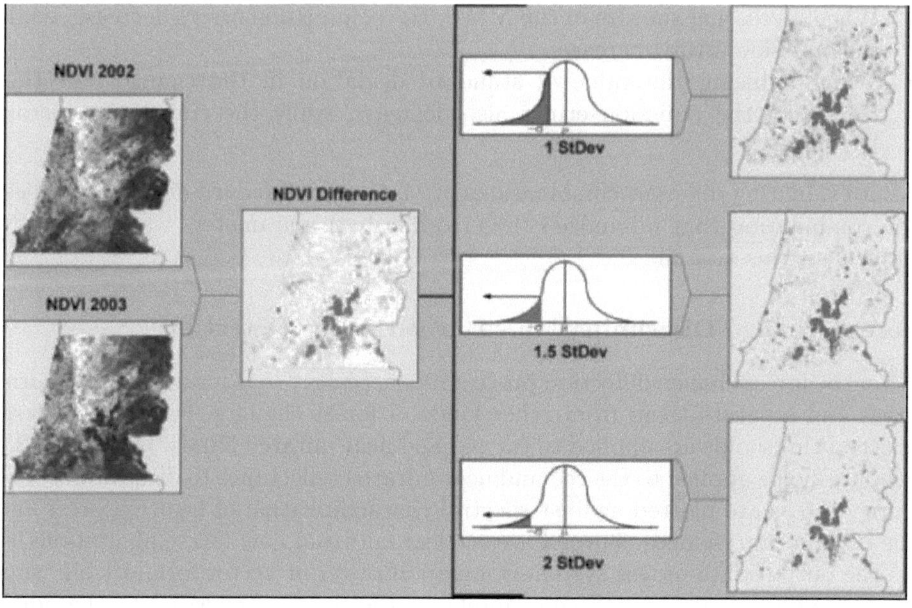

Fig. 12.4. Forest Change Detection Methodology example with different thresholds applied (Armas and Caetano, 2005a)

vegetation biomass). To discriminate the less significant differences on NDVI from the relevant ones (which correspond to real changes), a threshold is applied to the NDVI difference image.

In a first step, NDVI is calculated for the current day and subtracted of the maximum value composite (MVC) technique produced with the last 7 days NDVI. Then, it is applied a threshold based on the mean and standard deviation of the vegetation index difference image. Pixels with a value above a certain number of standard deviations to the mean value of the NDVI difference are considered to be changes. Since the decreases in the difference image are the ones intended to be founded, the standard deviation value is subtracted from the mean value of the vegetation index difference image. During this process it is applied an analysis mask based on land cover maps produced with Landsat TM imagery, in order to remove from the analysis changes occurring in urban, agriculture and water land covers. Figure 12.4 presents the influence of different values of standard deviation on the threshold applied in the change detection methodology.

After detecting the changes, a Minimum Mapping Unit (MMU) of 9 pixels (48 hectares) is applied. All changed areas smaller then the MMU are generalized (i.e., eliminated) before calculating the final product. Before implementing this system, several studies [1, 2, 3] were performed with MODIS imagery in order to evaluate the influence of the standard deviation applied in the change detection threshold, as well as in the MMU definition. Within these studies, the following was confirmed:

- When increasing the size of the MMU, the commission errors decrease, while the omission errors increase.
- When reducing the value of standard deviation in the change detection threshold, the omission errors also decrease, while the commission errors increase.

After evaluating different combinations of MMU and standard deviation values, the combination that minimized the errors in both parameters was selected to implement this system.

Forest Change Discrimination - Burned Areas Detection

After the forest change detection process it is necessary to discriminate the burnt areas and separate them from other kinds of forest changes. For that purpose spectral thresholds are applied to the red and near infrared bands. In Figure 12.5 the thresholds applied to the red and near infrared reflectance to discriminate the burned areas are plotted against spectral characterization of healthy forest and the main forest changes (burned areas, clear cuts and new forest plantations).

The output data of this system is a map in raster or vector format (.bil, .shp, .kmz). Figure 12.6 shows three examples of the product referring to the north western part of Portugal during the summer of 2006 (map layer superposed against MODIS images).

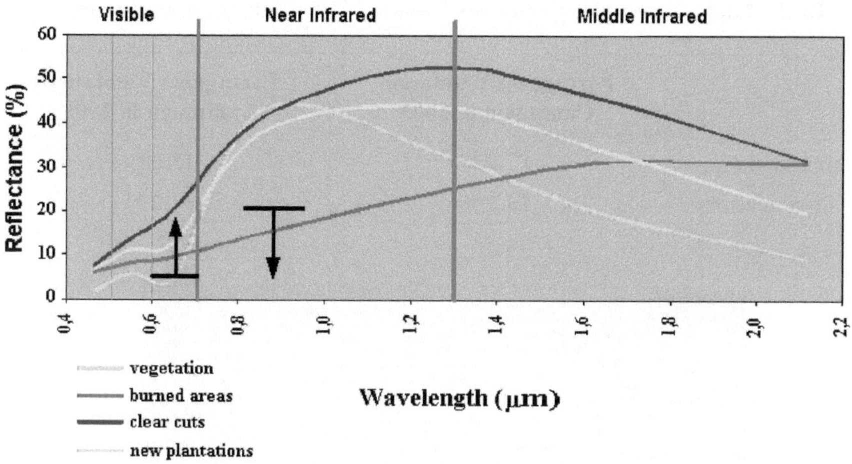

Fig. 12.5. Thresholds Applied to discriminate burned areas from other forest changes (adapted from [2])

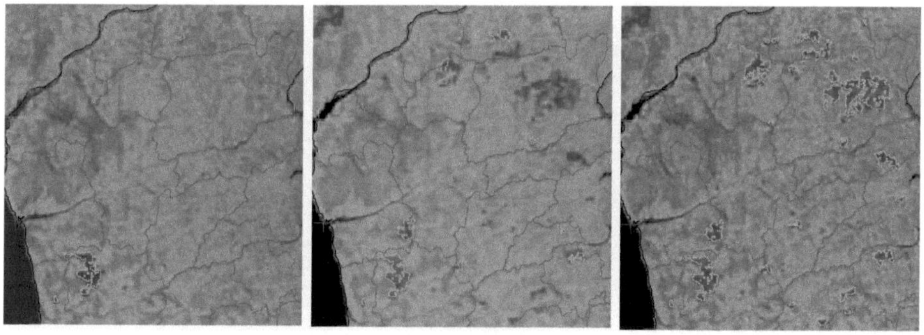

Fig. 12.6. Automatic Burn Scar Mapping results obtained for the year of 2006

12.3.4 Validation

This service started to be developed in 2003 and has been under constant improvements since then. During the forest fire season of 2006 it was operated in Portugal, for the first time, as a pre-operational service, achieving excellent results. This methodology has been validated during the years of 2003 and 2006 and, due to the inexistence of ground truth data, the service has been validated against reference data. For the year of 2003, the reference data was manually digitized over high spatial resolution imagery (SPOT-HRG), while for the year of 2006, the reference data was manually digitized over medium spatial resolution imagery (MODIS). The validation processes done until the year of 2007, have taken into account the quantification of omission errors and commission errors obtained for the burnt areas class. The omission error of the "burnt areas" indicates the percentage of

Table 12.4. Automatic Burn Scar Detection thematic accuracy values

	Portuguese Validation Campaign in 2003	Portuguese Validation Campaign in 2006
Commission errors	17,1 %	15,2 %
Omission errors	18,2 %	17,9 %

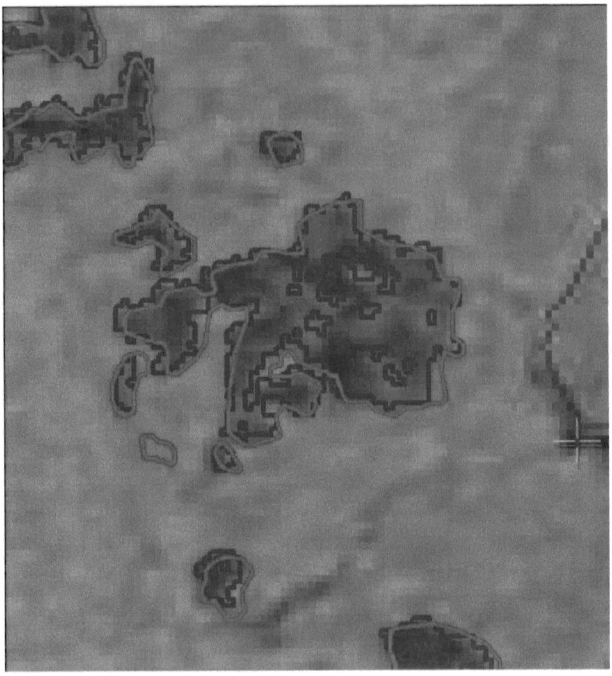

Fig. 12.7. Automatic detection (blue) plotted against reference data (red) superposed to MODIS imagery (Armas and Pelizzari, 2007)

changes that were not identified. The commission error of the "burnt areas" indicates the percentage of areas that were misclassified as "burnt areas".

Table 12.4 presents the accuracy values obtained during the years of 2003 and 2006 [2, 4].

As it can be seen, after the validation for the years of 2003 and 2006, similar accuracy results were obtained, indicating the system stability and operationality. Since the Omission and commission results obtained are very similar, it indicates that this system has reached a good compromise between all the parameters within the system. Figure 12.7 presents the results obtained for the year of 2006 by the system herein presented (in blue) against reference data manually digitized over MODIS imagery (in red).

The main advantages of this system are:

- The fully automatic process, eliminating the subjectivity of the human intervention;
- The automatic provision of island polygons not burned inside a large burned area.
- The clear definition between the burned and not burned areas.

The main limitation of this service is its unavailability whenever 50% of the source data (surface reflectance) has clouds or is acquired with large acquisition angles.

12.4 Conclusion

This chapter showed that computational intelligence is a valuable support to solve complex problems where traditional approaches are not sufficient. The case studies presented in this chapter and related to forest fires are based on remote sensing data fusion and automatic change detection in satellite images. The sensors used in these case studies are medium and high resolution space borne multispectrometers (MODIS and Landsat TM). The processing chains are fully automatic and daily maps are delivered throughout the entire fire season without intervention of human experts.

Future developments of earth observation based techniques, especially in the case of forest fires emergency management, will be driven by the following main factors:

- Increased data availability;
- Higher computational performance.

Whereas the pre-fire and post-fire phases are already well supported by currently available remote sensing missions, data availability, i.e. higher revisit frequency, is a key issue for an efficient monitoring of the fire outbreak (early warning) and evolution during the crisis. The ongoing public and commercial efforts to design and build constellations of small satellites bearing the same kind of sensors will be a decisive step forward for a continuous global surveillance of the forest environment. At the same time the usage of different platforms like unmanned aerial vehicles could complement observations from space and significantly improve the coverage of regions characterised by a high level of risk.

The huge data volume that such a system produces must be processed and analysed in a fully automatic way to guarantee timely maps provision. Computational intelligence plays here a major role and a strong interaction among the computer science community (image processing, high performance computing, computer vision) and the remote sensing data interpretation experts is required.

The results obtained by the usage of computational intelligence to analyse remote sensing data for forest fires emergency management are encouraging. This is true not only from the scientific point of view but also in terms of operational applications that can be already used to better prevent environmental loss as well as to rapidly and objectively monitor and assess the damages.

References

1. Armas, R., Caetano, M.: Scale effect on burned area assessment using earth observation data. In: Oluic, M. (ed.) New Strategies for European Remote Sensing, pp. 61–67. Mill Press, Rotterdam (2005a)
2. Armas, R., Caetano, M.: Mapping changes in forest cover using multi-scale MODIS imagery. In: Proceedings of the 31st International Symposium on Remote Sensing of Environment - Global Monitoring for Sustainability and Security, Saint Petersburg, Russian Federation, unpaginated CD-ROM, June 20-24, 2005 (2005b)
3. Armas, R., Caetano, M.: Cartografia de áreas ardidas com imagens multi-resolução do sensor MODIS. In: Casacae, J., Matos, J. (eds.) Cartografia e Geodesia 2005, Lidel, Lisboa, pp. 293–303 (2005c)
4. Armas, R., Pelizzari, A.: Automatic Burn Scar Mapping - Project Final Report, p. 37, Critical Software (2007)
5. Anderson, H.E.: Moisture and fine forest fuel response. In: Proceedings of the 8th Conference of Fire and Forest Meteorology, Detroit, Michigan, pp. 192–199 (1985)
6. Barbosa, P., San-Miguel-Ayanz, J., Schmuck, G.: Remote Sensing of forest fires in Southern Europe using IRS -Wifs and MODIS data. In: Proceedings SPIE Remote Sensing Symposium, Toulouse, France, September 18-21, 2001, pp. 165–173 (2001)
7. Barbosa, P., Ayanz, J., Martinez, B., Schmuck, G.: Burnt area mapping in southern Europe using IRS-WIFS. In: Proceedings IV International Conference Forest Fire Research, Luso, Portugal, November 18-23 (2002)
8. Burgan, R.E., Hartford, R.A.: Monitoring Vegetation Greenness with Satellite Data, Gen. Tech. Rep. INT-297. Ogden, UT: U.S. Department of Agriculture, Forest Service, Intermountain Forest and Range Experiment Station (1993)
9. Burgan, R.E., Klaver, R.W., Klaver, J.M.: Fuel Models and Fire Potential from Satellite and Surface Observations. International Journal of Wildland Fire 8(3), 159–170 (1998)
10. Caetano, M., Carrão, H., Freire, S.: Methodology for Fire Risk Mapping, PREM-FIRE Project (ESA 15720/01/I-SB) WP330 Technical Report, p. 40 (September 2002)
11. Caetano, M., Freire, S., Carrão, C.: Fire risk mapping by integration of dynamic and structural variables. In: Goossens, R. (ed.) Remote Sensing in Transition, pp. 319–326. Millpress, Rotterdam (2004)
12. Carrão, H., Freire, S., Caetano, M.: Fire Risk Mapping Using Satellite Imagery and Ancillary Data: Towards Operationally. In: Owe, M., d'Urso, G., Toulios, L. (eds.) Proceedings of SPIE, Remote Sensing for Agriculture, Ecosystems and Hidrology IV, vol. 4879, pp. 154–165 (2002)
13. CCRS, Fundamentals of Remote Sensing, Canadian Centre for Remote Sensing (2003), `http://ccrs.nrcan.gc.ca/resource/tutor/fundam/index_e.php` (consulted on May, 13th 2007)
14. Chuvieco, E., Congalton, R.G.: Application of Remote Sensing and Geographic Information Systems to Forest Fire Hazard Mapping. Remote Sensing of Environment 29, 147–159 (1989)
15. Chuvieco, E., Salas, F., Vega, C.: Remote Sensing and GIS for long-term Fire Risk Mapping. In: Chuvieco, E. (ed.) A review of remote sensing methods for the study of large wildland fires, pp. 91–108 (1997)
16. Daryaei, J.: Digital change detection using multi-scale wavelet transformation & neural network. Master thesis, International Institute for Aerospace survey and Earth Sciences, Enschede, Netherlands (2003)

17. Escuin, S., Navarro, R., Fernández, P.: Using Remote Sensing and GIS to assess wildfire damage throughout the Mediterranean. Earth Observation Magazine 11(5), 19–22 (2002)
18. Fosberg, M.A., Deeming, J.E.: Derivation of the 1- and 10-hour Timelag Fuel Moisture Calculations of Fire Danger. United States Department of Agriculture, Forest Service, Research Note RM-207, Rocky Mountain Forest And Range Experiment Station, Fort Collins, Colorado (1971)
19. Freire, S., Carrão, H., Caetano, M.: Produção de Cartografia de Risco de Incêndio Florestal com Recurso a Imagens de Satélite e Dados Auxiliares. In: Proceedings of ESIG 2002- VII Encontro de Utilizadores de Informação Geográfica, Oeiras, November 13-15 (2002) (unpaginated CD-ROM)
20. Gong, P.: Land Cover and Land Use Mapping and change dynamics modeling with remote sensing. In: Advanced Training Workshop on Landuse and Landcover change Study, Taipei, Taiwan (2002), http://www.gcc.ntu.edu.tw/workshop/lecturers_materials.htm (consulted on March 20, 2004)
21. Joint Research Centre. Pilot Projects on Forest Fires (2001), http://natural-hazards.aris.sai.jrc.it/fires/ (consulted on April 12, 2002)
22. Joint Research Centre. Forest Fires in Europe - 2003 Fire Campaign (2004), http://www.fire.uni-freiburg.de/programmes/eu-comission/JRC-Report-4-Forest-Fires-Europe-2003-Fire-Campaign.pdf (consulted on May 15, 2007)
23. Joint Research Centre, Forest Fires in Europe 2005, EUR 22312 EN (c) European Communities (2006)
24. Klaver, J., Klaver, R., Burgan, R.: Using GIS to assess forest fire hazard in the Mediterranean region of the U.S. In: Proceedings of 17th Annual ESRI Users Conference, San Diego, p. 10 (1997)
25. Koutsias, N., Karteris, M., Chuvieco, E.: The use of Intensity-Hue-Saturation Transformation of Lansdat-5 Thematic Mapper Data for Burned Land Mapping. Photogrammetric Engineering & Remote Sensing 66(7), 829–839 (2000)
26. Lopéz, A., San-Miguel-Ayanz, J., Burgan, R.: Integration of Satellite Sensor Data, Fuel Type maps and Meteorological Observations for Evaluation of Forest Fire Risk at the Pan-European Scale. International Journal of Remote Sensing 23(13), 2713–2719 (2002)
27. Mangana, P., Gonçalves, R., Caetano, M., Carrão, H., Pereira, C., Carreira, J., Viegas, D., Ribeiro, L., Galante, M.: Previsão de Incêndios Florestais. COTEC Portugal - Iniciativa sobre incêndios florestais. PREVINFL Project Final Report, p. 118 (2006)
28. Mitri, G., Gitas, I.: The development of an object-oriented classification model for operational burned area mapping on the Mediterranean Island of Thasos using Landsat TM images. In: Viegas, X. (ed.) Forest Fire Research & Wildland Fire Safety. Mill Press, Rotterdam (2002)
29. National Interagency Fire Center - NIFC (USA, 2007), Wildland Fire Statistics, http://www.nifc.gov/fire_info/fire_stats.htm
30. Recondo, C., Wozniak, E., Pérez-Morandeira, C.: Map of burnt zones in Astúrias in the period 1991-2001 created from Landsat-TM images. In: Proceedings of the 4th International workshop on Remote Sensing and GIS Applications to Forest Fire Management: Innovative concepts and methods in fire danger estimation (Ghent: EARSEL- European Association of Remote Sensing Laboratories), pp. 193–197 (2003)
31. Rothermel, R.C.: A mathematical model for predicting fire spread in wildland fuels. General Technical Report INT-115, USDA Forest Service, Intermountain Forest and Range Experiment Station (1972)

32. Rouse, J., Haas, R., Schell, J., Deering, D.: Monitoring Vegetation Systems in the Great Plains with ERTS. In: Proceedings of 3rd ERTS Symposium, vol. 1, pp. 48–62 (1973)
33. Salvador, R., Valeriano, J., Pons, X., Díaz-Delgado, R.: A semi-automatic methodology to detect fire scars in shrubs and evergreen forest with Landsat MSS time series. International Journal of Remote Sensing 21(4), 655–671 (2000)
34. Sunar, F., Özkan, C.: Forest fire analysis with remote sensing data. International Journal of Remote Sensing 22(12), 2265–2277 (2001)
35. Wulder, M.A., Franklin, S.E.: Remote Sensing of forest environments, Introduction. In: Remote Sensing of Forest Environments, pp. 3–12. Kluwer Acad. Publ., Dordrecht (2003)
36. Gong, P., Xu, B.: Remote Sensing of forests over time. In: Remote Sensing of Forest Environments, pp. 301–333. Kluwer Acad. Publ., Dordrecht (2003)

13

Automatic Preprocessing and Classification System for High Resolution Ultra and Hyperspectral Images

Abraham Prieto, Francisco Bellas, Fernando Lopez-Pena, and Richard J. Duro

Integrated Group for Engineering Research, Universidade da Coruña
Campus de Esteiro, 15403, Ferrol, A Coruña, Spain
{abprieto,fran,flop,richard}@udc.es

13.1 Introduction

Spectroradiometric imaging is based on the fact that all materials display a specific spectral signature when light impinges on them. That is, they reflect, absorb, and emit electromagnetic waves in a particular pattern depending on the wavelength, leading to a spectrum for the material that can be used to classify it. An imaging spectroradiometer captures images where the spectral information of every pixel is collected for a set of contiguous discrete spectral bands. It is customary to classify these instruments as ultraspectral, hyperspectral, multispectral, and panchromatic imagers. Typically, ultraspectral data sets are made up of thousands of spectral bands, while hyperspectral sets have hundreds, multispectral tens, and panchromatic sets just a few. Thus, each ultraspectral or hyperspectral image contains large amounts of data which can be viewed as a cube with two spatial and one spectral dimension. Regardless of the application of the images, the analysis methods employed must deal with large quantities of data efficiently [1]. Originally, imaging spectroradiometers were used as airborne or spaceborne remote sensing instruments. During the last decade, spectral imaging has become a common advanced tool for remote sensing of the surface of Earth. It is being used to collect crucial data about many activities of economic or scientific relevance [2] as, for instance, mineral exploitation [3], agriculture and forestry use [4], environmental pollution [5], ecosystem studies [6], assessment of natural resources [7], water resource management [8], monitoring of coastal zones [9] and many others. For all these remote sensing applications the hyperspectral images obtained usually contain mixtures of spectra in every pixel due to the poor spatial resolution of the images when taken a large distance from the target. Thus a single pixel could typically correspond to one hundred or more square meters of land. As a result, most analysis methods developed had as their main objective to provide the segmentation of the images in terms of the ratio of endmembers present in every pixel to improve the spatial discrimination of these systems when analyzing different types of land covers [10].

Recently, imaging spectroradiometers have started to be used in several ground based applications ranging from different types of field measurements

M. Graña and R.J. Duro (Eds.): Comput. Intel. for Remote Sensing, SCI 133, pp. 313–340, 2008.
springerlink.com

to laboratory or industrial purposes [11]. This chapter is devoted to this new type of applications for imaging spectroradiometers and more particularly for the industrial classification of real materials. For all these new types of applications, the images are taken close enough to the subject to obtain a very detailed view of it. Therefore, the problem does not always lie in the extraction of subpixel information, but rather, it can sometimes be formulated as a multipixel based classification problem. In such cases the grains of different components constituting the material can be distinguished in the image. Therefore a single pixel does not contain a representative enough set of spectra to characterize the whole material, and an appropriate spatial and spectral combination of information coming from different pixels is required to achieve a correct classification. As ultraspectral imagers are just in their beginning and have not yet been used for ground based applications, in the following we are going to talk only about hyperspectral images, but the methodology presented here is general and may also be applied to ultraspectral images.

For ground based applications hyperspectral systems are generally deployed following a scheme commonly used for any other standard imaging device and under well controlled conditions of luminance and position. This represents a great advantage with respect to remote sensing applications characterized by their intrinsic changeability and the presence of confounding factors, and makes it possible to simplify, improve and, in some cases, automate the analysis of the data. As labeling of ground truth images for training is quite easy under well controlled conditions, this method of calibration is usually preferred over radiative transfer based models. Thus, hyperspectral processing depends on how some experts classified the ground truth in order to be able to classify new samples. For instance, an expert will label some material as type x granite without specifying what features make it so. These features have to do with the spatial combination of the components of the granite sample (quartz, feldspar, mica and other minority components), each component represented by some features in its spectrum. The problem now is how to extract the spatial and spectral features required for the correct classification of material samples given some ground truth sample classifications provided by the experts when the spatial resolution of our hyperspectral system is high enough so that each grain of a single component of the material is larger than the pixel size.

Some work has been carried out by different authors on the integration of spatial and spectral information coming from hyperspectral imaging devices in order to improve the discrimination capabilities of the systems. Most of this work has been aimed, as stated before, at improving endmember extraction in remote sensing through the consideration of the surroundings of a pixel [12]. Only recently, have some authors presented approaches for the extraction of multipixel patterns from airborne hyperspectral sensors. An example of this approach can be found in [13]. The authors propose an autoregressive Gauss-Markov random field (GMRF) to model the predictability of a target pixel from neighboring pixels for the classification of spatial patterns of aircraft and residential areas using information provided by the HYDICE airborne sensor. Another example is the work by Van der Werff and col. [14] who have proposed an algorithm based on sequential Hough transforms to detect botanical and mineralogical alterations that result from natural seepage of carbon dioxide and light hydrocarbons. These halos can only be distinguished from the background by their specific spatial pattern.

Both cases have resorted to the hand tuning of their algorithms and parameters to be able to take into account the specific spatial features needed for the discriminations required in their particular problems. These cases share with the present one the need of searching for spatial structures within the image; however they are looking for specific common spectral information characterizing these structures while the problem stated here is to try to blend individual pixel spectral characteristics to obtain a common footprint of the structure.

To do this in this chapter we are going to describe and present some results using a neural network based classification system that is embedded within an automatic preprocessing stage used for automatically determining the optimal parameters for the dimensional reduction that would be required so as to obtain a classification that is in accordance with the labeling provided by the experts for the ground truth. Thus, the idea is to obtain the smallest possible description of the hyperspectral image that is consistent with a robust classification of the areas under consideration according to the labels provided by experts. Consequently, throughout the chapter, we will consider both, the dimensional reduction and the problem of establishing measurable indices for determining what is a robust and accurate classification, or, rather, how robust we want it to be for a given problem and thus be able to find an equilibrium between dimensional reduction and classification accuracy and robustness.

The chapter is structured as follows: section 13.2 describes the initial design of the system and the first results where the parameters of the preprocessing were hand tuned. Section 13.3 is devoted to the automation of the preprocessing stage and the results obtained over synthetic images. The final implementation of the system and its application to real images is described in section 13.4. Finally section 13.5 discusses some conclusions that can be derived from this work.

13.2 Experimental Set-Up

The hyperspectral sensor used in this work was developed in the framework of a research project for developing a complete, automatic, flexible, reliable, fast and precise real time detection and cartographic representation system for marine spills. The system is described elsewhere and its core is an instrument designed to be used from light aircraft or AUVs and was constructed by merging a lightweight hyperspectrometer with a series of positioning sensors for determining its geographical position and orientation. This push-broom type light imaging hyperspectrometer is an independent module developed to be used as a portable device on its own. Fig. 13.1 displays its configuration.

It was made of commercially available components allowing for low-cost fabrication. In its operation, the slit allows just one line of the subject image which

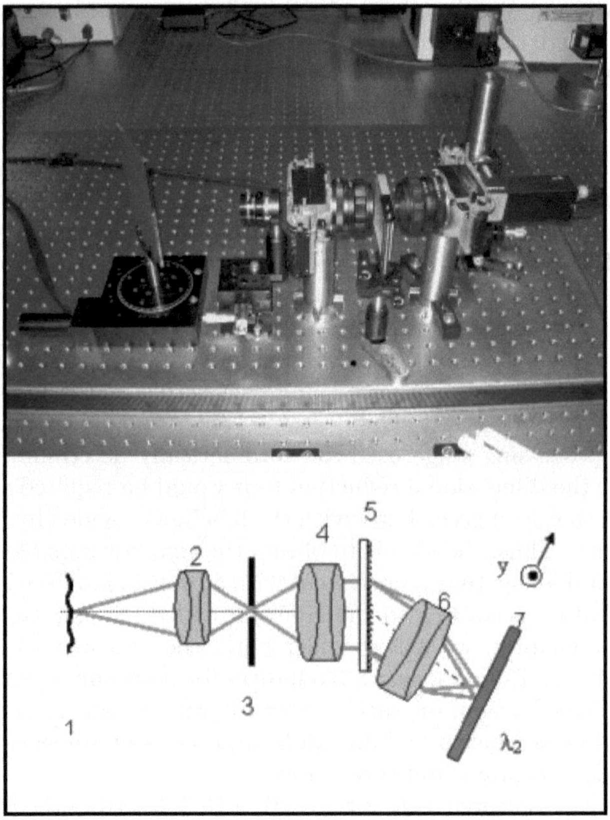

Fig. 13.1. Real hyperspectrometer (top) and schematics (bottom): (1) subject, (2)(4)(6) objectives, (3) slit, (5) grating, (7) CCD

is decomposed into a spatial dimension plus a spectral dimension in the image plane. A 12 bit CCD camera acquires these images corresponding to a section of the hyperspectral cube. The instrument has an acceptable signal to noise ratio for wavelengths in the range between 430 and 1000 nm. The measured spectral resolution during calibration tests was of 5 nm while spatial resolution was about 30 μm in the image plane. More details of the instrument can be found in [15].

A virtual instrument has been developed to implement the Artificial Neural Network (ANN) based processing module plus a user friendly graphical interface for visualization, control and post-processing. Fig. 13.2 displays a screenshot of the graphical interface, where four samples of different qualities of a given material are being analyzed. The color image of the subject is reconstructed from the hyperspectral data and represented on a frame. Another frame represents the spectrum of the sample at the position of the cursor on the color image.

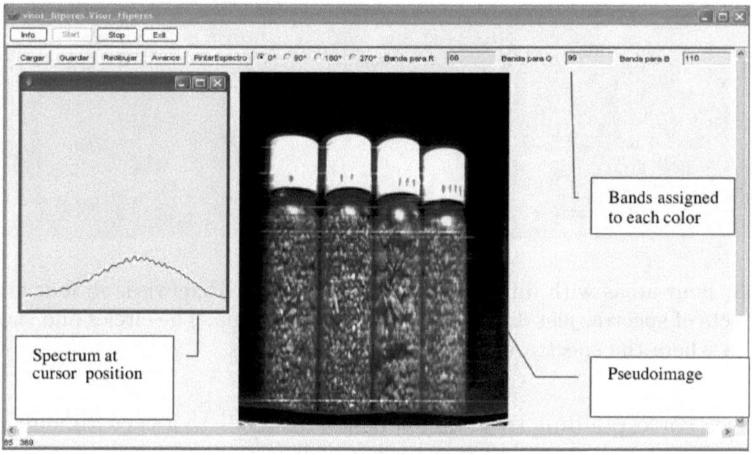

Fig. 13.2. Graphical interface developed for visualization of the hyperespectral cubes

13.3 Preprocessing Stage

The first step in the definition of a system for the automatic classification of materials using hyperspectral images came about as a consequence of a real task we needed to carry out: to perform quality control in an andalusite processing plant. Andalusite is an anhydrous, aluminum silicate mineral used in refractory applications both in bricks and in monolithic products. The main characteristics when determining the quality and uses for andalusite samples are their grain size and purity [16]. Thus, in order to classify the minerals going through the processing line some type of discrimination based on the optical characteristics induced by variations of these two parameters should be obtained from the hyperspectral images. To this end an ANN based hyperspectral classification system was developed to detect in real time and in a given section of the processing line the andalusite quality based on these two parameters. The key premises for this application were that the spectra of the andalusite changed with the level of impurities and with its granularity. ANNs are a good choice due to their proven supervised classification or categorization capabilities in general, and in the case of hyperspectral images [17] in particular.

For the classifier ANN to be trained, a training set including true input-output pairs must be provided. Inputs are usually spectra and target outputs are the classification category provided by the expert. If the area that has been labeled as belonging to a particular category is not spectrally homogeneous, that is, any of its pixels produce different spectra, some of which may even belong to other categories, the training process degrades, becoming impossible in some cases. One very good example of this can be found in the literature on the processing of the classical Indian Pines hyperspectral image [18] [19] regarding the areas classified as corn and soybean. In these areas the plants were very young and thus only represented a small percentage of the pixels, being earth most of the

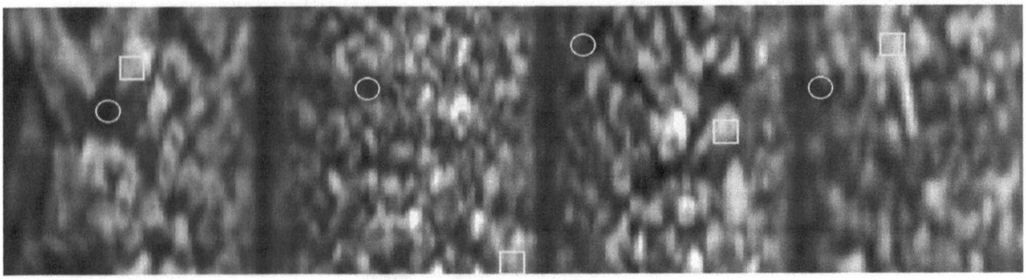

Fig. 13.3. Four areas with different types of andalusite. Pixelwise, all four areas have the same sets of spectra, just different distributions of them. The circles and the squares show areas where the spectra are the same.

ground area corresponding to a single pixel. This leads to a very difficult training problem, as the labels do not correspond to a given spectrum or a range of similar spectra and some pixels in both areas contain spectra that are the same but that are labeled differently.

The original solution for the problem in hand was to develop a preprocessing stage to determine the *size for the processing window* that will be displaced over the image and the *optimal combination of the parameters of the spectra* in the different pixels of the window so that all the windows within an area with a particular label are represented by a similar set of inputs. Thus, one is basically trying to obtain the smallest window that through some function that combines the spectra of its pixels provides a vision of the area that is spectrally homogeneous. For the classification itself, several supervised classification system using different types of Gaussian based ANNs and a multilayer perceptron were tested. Gaussian functions have been widely applied in ANN based image processing due to their noise filtering capabilities. A comparison of the results provided by three different types of these networks when applied to hyperspectral classification is presented in [20]. These types of networks are based on the standard radial basis function (RBF), the radial basis function with multiple deviations (RBFMD) and, finally, on the use of Gaussian synapses (GSBN). The first two types are structurally similar; the Gaussian function is applied over the inputs by using one parameter per synapse (center) and one parameter per neuron or per synapse in the hidden layer (deviation). The latter (GSBN) presents a multilayer perceptron structure, but replacing the simple weights used in the synapses by Gaussian functions in order to filter the inputs. The main difference between these three types of networks is the number of parameters to be trained, being the RBF the one requiring fewer parameters and the GSBN the one requiring most. This number of parameters determines the complexity of the learning process and the generalization capability of the system once trained.

These networks provide more complex and flexible decision boundaries, better adapted to the high dimensionality of the search space and the complexity of the subspace to discriminate in hyperespectral image classification problems. The application of these networks increases the number of parameters that must be trained and, consequently, the computational cost, but it should be compensated by the decrease of the minimum necessary network size and the increase in the speed of the network training stage while using fewer training samples. Thus, as an initial test, the andalusite images were processed in a pixelwise manner using these types of networks. The results obtained were quite poor, well below 40% classification accuracy. After analysis of the results it was concluded that the problem lies in the images themselves.

The andalusite samples all contained the same elements (as shown in Fig. 13.3 for four samples) the only difference between them was how they were mixed. Thus, pixel by pixel and with the resolution of the images available there was no way of discriminating between the four samples. The experts would say that the second sample from th left was andalusite with the right granularity and the appropriate proportion of andalusite to the other minerals present and the others had different proportions of andalusite with a different granularity. But experts were talking about an areawise appreciation and not a pixelwise one. In terms of training the neural networks, what was happening was that they were being presented with spectra corresponding to the same compound and told to train towards different labels. As a result, the networks ended up producing an average of the labels which led to very low classification accuracy. The conclusion of the study was that spatial or area wise information must be taken into account.

As a first approximation to test this hypothesis, instead of processing individual pixels, the images were divided into windows of different sizes, and it was determined that the smallest window size that allowed a correct classification was 20x20 pixels. Consequently, in order to train the networks the images were transformed in the spatial dimension by taking 20x20 pixel blocks (with their corresponding spectral dimension). These windows overlapped by 10 pixels.

Obviously, dividing the image into windows is useless if one does nothing to combine the information from the pixels that make it up. That is, one would like to obtain from the information present in the spectra corresponding to the window pixels some type of descriptor that uniquely defines them as a set. The most obvious approach would be to average the spectra for all the pixels. Averaging is a way of losing information present in the signals, and that is what happened, most areas ended up having average spectra that were so similar that it was very difficult to discriminate them. This can be appreciated for the spectra of Fig. 13.4 where the left side of each graph displays the corresponding average spectrum.

If the deviations of the different points with respect to the average spectrum (right side of the graphs of Fig. 13.4) are studied, it can be seen that the shape of these deviations is different for different samples. Consequently, the average deviation spectrum in the window together with the average spectrum were used as inputs to the networks. What is relevant here is that these descriptors are diverse for different classes and similar enough in the case of the same class for a network to be able to classify them easily.

One of the most important aspects of this work is that once this type of spatial integration preprocessing procedure is carried out, a simple multilayer perceptron with 144 inputs (corresponding to the different bands of the average

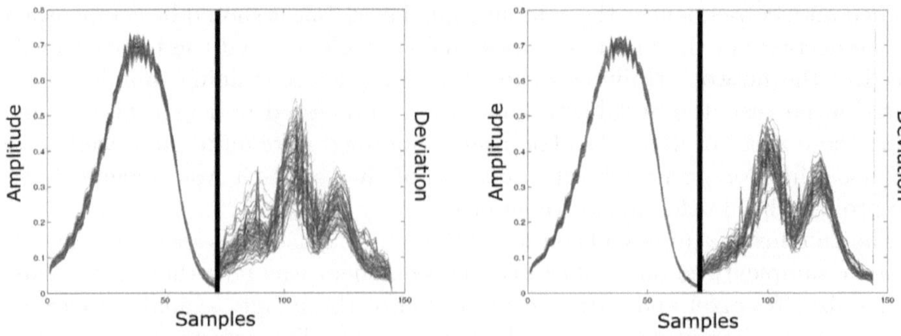

Fig. 13.4. Two different graphs showing the average spectrum (left part of each graph) and the rescaled deviation spectra (right part of each graph) for the correct andalusite composition and grain sample

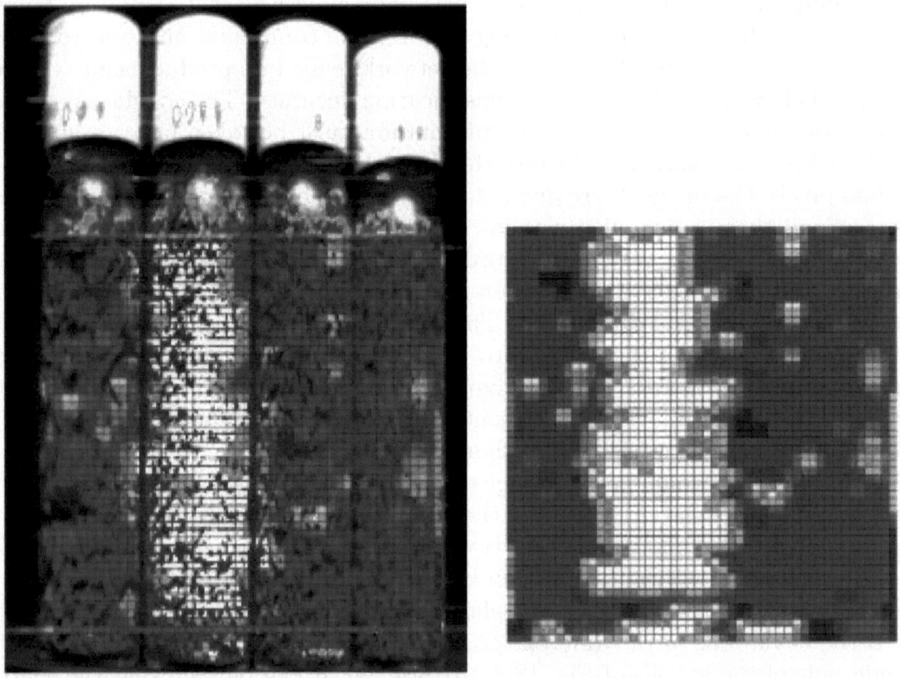

Fig. 13.5. Classification results of a test case having 4 samples of andalusite of different quality. The system is able to detect the right one (white) with 98% classification accuracy.

and deviation spectra), 2 hidden layers with 20 neurons each and one output could obtain on a test set of real images better than 98% classification accuracy.

The test case presented consists of four samples of andalusite of different quality extracted at four different points along a processing line. Only one of them (the second one from the left) is of the right purity and granularity. After training the systems as stated, the resulting network was tested on these samples. Fig. 13.5 displays an image where white color represents positive match for the appropriate andalusite grain and composition and dark grey color no match. This image clearly shows that the samples are classified correctly and there are only a few false negatives in the border of the bottle due to the reflections in the glass from the illuminating light source that induce false spectra.

As a conclusion for this initial study, the preprocessing procedure allows a large reduction in the complexity of the networks that make up the classifiers in addition to making the training procedure more reliable and faster. However, even though this first experiment can be taken as a proof of principle, it is clear that having to obtain the right window size and deciding what is discriminating for a given set of samples, that is, what parameters are to be taken into account by the network, is quite a tedious and not very methodological approach to the problem. Consequently, a way to automatically produce the same results was sought and is presented in the next section.

13.4 Automatic Procedure

The preprocessing stage presented in the previous section provided successful results but was very costly in terms of computational resources and in terms of user/researcher time, as the optimal windows size and processing parameters were basically hand tuned. In addition, the training of the ANN was a process that could not be started until the first stage had finished due to the fact that some parameters of the networks, such as the number of input neurons, depend on the parameters obtained in the preprocessing stage. At this point it was decided to automate the system in order to jointly obtain the optimal parameters for the preprocessing stage and the classification results provided by the ANN that uses those parameters.

The system core consists of three main interacting parts shown in the block diagram of Fig. 13.6. The left block represents the training dataset (hyperspectral cube). The second and third blocks constitute the first processing stage where the transformation of the portion of the raw data under consideration into the feature space description takes place through the determination of the window size and the dimensional transformation parameters. The fourth block corresponds to the ANN-based classifier and constitutes the second stage of the system, where the parts of the dataset are classified into the categories provided by the expert's labeling. Finally, the third stage is an adviser that, using the variation trends of the error along the training process, adjusts the parameters of the first stage so as to better adapt them to the production of the best possible classification. The next subsections provide a more detailed description of the three stages.

It is important to note here that the procedure is presented in the most general way, as it was thought out for problems of any dimension. In this line, as the first applications were designed for processing hiperespectral images, the initial matrix is a hyperespectral cube. The continuous dimension as contemplated in

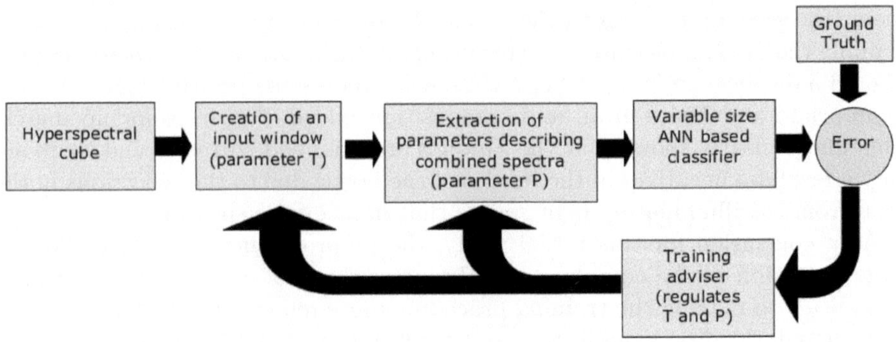

Fig. 13.6. Simple block diagram of the processing and classification system

the terminology employed is the spectral dimension and the discrete one is the spatial dimension. The distance measure used to create the data windows is the Euclidean distance in the area we are characterizing with the hyperspectrometer.

13.4.1 Data Window Size Selection and Dimensional Transformation of the Input Data

The first step is to transform the initial matrix for a given window of the input image into a set of parameters in a feature space that describe the spatial distribution of the vector components in the window. This basically consists in a transformation of the set of elements in the window into a new representation that can be taken incrementally, the more parameters that are considered the more precise the representation. The objective is to provide a natural way to choose a number of parameters that provides a good enough description for our purposes. These parameters are arranged in a given order so that the training procedure used in the following stages of the complete algorithm can choose how many of these parameters are really necessary for the ANNs to produce an appropriate classification. It is desirable to minimize the number of descriptors necessary to obtain the classification accuracy required so as to simplify the problem as much as possible. The following subsections provide a formal description of the 5 consecutive processes that take place during the dimensional transformation stage as well as the definition of the different terms used.

Initial conditions

1. The algorithm starts from N_c matrices of dimension MxB, this is, M vectors of B values each (Fig. 13.7). Each vector is used to represent a continuous function, denoted as a continuous dimension, and the different values of one element along the different vectors represent the discrete dimension. In the case of hyperspectral data the continuous function is the spectral representation of this pixel. All the vectors are grouped in a spatial dimension then obtaining a data cube.

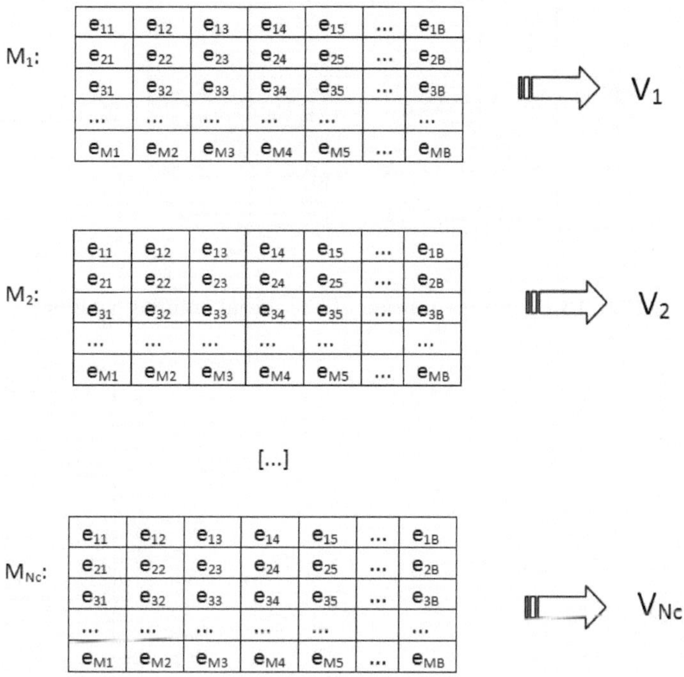

Fig. 13.7. Representation of the initial N_c matrices

2. Each of the N_c datasets must contain a numerical categorization of the relevant features to be extracted. Thus the training set is an $\Re^{MxB} \rightarrow \Re$ application.

The categorization could be based, for example, on the proportion of a given element or on the purity of the element. For instance, if we have three different types of granite A, B and C, the original data could have them labelled as 0.2, 0.5 and 0.8. This is basically an encoding of the target value categories.

Input matrix

Every step of the training process starts by considering a window of T vectors, extracted from one of the MxB matrices (Fig. 13.8). The data processing is performed on this window, which moves over the data set and which, in general, should be as small as possible without affecting the performance in the resulting classification capability. This is because the smaller the window size (parameter T), the larger the resolution the system can provide when analyzing data. The window data set is created based on a distance measurement, not necessarily in the same

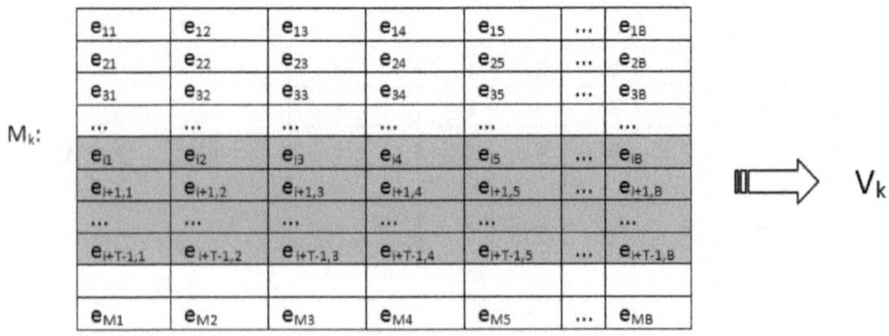

Fig. 13.8. Representation of the initial training window

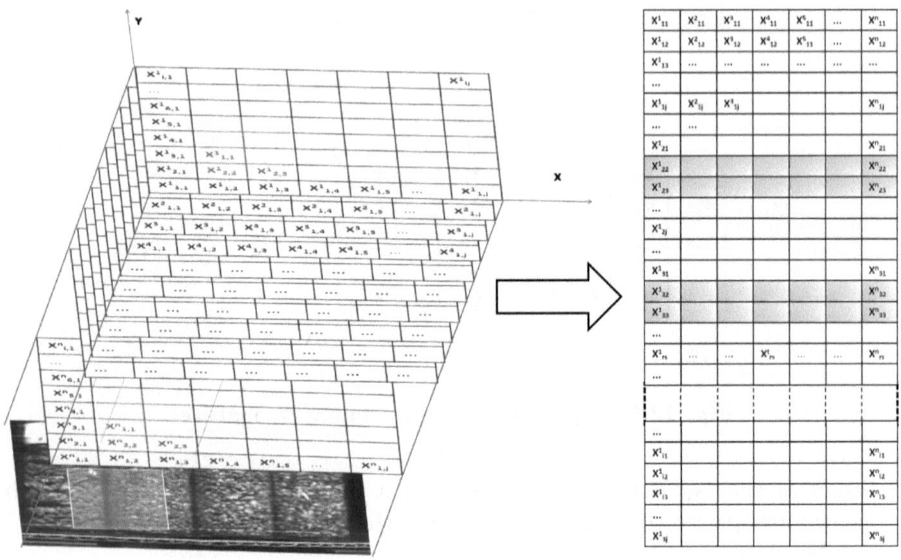

Fig. 13.9. Representation of the process over a hyperespectral image

order as in the original matrix. This way, the reduction of dimensionality is carried out over each dimension.

For the hyperspectral cube, the window selected corresponds to a rectangular area in the surface under analysis as shown in Fig. 13.9. Notice that after the transformation to the 2D matrix, the selected vectors are not necessarily correlative. They are just elements that are close according to a given distance metric.

Incremental vector description (spectral dimension)

To reduce the high dimensionality it is necessary to describe each vector in an incremental way. This is achieved by creating a new equivalent vector as follows:

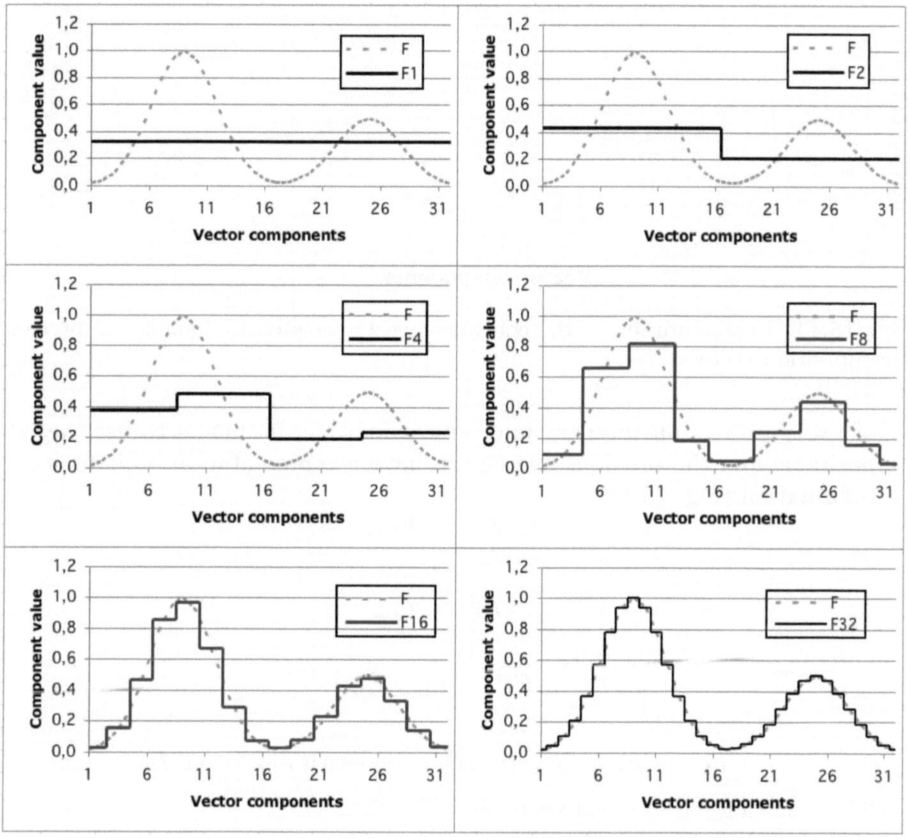

Fig. 13.10. Different accuracy levels for the representation of F with the projected vectors F_J

- Let F_J be a vector obtained from projecting the original vector of B elements into a new vector of J elements where $J \leq B$. This projection or resizing is obtained by dividing the original vector into J sub-vectors, of B/J elements and creating the new vector (F_J) using the average value of each subvector as its components.
- The coarsest approach would be to project the vector into a single band, this way it would be represented by a single value F_1: $[F_{1,1}]$, which corresponds to the average of the whole original vector. The next level of projection would lead to two values F_2: $[F_{1,2}, F_{2,2}]$ (corresponding to the average of two values of the original vector), the next by four and so on. This incremental process in represented in Fig. 13.10.
- In order to build an incremental procedure, the equivalent vector is created by joining the previous decomposition vectors but in a special way: when a new, finer vector is added, we only add the odd positions of the new vector to the representation (F^*). As each value of a given vector is in fact the average

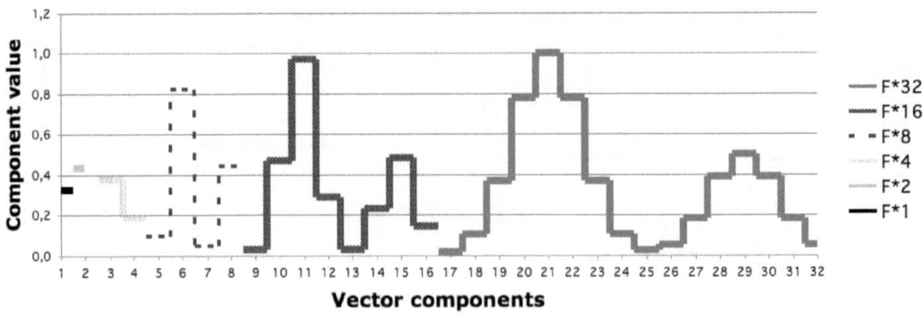

Fig. 13.11. Representation of the equivalent vector created by joining the previous decomposition vectors

of two positions on the next finer approximation, it suffices to use the odd positions and the previous average to have a representation without losing information (Fig. 13.11).

$$F_1{}^* = [F_{1,1}]$$

$$F_2{}^* = [F_{1,2}]$$

$$F_4{}^* = [F_{1,4},\ F_{3,4}]$$

$$F_n{}^* = [F_{1,2^n},\ F_{3,2^n},\ ...F_{2n-1,2^n}] \text{ with } n/2^n \in (1, B)$$

- The resulting incremental vector is:

$$[F_{1,1},\ F_{1,2},\ F_{1,4},\ F_{3,4},\ F_{1,8},\ F_{3,8},\ F_{5,8},\ F_{7,8}.....]$$

For the general case:

$$F_1 \cup F_2 \cup F_4 \cup F_8 \cup .... \cup F_n$$

- Notice that at each level of approximation the vector contains all the values of the previous level, which is consistent with an incremental description such as the one we seek. When description resolution is increased, descriptors are added to the descriptors corresponding to the lower resolution description.

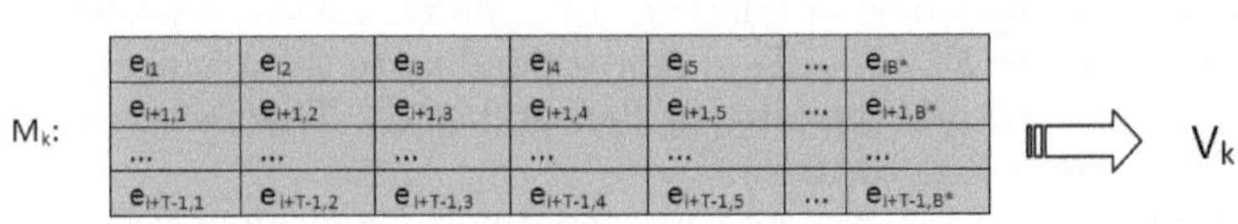

Fig. 13.12. Representation of the reduced matrix obtained after the spectral dimension reduction

- Finally as a result of this process a reduced matrix of TxB^* (being $B^* = 2n$) is obtained, with a value of B^* depending on the accuracy of the vector description (Fig. 13.12)

Incremental discrete dimension description (spatial dimension)

The objective is to create a description of the set of vectors, made up of several groups of similar vectors or vectors corresponding to the same category. Consequently, the description is oriented to the incremental extraction of the representative values for each group.

Let us define some concepts:

- $S(V, n)$ denotes the vector obtained by selecting a subvector of V such that this subvector has size n: $S(V, n) = S/S \subset V \cap dim(S) = n$
- The variability (Var) of a vector is defined as the difference between the largest and smallest values of the vector, this is: $Var(V) = max(V) - min(V)$
- The minimal subvector of size T of a vector V, denoted as $S^*(V, T)$, is defined as the subvector with the minimum vector variability in V of size n. This is: $S^*(V, n) = S(V, n)/min\,[Var(S(V, T))]$
- The difference vector (D) is defined as the vector resulting from removing the elements of a subvector from the original vector. In this case $D = Diff(V, S(V, n))$, and verifies $dim(D) = dim(V) - dim(S)$

From these definitions, the k partial subvectors of a vector V, ($Sp(V, k)$), may be defined as:

$$S^p(V,K) = S^*(D_{k-2}, T/k) \text{with} \begin{cases} D_0 = V; \\ D_n = Diff(D_{n-1}, S^*(D_{n-1}, T/k)) \text{ where } T{=}dim(V); \end{cases}$$

As a conclusion of the process of section 13.4.1 we have T vectors of B^* values each. And also, T values for each of the B^* positions of the TxB matrix. These T values are now represented by a vector V_b corresponding to each *b-th* band. In this step this vector is converted into another one with T incremental descriptors:

$$[S_1,\ S_2,\ S_3,\ S_4,\ S_5,\ .....S_n,\ .....S_T] \text{ with} \begin{cases} S_1 = O(V_b); \\ S_n = O(S^p(V_b, n)); \end{cases}$$

In Fig. 13.13, we show as an example, the $n/2$ size subvector selected (dark area) with the minimal variability. Similarly, on the left graph of Fig. 13.14, we represent the same example but where the minimal variability $n/3$ subvector is first selected and then extracted (right graph) to obtain a second vector (created by joining the remaining points) from which the second partial subvector is obtained (dark area of right graph).

After applying this stage to different sets of signals, we found the need to generalize the processing using operator $O(V)$. In the case of clearly different signals, operator O performs the mean of the elements of the vector, and in the case of very similar signals it produces the typical deviation.

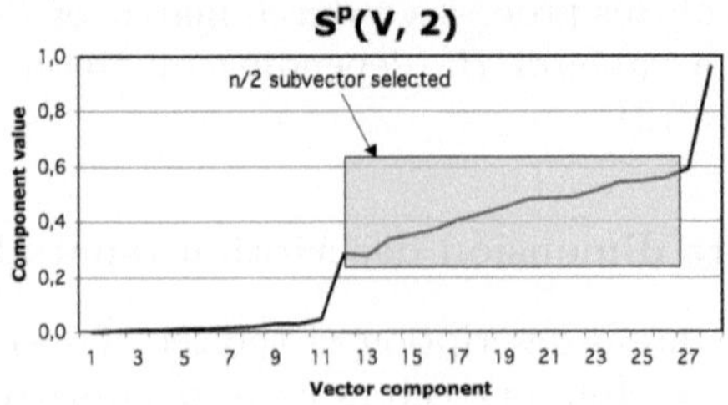

Fig. 13.13. Representation of the transformation of the vectors with incremental descriptors in the case of $S^p(V, 2)$

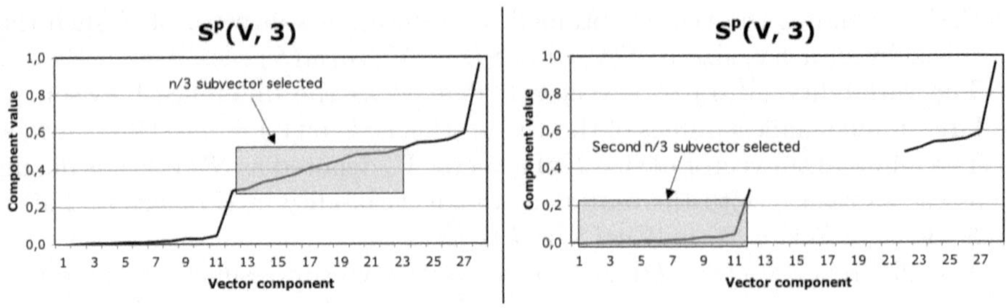

Fig. 13.14. Representation of the transformation of the vectors with incremental descriptors in the case of $S^p(V, 3)$

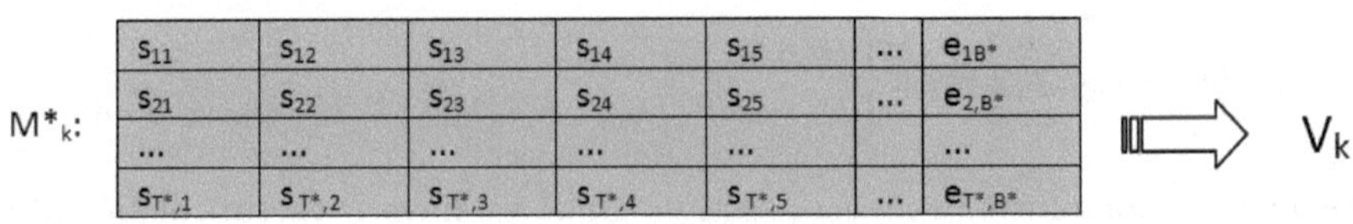

Fig. 13.15. Final incremental matrix representing the data contained in the window

As a result of this process, a T^*xB^* incremental matrix representing the data contained in the window is obtained (Fig. 13.15)

Two-dimensional sweep

The last step of this dimensional reduction or rather, a complementary step to the dimensional reduction process, is to incrementally sweep the TxB^* matrix to arrange the parameters that will be used as inputs to the ANNs in charge of classification in a given incremental information order. Obviously, the fewer parameters used the smaller the ANN as long as the required classification can be achieved.

First of all, the data are grouped in the continuous dimension (the B^* elements of each vector) into incremental blocks of $2n$ values as in the creation of the

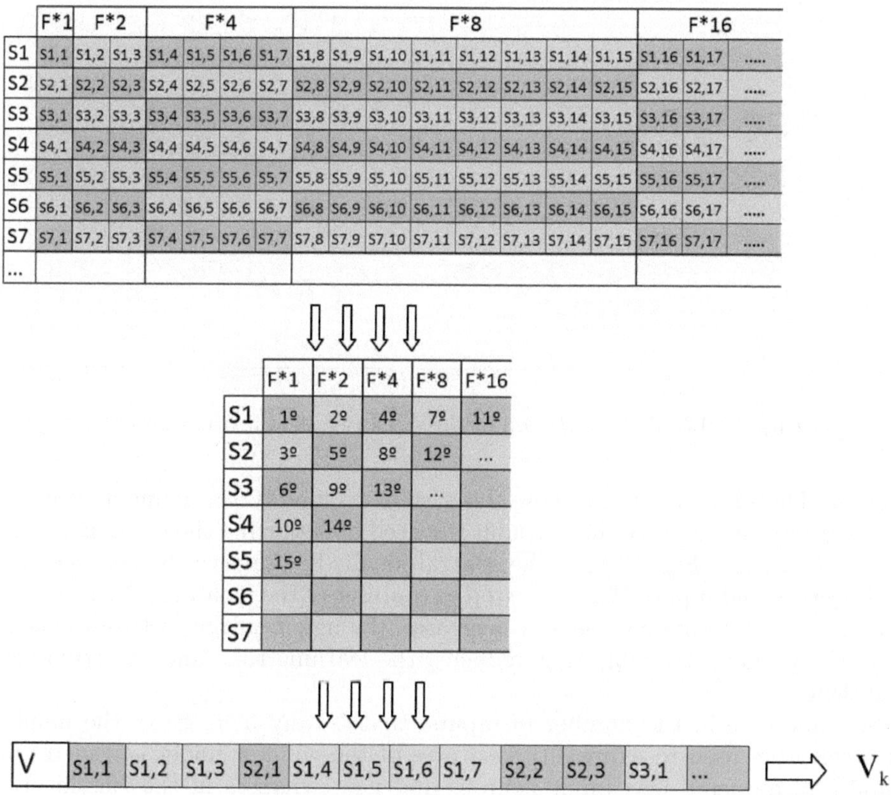

Fig. 13.16. Representation of the order followed for introducing or removing blocks of inputs on the two-dimensional sweep

incremental vector, thus, whenever a resolution increase is required, the number of descriptors are duplicated.

To carry out the sweep, an order for introducing or removing blocks of inputs is established when seeking to increase or decrease the number of descriptors used by the classifier. This order is shown in Fig. 13.16.

At this point a new parameter P is defined as the number of data blocks extracted from the T^*xB^* matrix, following the order shown in Fig. 13.16, to obtain a training vector for the classifier. The size of this vector varies with the value of P. A vector is obtained from the $M * k$ matrix.

Thus, as a result of this first stage, the initial data matrix with a very large amount of data has been reduced to a more manageable dimension through a reduction based on just two parameters, P and T.

13.4.2 Classification of the Categories through an ANN Structure

In order to perform the classification of the data into the categories provided by the expert, a modified RBF network was used as commented in the previous

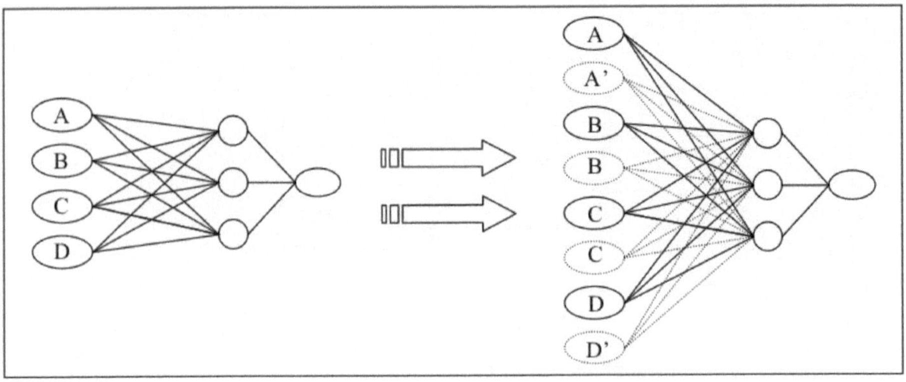

Fig. 13.17. Representation of the ANN size modification process

section. The objective is to allow the system to modify the number of inputs during the training procedure without the need of restarting the learning process from scratch (see Fig. 13.17). The general idea is to preserve the complete NN architecture and inputs of a given step as a subset of the next one. As a result, if the number of inputs increases or decreases, the new resulting network benefits from the previous training, thus reducing the learning time and smoothing the transients.

Any increase in the number of inputs, that is, any increase in the number of descriptors used to express the features of the window under consideration, implies introducing new input neurons and new synapses in the classification network. To avoid having to start training from scratch, the descriptors of the characteristics of the data window have been chosen so as to provide a gradually more detailed view of the data. as explained before, and whenever an increase of the number of parameters takes place (new inputs on the network), these parameters derive from previous lower resolution descriptors and, consequently, their influence can be smoothed by introducing neurons and weights for them that are the same as those of their originators. Nevertheless, in order to smooth even further the consequences of these modifications, several coefficients are introduced to gradually increase the effect of the new inputs on the hidden layers. These new attenuated inputs need to be trained in the same way as the old ones, and, consequently, the training algorithm must be modified in this direction. When the number of neurons is decreased, the redundant inputs and synapse connections are smoothly removed and the training is adapted to its new configuration. The rate of increase or decrease of the transition coefficients values has been regulated to reduce the sharp transitory increase of the error.

An added bonus of using this variable size network is that it is even possible to decrease the convergence time on high dimensional search spaces, as will be later shown on one of the examples.

13.4.3 On Line Tuning of the Window Size and Dimensional Transformation Parameters

The tuning or advisory module modifies the parameters of the first stage, that is, the determination of a processing window size (parameter T) and the description in terms of a set of parameters in feature space of the data matrix under consideration (parameter P), after evaluating the evolution of the error during the training phase. In order to perform this evaluation, it begins by extracting some descriptive parameters for the trends in time of the error obtained. This is done so as to provide a general approach for studying the average error level and its variability, these two parameters together with the window size, which is simultaneously an input and output for this module, are used to evaluate the global fitness function of the system.

It is important to note that this tuning or advisory system is operating in the classification space, that is, its quality is determined by the quality of the classification it obtains, and not in terms of other more classical parameters such as mean squared error or others. Thus, it appears that it would be very interesting to perform parameter tuning and the evaluation of the system in terms of parameters that are directly related to classification quality. In this line, some parameters and terms have been introduced and they will now be described:

- First of all, instead of the traditional MSE, a *category discrimination level* will be considered. It measures the proportion of misclassified samples using the number of categories to establish the boundaries between them. Therefore, every classification may be right or wrong.
- Another parameter considered here is what will be called *reliability*. It represents the ratio of discrimination values that are lower than a particular discrimination level. It is basically the stability of the value of the discrimination level. It is calculated using the ratio of values from a group of discrimination values that are lower than a particular discrimination level. Thus, reliability is based on a given discrimination level and on a given number of discrimination values, called the size of the use window. The use window represents the number of data used and considered as a unit to extract information from. That is, if we have an image with a spatial dimension of 40x40, the corresponding 1600 values are the use window, that is, the block of information used as a base for processing.
- An additional concept considered here is the analysis tool resolution; it is basically the window size (parameter T) used in the first stage of the system which is both an evaluation and a pre-processing parameter

Two more concepts derived from the previous ones are going to be introduced.

- The first one is the *reliable discrimination level*; it is defined as the maximum discrimination level obtained with a given reliability from a set of discrimination level data. This is basically the percentile reliability in the discrimination level data set. As a consequence, two parameters are merged into a single one

Fig. 13.18. Desired and actual discrimination areas S^* and S over an image

which establishes the required reliability level, and smoothes the search process. This is necessary, as using simple reliability may result in very bumpy functions.

- The *spatial discrimination level* is obtained based on the input data set size (T), and the required size (T^*) and using a geometric analogy for weighing the correct or incorrect classification.
  - Let S be the discrimination area the system is using in the first stage, that is, the smallest window it can classify (T) (Fig. 13.18)
  - let S^* be the discrimination area corresponding to the required windows size, that is, the maximum of the minimum area for which the tool must be able to perform an analysis.(Fig. 13.18)

The contribution to the resulting classification value obtained for S provided by S^* is S/S^*. On the other hand, the contribution to any other category is $(S - S^*)/S$. Taking into account that the area S^* will be classified with the same classification label as S and weighting the proportion of correct classifications with a 1, and the proportion of misclassification with $1/N_c$, which is the probability of coinciding with category of S^*, we have what we call the Spatial Discrimination Level:

$$P = min\left[\frac{S}{S^*} \cdot 1 + \frac{S - S^*}{S} \cdot \frac{1}{N_c}, 1\right]$$

This estimated value of correct classifications must be substituted by the real discrimination level D (as it is obviously not always 1). Consequently, the global reliable discrimination level will be:

$$P_G = D_r \cdot P$$

Thus, when the spatial resolution reaches the desired level, its value becomes one and its contribution to the global discrimination level becomes zero. Further increasing it is only possible by increasing the discrimination level of the classifier module. The user will stop the search when this global discrimination reaches the desired level. Notice that evaluation quality is influenced by three user parameters: the use window, the reliability and the resolution required.

Based on the previous evaluation, a search algorithm is used to guide the modification of the dimensional reduction parameters, that is, the descriptors employed to provide a characterization of the window in terms of feature space values. In this case the search algorithm is structured as a set of rules. It uses a random initial search direction and maintains it while the global error level decreases. If the error level increases, a new random advance direction is selected. In addition, in order to add some extra exploratory capabilities, a small probability of changing the direction selected for the algorithm is added. This simple search method appears to be enough for the experiments presented in the following sections and is surely enough to display the basic capabilities of the approach proposed in this paper.

13.4.4 Experimental Results: Synthetic Images

In the first experiment to test the self-adjustable classification system, three different spectral data groups were created by combining in different ratios the synthetic spectra of the set. In this particular test case the ratio values used are 0.20, 0.30 and 0.52. The three groups are stored as three cubes of 50x250 values in the spatial dimension and 128 values for each spectrum.

Spatially, these three groups represent three windows in which the different spectra are combined using a 10x10 regular mosaic pattern with three different appearance proportions (20%, 30% and 52%) of one of the spectra, as shown on Fig. 13.19. The mosaic ensures that the minimum window size which maintains the same spectral proportions is a 10x10 window. Below this size the deviation of the density grows as the size decreases, above this value the deviation remains almost constant as displayed in Fig. 13.20. This figure shows the standard deviation of the different density values obtained by moving the analysis window over the image for the different values of window size for the three different mosaics represented in Fig. 13.19.

The adaptive system tries to extract the actual ratio of one of the spectra trough the analysis of a data window. After a given number of training iterations

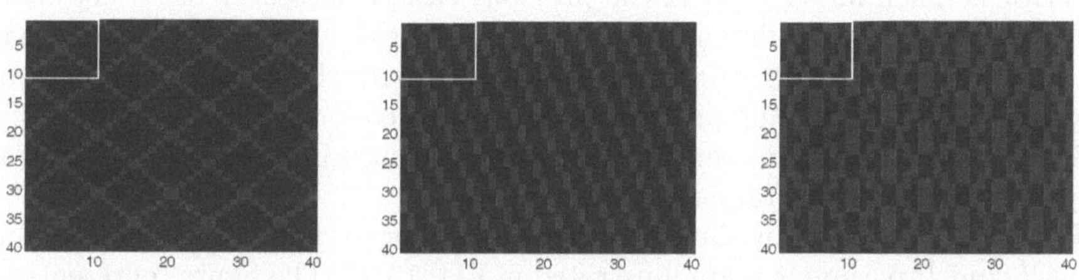

Fig. 13.19. Different spectra are combined using a 10x10 regular mosaic pattern with three different appearance proportions (20%, 30% and 52%) of one of the spectra for the synthetic training set

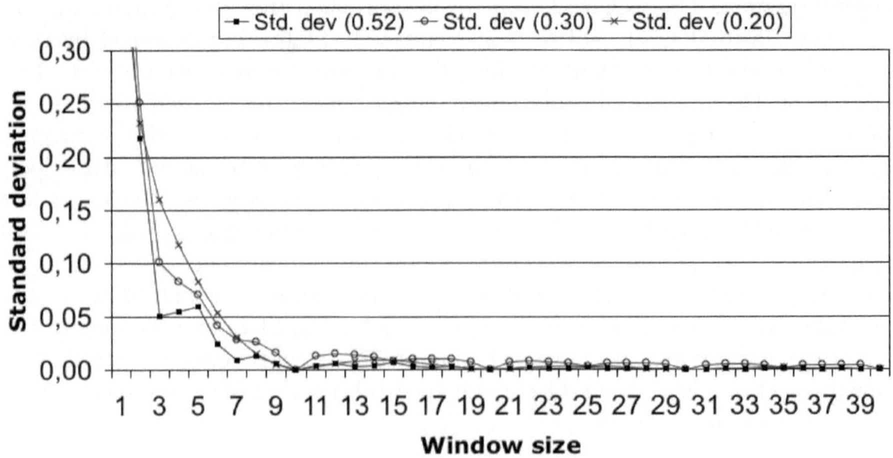

Fig. 13.20. Standard deviation on synthetic mosaic density vs the window size used

(900 in this case) two parameters are modified to study their influence on the error level; these two parameters are the windows size (T), and the number of descriptors (P). A transient time (100 iterations), which allows the system to reach a stable state before starting to evaluate the error level again, is also considered. Based on this, the advisory system decides new values for these two parameters and adds this information to its learning system in order to improve the decision rules for future actuations. Stability, average value and trend of the error evolution are used as fitness parameters by the advisory system in this training process. The strategy of starting with a coarser description of the spectra and progressively moving towards a finer one is illustrated in Fig. 13.21. The dashed line in this figure represents the error obtained by following this strategy, the line with markers represents the window size. We can see that with the coarser description the error drops rapidly because the search space is very small; when the window size is increased, the new inputs added to the ANN produce an error disturbance, but it is not very large and the error drops very fast in every case. This figure also represents the error when a direct strategy is used (solid line) instead of the incremental one. It can be appreciated that the error is much larger than in the previous case; it also decreases very slowly and with a lot of large fluctuations. In this case the ANN starts directly with 128 inputs corresponding to the spectral bands used; the search space becomes very large and the training process is not able to cope with it adequately. This indicates that using an incremental strategy can also improve the convergence properties of the ANN training algorithm.

The results obtained by the rule-based algorithm to choose the optimal values of the T and P parameters are illustrated in Fig. 13.22. The figure represents the evolution of the values assigned to these parameters as well as the errors obtained every iteration, and the average error obtained at each stage. We can see that this methodology, even though it is very simple, appears to work satisfactorily.

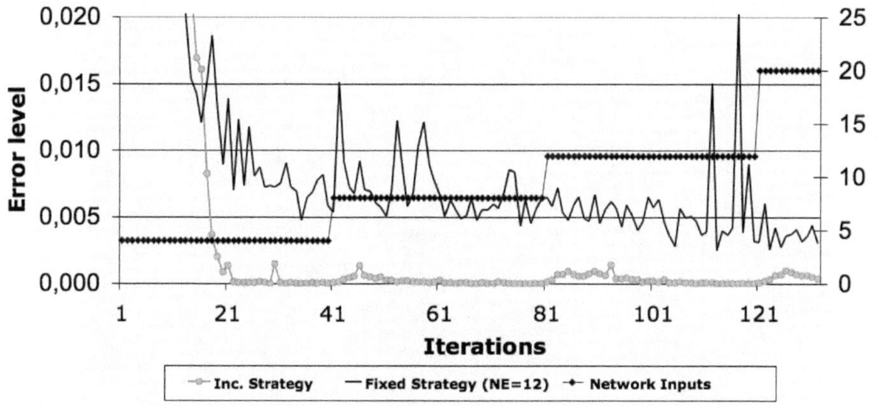

Fig. 13.21. Errors obtained in terms of MSE during training when using the incremental refinement strategy and a fixed strategy

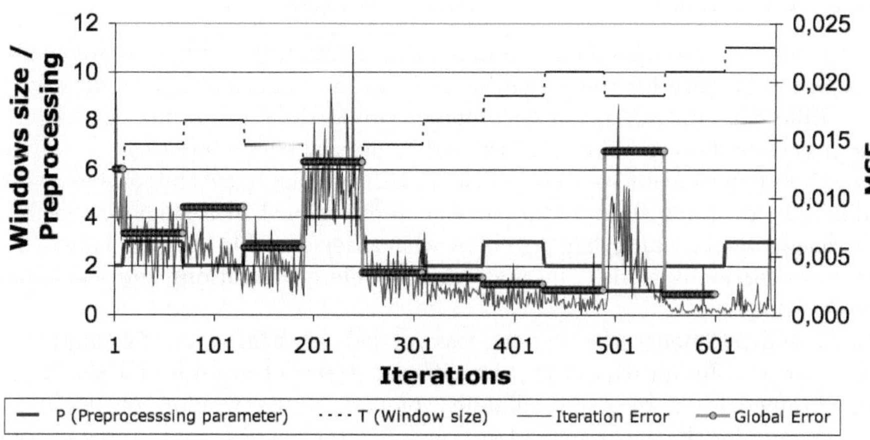

Fig. 13.22. Evolution of parameters T and P and training error in terms of MSE during a training process

The figure shows how the algorithm starts with a value for T of 6 pixels and for P of 2. This means 8 inputs to the network and as the procedure progresses and the error value decreases, parameter T tends to the optimal value of 10, and the number of parameters are maintained almost in the same level, some variations in these two parameters make the global error increase and to counteract this effect, the system inverts the search direction.

Fig. 13.22 also shows how the adviser isn't always following the expected tendency. For example, in iteration 300 the error level decreases with respect to the previous step as a consequence of an increase in T and a decrease in P. Unexpectedly, in the next iteration parameter T is increased. This is a consequence of the random exploratory component of the adviser.

Fig. 13.23. Samples of granite used for creating the real data

13.4.5 Experimental Results: Granite Images

The second test case, based on real mineral samples, was implemented considering a particular case for the classification of granite in a real industrial setting. Three different minerals occur in different proportions in granite, giving each type of granite its own color, texture and structural characteristics.

For these experiments we were provided with samples of this mineral extracted at different locations (Fig. 13.23), which were scanned in a single image by a hyperspectrometer which delivers cubes with a depth of 144 spectral bands. The tests were carried out with the materials in different positions and the results were equivalent.

In these experiments, the training was carried out using a set of ground truth labels provided by an expert for the different types of granite and the results were obtained for a required reliability of 85%, a use window of 10x10 and several values for the optimal window size. After the training and tuning process, the classifications obtained for the test sets were very satisfactory, leading to successful classifications with the required parameters.

In Fig. 13.24, a search process for which an optimal window size of 10 should be achieved is depicted. The left axis corresponds to the four different discrimination levels represented. The right axis corresponds to the values of parameters P and T. In the first 100 steps the low initial values of T and P (4 and 2 respectively) make the discrimination error levels very high. Take into account that the discrimination error values show the percentage of misclassified samples, this is, the lower the better. After this initial stage the value of T increases to a value slightly higher than the required window size and P also increases a little bit. As a consequence the discrimination improves to an average misclassification value of 1% and the global discrimination varies between values of 10 and 20% due to variations in the window size. In the last stage the increase in the input parameters and a small decrease of the window size to the optimal value makes the global discrimination error reach lower values. The minimum global discrimination error level obtained

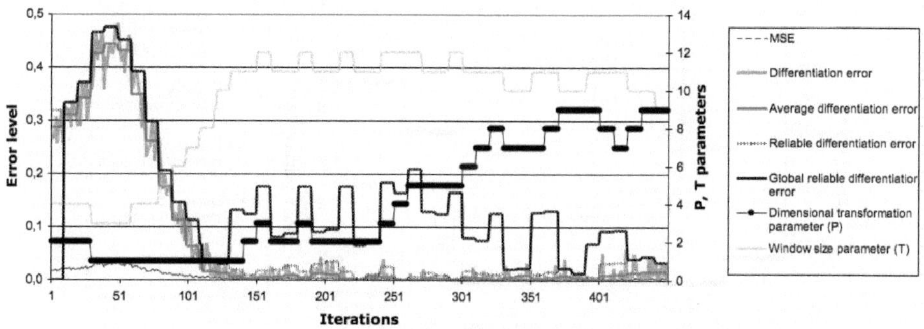

Fig. 13.24. Evolution results with granite samples

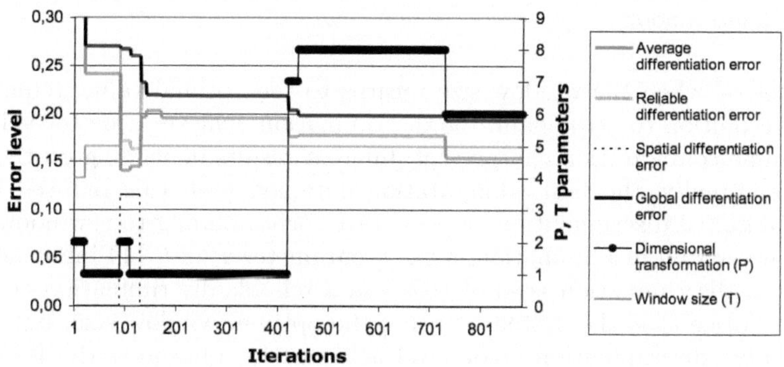

Fig. 13.25. Best-solution representation of the evolution results with granite samples for a 10 x 10 use window

is 1.3% corresponding to a window size of 10, a spatial discrimination error level of 1.0, a parameter P of 9 (56 input neurons), an average discrimination error level of 0.42% and a reliable discrimination error level of 1.3%. Notice that this is a search process, and the exploratory component creates fluctuations trying to continue the search after finding good solutions in order to determine if better solutions are available.

A clearer way of representing this search is using the best values obtained in each step. Fig. 13.25 displays this new representation for the search process. Notice that in this figure, due to the larger scale in the x-axis, the reliable and the average discrimination error, which are very close, are not discriminated. In Fig. 13.26 we have represented the search process obtained varying the optimal window size to 5x5 in order to study the effect of a requirement for a higher resolution. In this case, the initial conditions are the same as in the previous figure. The classification with such a reduced window size is more difficult and, as a consequence, the discrimination error values are higher. In step 100, the system increases the window size obtaining a better reliable discrimination error level, but only a slight improvement in the global discrimination error level.

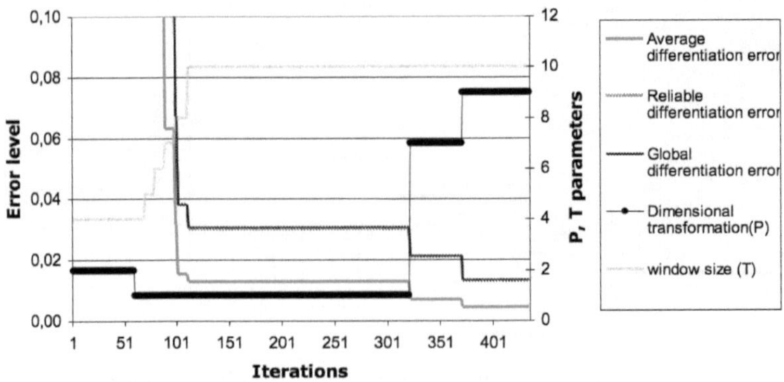

Fig. 13.26. Best-solution representation of the evolution results with granite samples for a 5 x 5 use window

Nevertheless, when the window size returns to the optimal value, reducing the spatial resolution to zero again, the discrimination achieves a better value, this is, the same configuration can present different results depending on the search sequence. Finally, the best configuration obtained with this reduced window reaches a global discrimination error of 19% corresponding to a window size of 5, with no spatial discrimination error, a parameter P of 6 (44 input neurons), an average discrimination level of 16% and a reliable discrimination error level of 19%. Notice that the system is using the optimal window size, but not the best reliable discrimination error level achieved for obtaining the best global discrimination error.

13.5 Conclusion

This chapter describes the implementation stages and the solutions followed in the creation of an ANN based hyperspectral processing system specifically developed to be used in ground based high resolution applications, particularly in performing classification and quality control tasks of materials in industrial processes. In these applications the spectral analysis must be carried out over a set of several neighboring pixels in order to obtain a sample representing the whole characteristics of the material under analysis. In addition, the spectral resolution may be decreased in order to reduce the computing power needed to perform the analysis. The determination of the optimal spatial window size, as well as a reduction of the spectral bands and determination of an optimal spatio-spectral description, is addressed in an automated way through the creation of a training algorithm that using a training advisor regulates just two parameters related with these terms. The advisor automatically adjusts these parameters as training takes place. The algorithm used for the determination of the spectral bands starts from a coarse spectrum and refines it in a smooth and progressive manner that permits a better training of the system. This is reflected in the gradual growth of the neural network used to process it. The algorithm increases the size of the NN preserving most of the training performed in the previous step so as to reduce the computational cost. The algorithm is studied on a synthetic test set and then adapted to be used in real samples and configured by the real user. It is finally applied to a real test case considering samples of granite and obtaining the desired results.

References

1. Chang, C.: Hyperspectral Data Exploitation: Theory and Applications. Wiley-Interscience, Chichester (2007)
2. Ballinger, D.: Space-based hyperspectral imaging and its role in the commercial world. In: Aerospace Conference, IEEE Proceedings, vol. 2, pp. 915–923 (2001)
3. Rajesh, H.M.: Application of remote sensing and GIS in mineral resource mapping - An overview. Journal of Mineralogical and Petrological Sciences 99(3), 83–103 (2004)
4. Boyd, D.S., Danson, F.M.: Satellite remote sensing of forest resources: three decades of research development. Progress in Physical Geography 29(1), 1–26 (2005)
5. Kooistra, L., Salas, E.A.L., Clevers, J.G.P.W., Wehrens, R., Leuven, R.S.E.W., Nienhuis, P.H., Buydens, L.M.C.: Exploring field vegetation reflectance as an indicator of soil contamination in river floodplains. Environmental Pollution 127(2), 281–290 (2004)
6. Ustin, S.L., Roberts, D.A., Gamon, J.A., Asner, G.P., Green, R.O.: Using imaging spectroscopy to study ecosystem processes and properties. Bioscience 54, 523–534 (2004)
7. Rochon, G.L., Johannsen, C.J., Landgrebe, D.A., Engel, B.A., Harbor, J.M., Majumder, S., Biehl, L.L.: Remote sensing as a tool for achieving and monitoring progress toward sustainability. Clean Technologies and Environmental Policy 5(3-4), 310–316 (2003)
8. Masoner, J.R., Mladinich, C.S., Konduris, A.M., Smith, S.J.: Comparison of Irrigation Water Use Estimates Calculated From Remotely Sensed Irrigated Acres and State Reported Irrigated Acres in the Lake Altus Drainage Basin, Oklahoma and Texas, 2000 Growing Season. In: U.S. Geological Survey. Water-Resources Investigations Report 03-4155 (2003)
9. Miller, R.L., del Castillo, C.E., Mckee, B.A.: Remote Sensing of Coastal Aquatic Environments. Springer, Netherlands (2006)
10. Landgrebe, D.A.: Signal Theory Methods in Multispectral Remote Sensing. John Wiley and Sons, Chichester (2003)
11. Grahn, H., Geladi, P.: Techniques and Applications of Hyperspectral Image Analysis. John Wiley and Sons Inc., Chichester (2007)
12. Plaza, A., Martinez, P., Perez, R., Plaza, J.: Spatial/spectral endmember extraction by multidimensional morphological operations. IEEE transactions on geoscience and remote sensing (IEEE trans. geosci. remote sens.) 40(9), 2025–2041 (2002)
13. Smartt, H.A.J., Tyo, S.: Classification of hyperspectral spatial/spectral patterns using Gauss-Markov random fields. In: Proceedings of SPIE, Algorithms and Technologies for Multispectral, Hyperspectral, and Ultraspectral Imagery XII, vol. 6233 (2006)

14. Van der Werff, H.M.A., Bakker, W.H., Van der Meer, F.D., Siderius, W.: Combining spectral signals and spatial patterns using multiple Hough transforms: An application for detection of natural gas seepages. Computers and Geosciences 32(9), 1334–1343 (2006)
15. Lopez-Pena, F., Duro, R.J.: A Hyperspectral Based Multisensor System for Marine Oil Spill Detection, Analysis and Tracking. In: Negoita, M.G., Howlett, R.J., Jain, L.C. (eds.) KES 2004. LNCS (LNAI), vol. 3213, pp. 669–676. Springer, Heidelberg (2004)
16. McCracken, W.H., Kendall, T.: Andalusite review 1995. Industrial Minerals (346), 53–54 (1996)
17. Atkinson, P.M., Tatnall, A.R.: Introduction: neural networks in remote sensing. International Journal of Remote Sensing 18(4), 699–709 (1997)
18. Landgrebe, D.: Indian Pines AVIRIS Hyperspectral Reflectance Data: 92av3c (1992), `http://makalu.jpl.nasa.gov/`
19. Guatieri, J.A., Bechdol, M., Chettri, S., Robinson, J.W., Garegnani, J., Vermeulen, A., Antonille, S.: From Spectra to Cassification. In: 1st International Symposium on Hyperspectral Analysis (2000)
20. Prieto, A., Bellas, F., Duro, R.J., Lopez-Peña, F.: A Comparison of Gaussian Based ANNs for the Classification of Multidimensional Hyperspectral Signals. In: Cabestany, J., Prieto, A.G., Sandoval, F. (eds.) IWANN 2005. LNCS, vol. 3512, pp. 829–836. Springer, Heidelberg (2005)

14

Using Gaussian Synapse ANNs for Hyperspectral Image Segmentation and Endmember Extraction

R.J. Duro, F. Lopez-Pena, and J.L. Crespo

Integrated Group for Engineering Research
Universidade da Coruña (Spain)
{richard,flop,jcm}@udc.es
http://www.gii.udc.es

14.1 Introduction

Remote sensing of the earth is becoming a commonplace technology which is growing driven by the combined advances in sensor technologies and image and data processing.

In the near future the international proliferation of remote sensing devices will require an ever increasing number of high–resolution systems with lightweight instruments for application in small satellites and light planes, as well as a reduction of costs. One important tool used in remote sensing is imaging spectroscopy, also known as multi, hyper or ultraspectral remote sensing. It consists in the acquisition of images where for each spatial resolution element a part of the electromagnetic spectrum sampled at different rates is measured [1][2].

In the case of hyperspectral remote sensing images, the data for each spatial element is acquired with high spectral resolution over the electromagnetic spectrum in the 400–2500 nm (visible to near infrared) interval. It is quite common to use 50 to 250 spectral bands of bandwidths in the 5 to 20 nm range. Traditional multispectral data sets usually comprise 5 to 10 bands of relatively large bandwidths (70–400 nm). Finally, ultraspectral imaging, which is an emerging technology that has not yet been applied to its full extent, divides the spectral range in thousands of bands. The large amount of information provided by ultra or hyperspectral imaging permits a detailed description of the spectral signature, thus greatly improving the ability to detect and identify individual materials or classes with respect to other remote sensing techniques. On the other hand, the analysis and processing of the large amounts of data involved requires new processing methods and algorithms [3].

In terms of classification, there are two options for obtaining the composition from the images: The spectra may be compared to well known spectral databases (usually obtained from classical spectroscopy laboratory experiments), or a set of reference spectra (endmembers) can be extracted (labelled by experts) from the remote sensing data, and used to classify the rest of the data.

M. Graña and R.J. Duro (Eds.): Comput. Intel. for Remote Sensing, SCI 133, pp. 341–362, 2008.
springerlink.com

Most of the approximations to hyperspectral analysis come from the well known and established techniques of laboratory spectroscopy. But these approximations weren't developed taking into account the fact that each one of the image pixels provided by remote hyperspectral sensing corresponds to a large target area as compared to the size of the regions covered by each sensing element in classical laboratory spectra. Consequently, in general, the spectral information that requires analysis comprises a number of spectral contributions from different components found in the same area and thus mixed into the same pixel and includes the effects of whatever is between target and sensor.

The effects induced by atmospheric agents present between the target and the sensor lead to problems that are common to all remote sensing techniques. The influence of the atmosphere may be divided into two groups of effects: Those that are spatially and temporally constant and those that are not. The first category includes the absorption and scattering by CO_2, N_2, CH_4 and O_2 (well mixed in the atmosphere), and the second, those elements that could vary in certain circumstances (like water vapor, ozone and aerosols –dust, water droplets and haze–). To eliminate such influences, it is necessary to transform the measured radiances into reflectances [2]. There are two possible ways to obtain such a transformation: through the implementation of radiative transfer models or by using ground truth.

Thus for processing the data, in the classical approach, the first step to analyze the composition of a hyperspectral cube is to perform artifact correction. That is, to clean the signal of contributions not related with the target. These contributions correspond to whatever is between the target and the sensors. In the list of possible artifacts, the main contribution is usually due to the effect of the atmosphere (for example, the well known absorption bands of water vapour). However, the components of the atmosphere that affect a particular region being imaged could be very different across time and space. Therefore it is not only a matter of removing known contributions, but of first characterizing the particular conditions.

Several atmospheric characterization models have been proposed and developed. For instance [4] proposes a hybrid solution that uses image and spectral scene data plus radiative propagation models in order to obtain an appropriate characterization. Like other hybrid approximations, standard images (in the sense of being described by RGB or equivalent vectors) are used as geometric references. Another model of the atmosphere is presented in [5]. In this study, there is a list of artifacts (H_2O, CO_2, N_2, CH_4, O_2, O_3, and aerosol contributions) that is known beforehand. The authors build and implement a local basis model for measuring and for atmospheric correction only for water vapour, aerosols and ozone. They assume that other gases are well mixed in the atmosphere.

Yamazaki presented in 1997 his approach for estimating the atmospheric optical thickness at the time of observation directly from the satellite data itself [6]. It starts from an atmospheric model based on a one–dimensional inhomogeneous atmosphere model and a horizontally non–uniform albedo distribution. The author found good results in the comparison between measured and estimated values for TM 3 and 4 bands of LandSat TM.

Most of the correction models proposed were developed for particular spatial or environmental conditions, or take into account the characteristics of a given sensor. Thome et al. present in [7] an atmospheric correction algorithm to be used in ASTER (Advanced Spaceborne Thermal Emission and Reflection Radiometer); their method considers both pixel information (via the comparison of data with a predefined lookup table) and adjacency considerations. In [8], neural network techniques are used for estimating column vapour content using data collected from AVIRIS (Airborne Visible / Infrared Imaging Spectrometer). This allows the generation of an algorithm that can be used to correct the images.

Notwithstanding the work performed by all of these and other authors, using radiative transfer models is seldom satisfactory as the information required on atmospheric conditions is hard to obtain and it changes in time and space. Thus, the reflectance accuracy is limited by the simulation of the atmosphere, which turns out to be a hard and resource consuming task. In this line other approaches have been proposed. These approaches are more local but provide ways of carrying out the task without having to resort to hard to obtain atmospheric models. One approach is to measure the reflectance in selected ground control points at the time the image is remotely sensed and from these measurements obtain an estimation for the calibration of the whole image. Obviously, this approach is not always practical as, in many cases, access to these selected control points is not feasible. In fact, if the objective of the sensing run is to classify whatever is in the image and not to try to compare the resulting spectra to standardized spectral databases, the need for these atmospheric models that allow us (in the best of circumstances) to obtain the exact spectra of the target covers may be waived and the process may be reduced to a classification problem where the objective is to determine what class or classes a pixel belongs to and to what extent, and to obtain a method for labeling the different classes present. The most common technique used in this direction is for an expert to provide in situ classifications of some targets instead of measuring their reflectances [9] and using the information on the image for these targets as a reference to establish a classification strategy for the rest of the pixels. This is one of the most common approaches used in remote sensing applications and the one considered in this chapter. It is called the ground truth based approach and basically uses these known target pixels and their spectra to try to construct some type of classifier that classifies the rest of the pixels in terms of these classes. It is thus a process of determining how similar or dissimilar the spectra are to each other and establishing some type of classification decision boundaries.

As commented before, remote sensing spectroscopy is an evolution of previous knowledge in laboratory spectroscopy. However, for laboratory spectroscopy the concept of pixel does not exist, at least in the same terms as for remote sensing spectroscopy, as a pixel is related to the spatial characterization of data. Consequently, many techniques from laboratory spectroscopy have been directly applied to the pixel by pixel classification of multi or hyperspectral images, without taking into account the spatial dimension of the images. An example of a simple, but very effective linear interpolation based method for establishing comparisons between a sensed spectrum and a reference one is developed in [10]. In this case, the objective is to extract the position value for spectra absorption bands (values that are the spectral "signature" for the different materials).

Even though it would appear that neighbourhood considerations could be a very useful tool for improving the results of the analysis, this type of solutions have not been widely implemented. The main reason is related to the spatial resolution: hyperspectral images taken from a satellite platform (and even in the case of high altitude planes) comprise large target areas (easily 300 square meters or more), in these conditions, the use of adjacency information would probably not be very useful, because the composition of the different regions covered by the sensor could be quite different from one pixel to the surrounding ones. In, fact, the problem of remote sensing images is usually the opposite; more than one type of cover may be included in a single pixel. Each one of these primitive components is called an endmember and what is usually sought is to try to determine the proportions of endmembers within a pixel.

Summarizing what we have said until now, the objective of any analysis or classification method destined to the processing of hyperspectral images is to be able to identify which pixels contain different spectrally distinct materials (endmembers) and in what proportions. The cause of the mixture of several material spectra into a hyperspectral pixel spectrum may be that the area whose radiance measurement corresponds to one image pixel encompasses different land covers. This work assumes a linear mixture model, in which several basic materials (endmembers) are combined according to some abundance coefficients at each image pixel. Through their spatial distribution, the abundance coefficients may be visualized as abundance images, which provide a description of the spatial distribution of the material.

A number of methods based on statistical theory or using filtering or correlations have been applied to the analysis of these data sets by different authors with the objective of improving the classification results [11]. Some of these methods are compared using classification performance in [12]. However, the first question that must be addressed now is how to determine how many and what endmembers may be present in each pixel of the image.

There are two main approaches to this problem: by hand or automatically. That is, a set of endmembers could be extracted from some spectral database or previously existing images (they could be, for example a well known subset of the image) and established by the operator as the endmembers to be sought on the image. They could also be determined in an automatic manner: some spectra are selected due to certain characteristics and are used to evaluate the rest of the data.

Some examples of the first case can be found in [13] where the adequate coefficients for a neural network that will estimate the different contributions of preselected endmembers are obtained by means of a Genetic Programming technique. Here the authors make use of a hand selected set of spectra of known composition. Shouten [14] uses a previously chosen set in order to configure a NN for executing fuzzy classification functions. Finally, in [15] we find another example of hand selection of candidate endmembers to establish a composition and classification system.

Classical methods are usually manual: the designer uses information (having to do with information theory considerations of the possible candidates) to choose a set of endmembers. This is, for example, the base of very well known methods like PPI [33] that is part of the most usual computer analysis development tools. [16] contains a brief but exhaustive review of these well established options before presenting their own approach. This approach shows a characteristic that is the main idea in some of the papers in this area: the consideration of the problem of determining adequate endmembers as a geometric problem, due to the fact that the endmembers may be seen as a simplex in a multidimensional space. Bajorski [17] is another author that follows this research path.

From an unsupervised and automatic point of view, the approach to the problem of extracting endmembers is to consider it a particular case of the blind source separation problem. This is the aim of Bayliss [18] who use ICA to select the most appropriate endmember set. These techniques are unsupervised, so it is not strange that there are attempts to use SOMs to achieve this end. The work of Aguilar et al. [19] is a good example of this trend. Other authors try to use neighbourhood pixel information in order to improve the search for endmembers. For example, Greiwe et al. [20] use a hybrid method, where high spatial resolution images are used in order to determine with high accuracy the regions in which endmembers could be located. Plaza and col. [21] use morphological operators to establish similarities between pixels, using this information to improve endmember selection.

Whatever the approach for classifying pixels, when a single cover is assumed for each pixel, or for obtaining the proportions of endmembers when mixtures occur, the basic idea is classical, if you concentrate only on what is relevant the classification becomes much more robust and efficient. This is the approach followed in the work presented here: An Artificial Neural Network architecture and training algorithm that implement an automatic procedure to concentrate on what is relevant and ignore what is not straight from the training set was designed and tested on images where the pixels contained mixtures of contributions from different endmembers.

From this point of view, ANNs are a very suitable and effective alternative to deal with spectral image analysis difficulties [5] [22]. Some reviews may be found, such as [26], on the applications of ANNs to this area. In the case of ANNs trained using supervised training algorithms, the main drawback to performing the analysis and classification of hyperspectral remote sensing data is the difficulty in obtaining labeled sample data that can be trusted to the point of using it for quantitative evaluation. The scarcity of ground truth data has been recognized and specific training strategies have been devised to cope with this handicap [23], [24]. In fact, in an ideal world, we would be able to obtain a training procedure that produces good classifiers from very small training sets. This problem becomes even more pronounced when the pixels in the image correspond to combinations of materials, that is,when facing "the spectral unmixing" problem [25]. Ideally, hyperspectral images may allow the identification of single pixel materials. However as these pixels are frequently combinations of materials, it is often a plus to be able to decompose each pixels spectrum into its constituent material spectra. The computation of the abundance coefficients given a pixel spectrum and a set of endmembers is what is termed the unmixing procedure. If the endmembers are given, the unmixing procedure is equivalent to the parallel detection of the spectral features represented by the endmembers. This is the approach that was followed here. A new neural network structure based on Gaussian Synapses, called Gaussian Synapse Artificial Neural Network (GSANN) and its associated training algorithm that can obtain very good classifiers from small training sets in a very small number of training epochs was used on these multidimensional data sets. The approach was initially applied to the classification of pixels, assuming a single material per pixel, and their assignment to their different categories within hyperspectral images without any atmospheric correction. As it was shown to be very efficient in this type of classification, the method was extended in order to ascertain the quantitative proportion of each endmember present in each pixel. Thus, the strategy was used to perform a spectral unmixing procedure trying to determine the proportions of different endmembers the methodology is capable of extracting from real and synthetic images. The emphasis is on a pixel by pixel classification without using any neighborhood information in order to test the appropriateness of using a specific backpropagation training algorithm for these tasks.

The use of Gaussian and other radial functions in ANNs is not new [26]; but, so far and to the best knowledge of the authors, they are restricted to their use as basis or activation functions thus, basically, constraining their application to a linear regression scheme through the optimization of the values of the weights. Our approach introduces Gaussian functions in the synapses of the networks and leaves the activation functions as sigmoids. As a consequence, the training procedure is not a simple regression scheme for adapting the weights but a conceptually more interesting solution that not only preserves the benefits of using Gaussian functions, but permits shaping them automatically and independently for each of the dimensions by modifying their parameters during the training process. The results are not RBF functions, but a sort of skewed RBFs where the width of the Gaussian is different for each dimension. The approach combines the effectiveness of synapses that are able to filter out or enhance areas of the input space with that of multiple sigmoid based activation functions in the nodes.

In the following sections, we will describe the network, provide a brief overview of its training procedure, see how it performs on multispectral images and then test its spectral unmixing abilities on benchmark synthetic hyperspectral cubes and finally on a real hyperspectral image.

14.2 The Network and Training Algorithm

The segmentation of multi, hyper or ultraspectral images is basically a decision process in which the correlation of different value intervals for each of the bands participating is the discriminating element between classes, thus, we have considered that the best way to perform this discrimination is acting directly on the values circulating throughout the network by the inclusion of trainable Gaussian functions in the synapses. This approach permits each synapse to learn to select the signal intervals that are relevant to the classification being performed in each node and to ignore the rest of the intervals without having to introduce a lot of nodes and connections in the network just to compensate irrelevant information. The architecture employed in this type of networks is very similar to the classical Multi Layer Perceptron. In fact, the activation functions of the nodes are simple sigmoids. The only difference is that each synaptic connection implements a Gaussian function determined by three parameters: its center, its amplitude and its variance:

$$g(x) = A * e^{B(x-C)^2}$$

To train this structure, an extension of the backpropagation algorithm, called Gaussian Synapse Backpropagation (GSBP) has been developed [27]. This algorithm works very much like general Backpropagation, by means of the calculation of the gradients of the error with respect to each one of the three parameters of the Gaussian functions in each weight and modifying them accordingly. Thus, if T_k denotes the desired target for output k and O_k the output actually produced by neuron k the outputs of the neurons in the hidden and output layers are:

$$\begin{cases} O_k = F(\sum_j h_j * A_{jk} * e^{B_{jk}(h_j-C_{jk})^2}) = F(O_{Net_k}) \\ \\ h_j = F(\sum_i I_i * A_{ij} * e^{B_{ij}(I_i-C_{ij})^2}) = F(h_{Net_j}) \end{cases} \tag{14.1}$$

Being F the activation function of a neuron, h_j the output of hidden neuron j, I_i input i and the sum is carried out over all the inputs to the neuron. We denote as O_{Net} and h_{Net} the values that input the activation function of a node.

Consequently, the gradients with respect to A_{jk}, B_{jk}, and C_{jk} we obtain:

$$\begin{cases} \frac{\partial E_{tot}}{\partial A_{jk}} = h_j(O_k - T_k)F'(O_{Net_k})e^{B_{jk}(h_j-C_{jk})^2} \\ \\ \frac{\partial E_{tot}}{\partial B_{jk}} = h_j(O_k - T_k)F'(O_{Net_k})A_{jk}(h_j - C_{jk})^2 e^{B_{jk}(h_j-C_{jk})^2} \\ \\ \frac{\partial E_{tot}}{\partial C_{jk}} = -2h_j A_{jk} B_{jk}(O_k - T_k)F'(O_{Net_k})(h_j - C_{jk})e^{B_{jk}(h_j-C_{jk})^2} \end{cases} \tag{14.2}$$

For the hidden layer if we denote the variation of the error with respect to the net value of the neuron as:

$$\Theta = \frac{\partial E_{tot}}{\partial h_{Net_j}} = \frac{\partial E_{tot}}{\partial h_j}\frac{\partial h_j}{\partial h_{Net_j}} = \frac{\partial E_{tot}}{\partial h_j}F'(h_{Net_j}) \tag{14.3}$$

The variation of the error with respect to A_{ij}, B_{ij} and C_{ij} is:

$$\begin{cases} \frac{\partial E_{tot}}{\partial A_{ij}} = I_i \Theta_j e^{B_{ij}(I_i - C_{ij})^2} \\ \\ \frac{\partial E_{tot}}{\partial B_{ij}} = I_i \Theta_j A_{ij} (I_i - C_{ij})^2 e^{B_{ij}(I_i - C_{ij})^2} \\ \\ \frac{\partial E_{tot}}{\partial C_{ij}} = -2 I_i \Theta_j A_{ij} B_{ij} (I_i - C_{ij}) e^{B_{ij}(I_i - C_{ij})^2} \end{cases} \tag{14.4}$$

Once the gradients have been obtained, the updating of the different parameters every epoch of the training process is straightforward and can be carried out in the same way as traditional backpropagation.

As mentioned earlier, the proposed network has some resemblances to the classical Radial Basis Functions as both make use of Gaussian functions (RBF), but in the present case these functions are applied in the synapses while RBF use then as activation functions. Obviously, the performances of both types of networks will be quite different as shown by Prieto et al. [28]. Those authors perform a comparison between these types of networks and a modified version of the RBF. This comparison is carried out on terms of performance when analyzing a set of benchmark synthetic hyperspectral images. Results show that GSBP converge to good classification results in less iterations than the two RBF ones, making GSANN a good choice for this type of task at the cost of being more sensitive to overtraining.

14.3 Segmentation System

The segmentation system used in this work can be viewed in figure 14.1 and it is made up of a set of NN based detectors followed by a decision module. The set of GSANN detectors work in parallel over the spectral dimension of each pixel of the image. A specific detector is implemented for every category; therefore the set of detectors produce a detection probability vector for every pixel associated with the categories (endmembers) they have been trained for. Obviously, by working in this way any pixel may be assigned a detection probability by two or more detectors.

The occurrence of a multiple probability assignment to a pixel may be caused by a variety of factors: a category having non discriminant training sets, two or more categories having very similar spectra, mixtures of categories within the same pixel (taking into account that depending on the altitude of the flight and the spatial resolution of the instrument a pixel may represent very large areas), noise, etc. Thus, after going through the detectors, each pixel is characterized by a detection probability vector and the way this detection vector is used thereafter will depend on the application in progress. Consequently, to decide on the final label assigned to the pixel, all the detectors send their information to a final decision module which will produce a final decision in agreement with the user requests. For instance, the decision module may be trained to choose the most

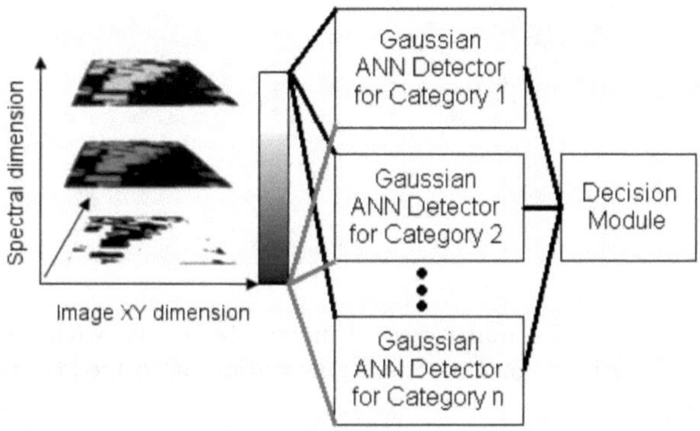

Fig. 14.1. Structure of the detector based segmentation system. The spectral cube is scanned in the x–y dimension and each spectrum is processed by the different detectors in parallel. The decision module constructs the final detection image.

likely category for a pixel or to assign combinations of detections to different new categories so that the final image indicates where there is doubt, but it can even prioritize some types of detections when searching for particular objectives such as minerals. In fact, this way of doing things allows for a very strict classification of pixels on one end or for proportional detections in the sense of spectrally unmixing pixels in the hyperspectral image through the probabilities provided by each detector. In this work we always assume linearity in the mixtures as do many other authors [29].

14.4 Some Experimental Results

This section discuses the application of this type of networks to process different imaging spectrometry problems. First we consider an example that contemplates multispectral images in order to display the capabilities of the Gaussian based networks for the correct classification of classes. Here each pixel is assumed to contain only one endmember. In the next subsection the problem is made more complicated through the introduction of hyperspectral images. In a first test case these are synthetic images so as to test the endmember extraction capabilities of the network. After this the system is applied to a real hyperspectral image often used as a benchmark in this type of work, the Indian pines image, where we will see how it performs even in real conditions.

14.4.1 Application to Multispectral Images

A multispectral image with 14 bands has been considered as a first application of these networks. This image has been obtained by the Center for Ecological Research and Forestry Applications, of the Universitat Autonoma de Barcelona in Spain. It was acquired by a CASI (Compact Airbone Spectrographic Imager) sensor using a general configuration of 14 bands, that is, each pixel is represented

Fig. 14.2. Segmentation of a multispectral image (14 bands) with four labeled categories. Top: Ground truth image. Bottom: segmentation performed by a set of Gaussian synapse networks.

by a vector of 14 8–bpp gray scale values. The original image has a size of 710x4558 pixels and covers a large area including vegetation, cultivated land regions and urban areas. Out of this set, a representative subsample consisting of a smaller region of 448x2048 pixels has been extracted for processing. In this case the image was not preprocessed to eliminate atmospheric effects at all. The ground truth for the image, displayed on the top part of figure 14.2, consisted of five different categories, four of which were used in the experiments, the fifth (black in the ground truth image) was too small to allow us to obtain statistically relevant training and test sets. The bottom part of the figure shows the results obtained by a set of four GSANN s consisting of 14 inputs, two hidden layers with 6 nodes each and an output for the detection of each category.

The networks were trained on only 50 pixels of each category for 10 epochs. The results presented in the bottom part correspond to pixels the networks had never seen before.

The results are very good, all the wooden and field areas are correctly classified and there are very few misclassifications. It must be pointed out that wooden areas (dark green and blue for evergreen and deciduous trees) are visually very similar; they just correspond to different types of trees and bushes, which in certain areas intermingle, especially in borders between zones. The areas labeled light green correspond to meadows and crops. In addition, the urban areas, which were labeled as solid gray areas in the ground truth image, were detected as what they really are: houses intermingled with trees and grassy areas. This particular result provides a measure of confidence of what the classifiers are doing. The problem of dubious or ambiguous ground truth is a common problem when trying to use supervised learning on multi or hyperspectral images. Mislabeled points in the training set may cause havoc on the results when using many types of networks. In the case of gaussian synapse based networks this problem is reduced, due to the intrinsic filtering operations they perform. This will become evident in the next example when considering a hyperspectral image. In addition, this problem of wrongly or ambiguously labeled ground truth is not only a very common problem when training, but it is also a big problem in order to provide a good statistical analysis of the results of the classifier systems in these domains. In fact, using a confusion matrix to present the results, even though it provides

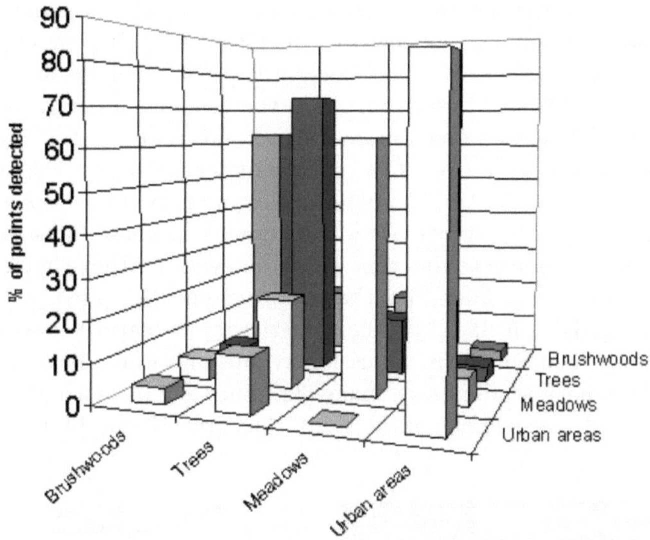

Fig. 14.3. Confusion matrix for the four categories considered in the CREAF multispectral image, the percentage of points assigned is represented as a bar

an indication of the operation of the system, is quite misleading in terms of absolute values due to the fact that when a misclassification occurs it is not easy to determine if it is due to the classifier or to the mislabeling of the ground truth. This is especially so in regions covering borders between classes. In the present case of the CREAF multispectral image, the results of the confusion matrix are presented in figure 14.3.

14.4.2 Application to Hyperspectral Images

In this section we carry out three experiments with synthetic and real images. In the first test case the network is running over the Indian Pines image much in the same way as it was in the previous section with the multispectral image.

The spectra used for this work correspond to the Indian Pines 1992 image obtained by the Airborne Visible/Infrared Imaging Spectrometer (AVIRIS) developed by NASA JPL which has 224 contiguous spectral channels covering a spectral region from 0.4 to 2.5 micron in 10 nm steps. The image used is a 145 by 145 pixel image with 220 spectral bands that contains a distribution of two–thirds of agricultural land and one–third of forest and other elements, namely two highways, a rail line, some houses, and smaller roads. The ground truth available for this image [30] designates 16 not mutually exclusive classes. This scene has been studied by Tadjudin and Landgrebe [31], and also by Gualtieri et al. [5].

Instead of the atmospheric correction model used by these authors, we have started with a very simple preprocessing stage consisting in subtracting the average spectrum of the pixels in the whole image in order to eliminate offsets.

One of the main characteristics of this image is that the labeling was carried out in a blockwise manner. This leads to the effect of having many pixels corresponding to features other than the block area label being taken as part of the block. Obviously, this will induce errors in the training process and will make the generation of detectors much more difficult. Also, due to the altitude of the flight and the spatial resolution of the spectrometer, each pixel corresponds to a very large area (more than 200 sq. meters) thus increasing the probability of the pixel corresponding to mixtures. As a consequence of these facts, by analyzing the spectra corresponding to different pixels one will see that they can be quite complex and misleading. Therefore, the use of systems that incorporate the ability to obtain highly non linear divisions of the input feature space is required. This is the context in which the Gaussian synapse networks have been used.

Some authors have studied this image [32], but they have usually constrained their efforts to sub areas in one of the sides, thus greatly simplifying the problem

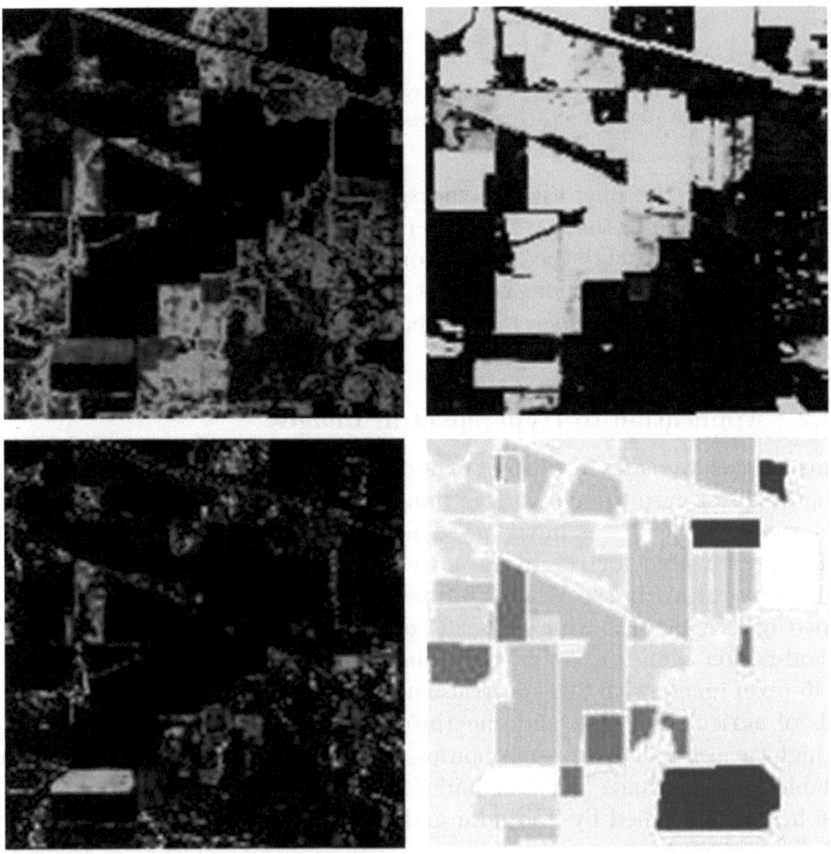

Fig. 14.4. Information provided by three of the detectors over the whole image. Top left: Grass pasture. Top Right: Soybean. Bottom left: Wheat. Bottom right: Ground truth.

as fewer spurious or confusing spectra were present. In this paper, the aim is to show how Gaussian Synapse based Networks are very easy to train to perform good segmentations of hyperspectral images in a computationally very efficient manner even when the labeling is misleading and partial. The objective is not to obtain 100% classification efficiency in classical image segmentation terms, as this would be impossible over the whole image, and probably useless if we constrain our efforts to a very small delimited area. The objective is to obtain a system that provides useful, reliable and interpretable information on what is on the ground without too much computing cost. Therefore this work has considered the labeling of the whole image. The objective was to obtain a maximum likelihood segmentation scheme, that is, a segmentation that would work as a winner take all for each pixel depending on its spectral features and without considering any neighborhood information. The final label of a pixel would be taken as corresponding to the category having the highest probability.

Seven networks were built in order to detect the categories in the Indian Pines image. The detectors were trained for: Soybean, Corn, Grass–Pasture, Grass–Trees, Hay–windrowed, Wheat and Woods. The different types of soybean and corn that were present in original image were grouped because the only difference between the types is the amount of weeds. The image remaining categories were not used because there were an insufficient number of pixels to train and adequately test a network for any of them.

The training sets consisted of different numbers of pixels corresponding to the different categories (Soybean, 220; Corn, 350; GP, 220; GT, 292; HW, 320; Wheat, 130; Woods, 450). These training points were extracted from certain regions of the image, but the tests were carried out over the whole image in order to prove the generalization capabilities of the networks. In fact only less than 1% of the points of the image were used as training pixels.

Due to the non–physical character of the preprocessing stage it was decided to use the whole spectrum and let the network decide what was relevant. The networks initially consist of 220 inputs, corresponding to the spectral bands of AVIRIS. There are two hidden layers, with 18 nodes each, and one output layer corresponding to the presence or absence of the category. The training process consisted of 50 epochs in every case, with the same values for the training coefficients: 0.5 for amplitude training, 2 for variance training, and 0.5 for center training.

Table 14.1. Confusion matrix (percent) for the part of the Indian Pines image where there is information on ground truth

	Corn Soybean	Grass Trees	Grass Pasture	Hay	Wheat	Woods
Corn Soybean	**93.5**	0.7	1.6	1.7	0.3	2.1
Grass Trees	0.0	**83.6**	14.4	2.1	0.0	0.0
Grass Pasture	15.6	7.2	**66.8**	6.5	1.3	2.6
Hay	0.0	0.0	0.0	**100.0**	0.0	0.0
Wheat	0.0	0.0	8.0	0.0	**92.0**	0.0
Woods	0.0	0.0	0.0	0.1	0.0	**99.9**

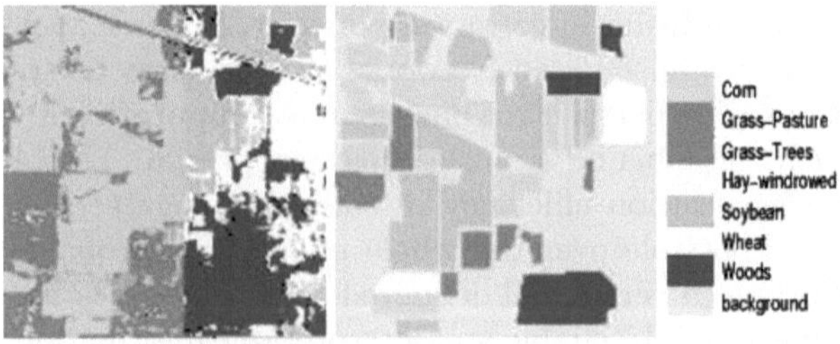

Fig. 14.5. From right to left. Color code of the categories, ground truth obtained in the literature for this categories and result of GSANN networks with a final maximum likelihood decision module in segmenting the image.

Before going into the classification results it is necessary to point out that these networks are quite small as compared to the regular multilayer perceptrons that would be required in order to perform a similar task. These networks only contain 4302 synapses and thus 12906 parameters to train. Take into account that for 220 inputs a regular MLP would need more than 100 neurons per hidden layer in order to have enough degrees of freedom to be able to perform this type of processing, leading to more than 32000 parameters. In addition, with such a huge number of parameters and to adequately train a regular MLP a very slow training procedure with many training samples would be required. Some tests have been performed by using this procedure, but every time they have ended up in local minima and so it was impossible to obtain a reliable classifier using a classical MLP on the whole image. On the contrary, the GSBP detection scheme got good results by requiring only 50 epochs having between 200 and 400 training samples each. Thus, the overhead introduced by having to train three parameters per synapse is largely compensated by the achieved efficiency, as well as by the reduction in the number of synapses and the number of training epochs required induced by the fact that some non linear processing and filtering is carried out through the use of the Gaussian synapses.

For the purpose of providing an idea of the operation of the individual networks, figure 14.4 displays the outputs of three of the detectors after scanning the hyperspectral image. The detectors correspond to the categories of grass pasture, soybean and wheat. There are pixels that are detected to different degrees by more than one of the detectors. Figure 14.5 depicts the results after going through the final decision module designed to take a maximum likelihood decision that does not take into account any neighborhood information. The ground–truth image in the NASA–JPL set for Indian Pines is also provided. It should be notice that the ground truth provided had enormous areas without any labels. The final labeling obtained by our system is very good and is consistent with the results and indications found by other authors [25][31] concerning these dark areas. In addition, the ground truth is not detailed in the sense that there are elements within the labeled zones which were not indicated, such as roads, pathways, trees within other types of areas, etc. Nevertheless, the implemented segmentation system consistently detects all these features. The network performs a very good classification of the whole image including all of the image regions that were no used at all in the training process. The only problem in the classification is the double label obtained in corn and soybean regions, where both detectors indicate the presence of their species. This happens because the image was taken when the plants were only three or four weeks old and only occupied 5% of the areas labeled as soybean and corn, being the remaining 95% bare soil with some weeds in both cases. In addition, corn and soybean plants at this age are very similar from a spectroscopic point of view. As a result, when one of these corn or soybean areas is analyzed by the system both detectors become active. For this reason, in the confusion matrix both categories have being grouped. This matrix, shown in Table 1, provides levels of detection of more than 85% for every category except for grass pasture, which presents a level of around 66% presenting some confusion with corn/soybean and grass tress. These confusion levels are coherent with what was mentioned earlier on what is really on the ground (earth with some weeds). Grass and wheat classes present a similar problem and, consequently, a double labeling is also obtained in some regions.

14.4.3 Endmember Extraction

The next steep in testing thi GSBP based classification system is trying to obtain more information in terms of subpixel endmember extraction. Firstly, a set of

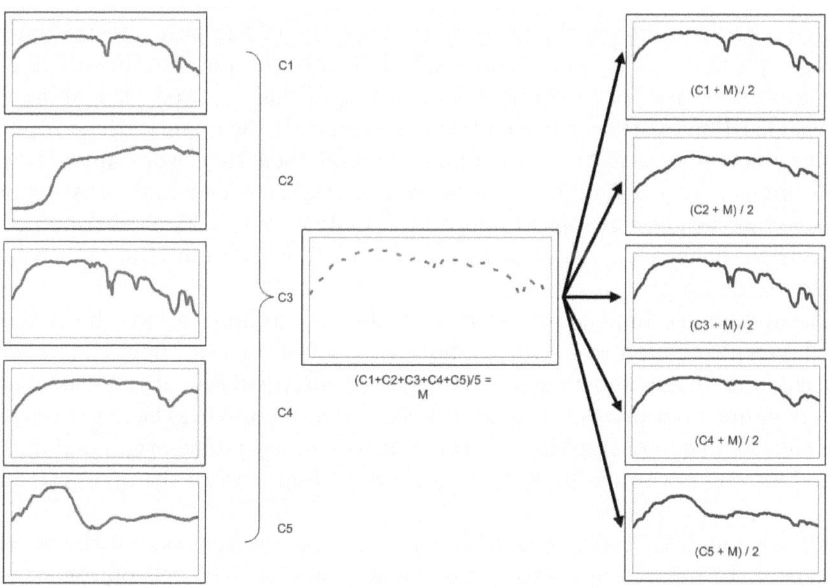

Fig. 14.6. The 11 element set used for the training of the detectors: The five spectra of the synthetic endmembers (left), the mean spectrum (center), five combination spectra (right)

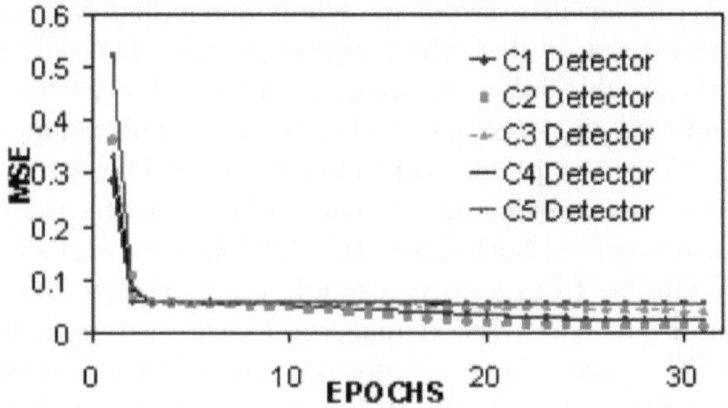

Fig. 14.7. Training error graph for each of the five detectors used in the experiment

synthetic images have been used to assure a test under the most controlled conditions. On one hand this will serve to check the capability of Gaussian synapse network in recognizing patterns, on the other it will explore its ability in discriminating and identifying endmember mixtures. The synthetic images used in these tests have been generated by the group of Prof. Graña [29] using five different endmembers. These images contain 64 by 64 pixels each characterized by 224 spectral bands. The spectra for the pixels were obtained by combinations of the different endmember spectra. A very limited set of spectra has been taken as training set, as shown in figure 14.6. This set has only 11 members, corresponding to the 5 pure endmembers plus the arithmetic mean of the five and the five spectra corresponding to the average of every one of the pure spectra with the average spectrum. It is quite significant that only the pure endmembers and a very limited number of arbitrary linear combinations are used for training.

Five GSBP networks have been trained to identify the quantitative proportion of endmembers present in the training set and then they were applied to the whole image. The training of the networks was very fast and convergent, as displayed in the error graphs of figure 14.7. Only about 30 steps of training were necessary to achieve very good results and just after less than 10 steps the results were quite acceptable.

The abundance images provided with the source images have been used to test the resulting trained networks. For every set of images, there is a set of five greyscale jpg images showing, for every pixel from the data cube, the proportion of each endmember present. So the test consisted in applying the networks to the data cubes and comparing the generated output image to the original abundance data. This comparison is presented in figure 14.8 for two of the synthetic image sets.

But, although it could be said that the images are evidence of the good behaviour of the nets, in order to assess the accuracy of the levels obtained, which correspond to the proportion of the endmember in the pixels throughout the image, in figure 14.9 a comparison of the histogram of the images corresponding to one of the categories is presented. The remarkable similarity between the histogram corresponding to the original abundance image and that of the one

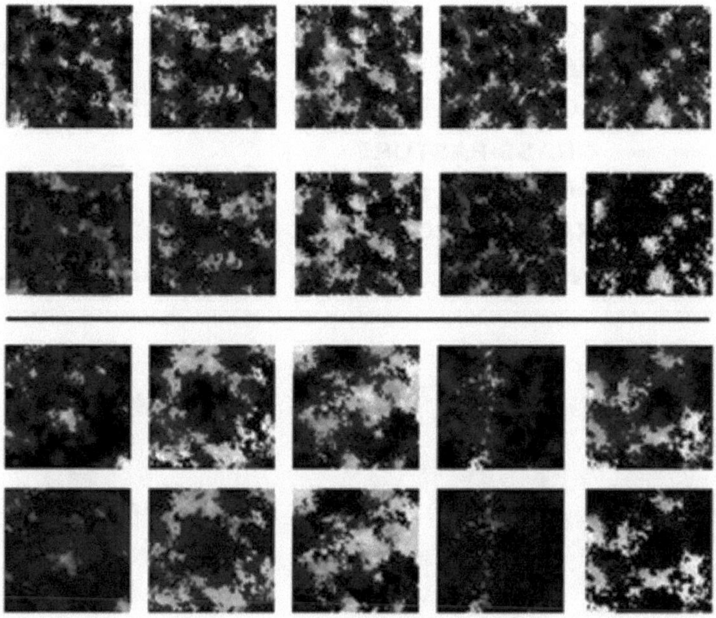

Fig. 14.8. Two different abundance images; for each one in the upper row we can see the original abundances for each of the 5 categories. The bottom row for each image is the abundance image generated by the ANNs.

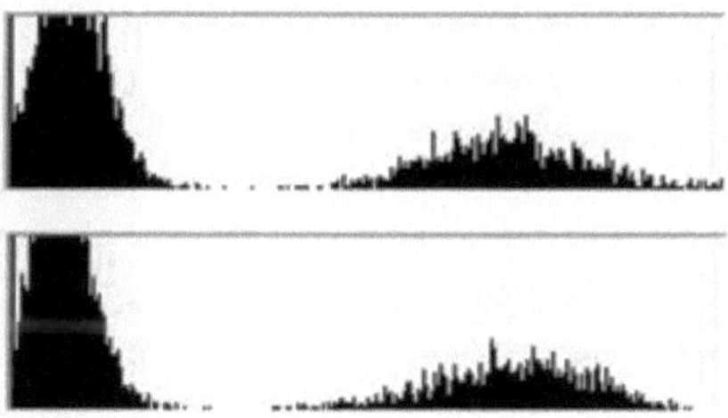

Fig. 14.9. Top: Original Abundance Histogram for C2. Bottom: GNet Generated Abundance Histogram. We can see that, although they are not identical, the number of levels is very similar in both cases. This provides an indication of the generalization ability in terms of abundance level of a net trained with a very limited number of combinations.

reconstructed by the Gaussian Synapse Network provides a good indication of the power of the GSANN for extracting spectral proportions from a pixel.

As a next step some more tests have been performed on the very well known real hyperspectral Indian Pines cube [30] in order to complement the results from the analysis of controlled synthetic images, and verify how well GSANN are adapted to the task of discriminating the presence of those marginal elements within the

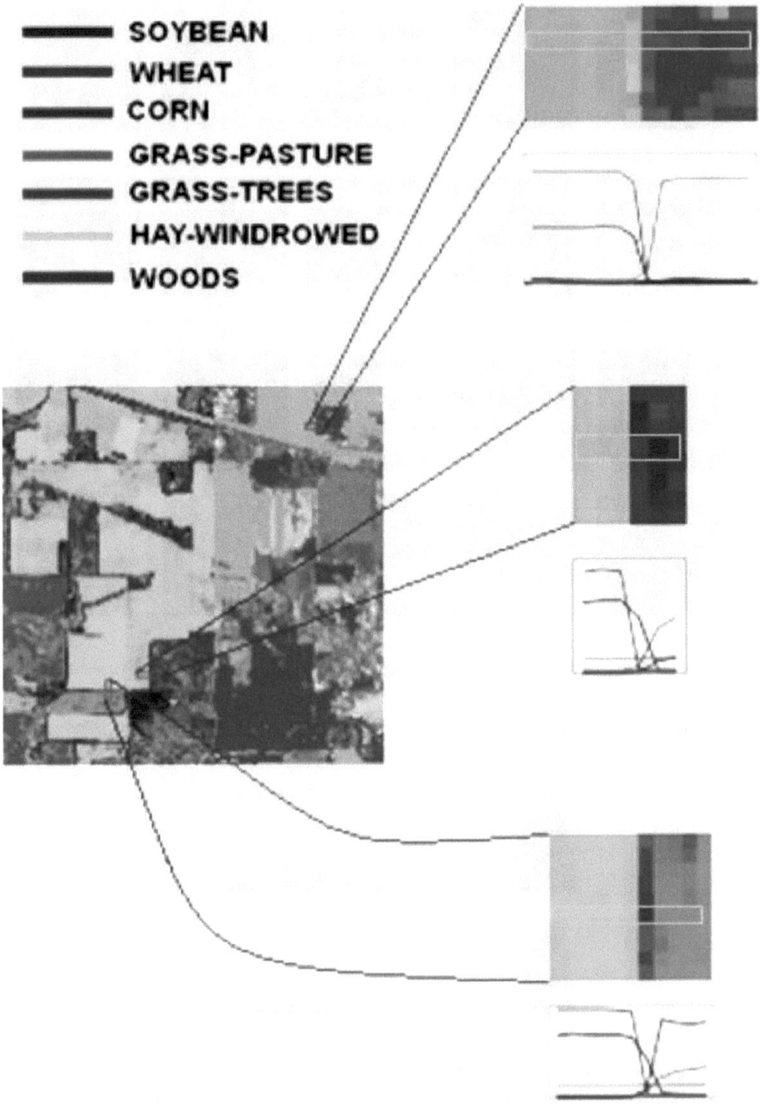

Fig. 14.10. Spectral mixtures using the networks in three transition zones of the Indian Pines image

pixels. From this image we have extracted groups of pixels representing transitions between areas with different labels in the hope that the intermediate pixels contain mixtures of elements from both sides of the transition. The networks were trained using pixels inside each labelled area and a few linear combinations of the spectra and then applied to the detection of proportions in the transitions. As no ground truth was available on the proportions of elements present on these pixels, some

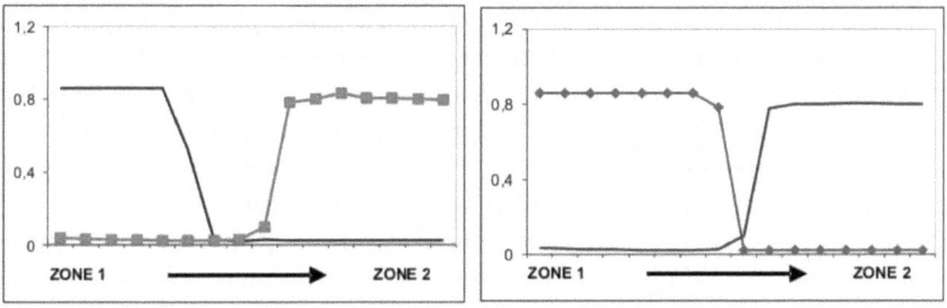

Fig. 14.11. Behavior of the detectors of two given endmembers through a brusque transition for a series of pixels obtained from the Indian Pines image (left) and for the synthetically generated pixels (right)

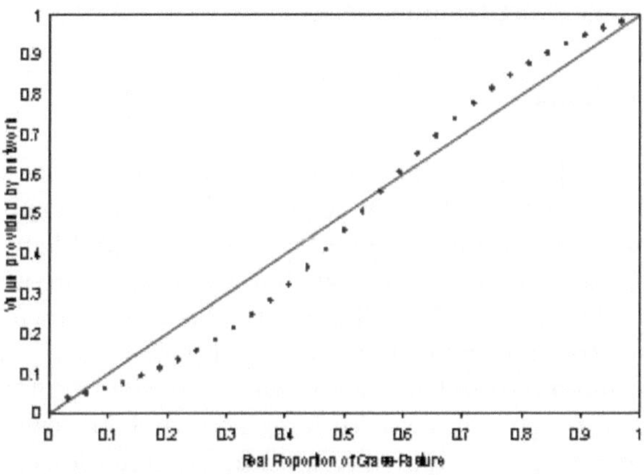

Fig. 14.12. Prediction by Gaussian Synapse Networks of the percentage of an endmember present in synthetically generated pixels as compared to the real values. Perfect results are represented by the black line.

artificial sets of pixels starting from the same endmembers were also generated to simulate the same type of brusque transition so as to provide a validation set for the results obtained from the real image. The results from running the detectors generated earlier over the transition areas selected can be seen in figure 14.10. These graphs clearly display a transition between categories where two or three pixels in the middle contain mixtures of the elements present on the sides.

As commented before, to check the validity of these results, we generated a synthetic set of pixels through the transitions with known proportions and compared them to the real ones obtained from the image in terms of the proportions produced by the GSBP networks. The evolution of the proportion of endmembers for these two cases along the pixel line can be observed in figure 14.11. These results are quite encouraging, as the detection on the real image closely follows that of the synthetic one with well known proportions.

To better decide how accurately these networks could predict proportions, and detect marginal elements a new transition is generated in which the evolution of proportions from one endmember to the other is linear. Figure 14.12 shows the results obtained by applying the detectors in analyzing this line of pixels. This figure displays the percentage of a given endmember (solid line) present in each pixel and the values provided by this endmember detector. It can be observed that the agreement between these values is quite close and that even low percentages of the endmember present in the image are easily detected. These results show how the GSANN based segmentation system is able to detect the presence of endmembers when their proportions are as small as 3% of the pixel surface. These results, coupled with those obtained using the synthetic images, confirm the adequacy of GSANN for the determination of quantitative endmember mixture results after training with very limited numbers of labelled samples and for very short training periods.

14.5 Conclusions

This chapter is aimed at providing an introduction to the use of GSANN and its GSBP training algorithm for the segmentation of hyperspectral images as well as explaining their use in the spectral unmixing problem, particularly in cases when the endmembers are known but some of them may appear in a pixel with a very low proportion. These networks have shown great potential for obtaining detectors for endmembers fast and with a very small training set. In the spectral unmixing problem they prove to be able to obtain very precise abundance images using very small training sets that are usually made up of the pure endmember spectra and a few linear combinations of them. In fact in this work we have thoroughly tested this approach on a set of benchmark hyperspectral images and confirmed the results on the well known Indian Pines image. The promising results presented here suggest that the combination of this strategy with an unsupervised method for the extraction of the endmembers would be a very promising path to follow in order to produce a system that can automatically obtain a segmentation of an unknown hyperspectral cube with unknown endmembers.

Acknowledgements

This work was funded by the Xunta de Galicia "Medio natural e desenvolvemento sostibel" Program.

References

1. Wiemker, R., Speck, A., Kulbach, D., Spitzer, H., Bienlein, J.: Unsupervised Robust Change Detection on Multispectral Imagery Using Spectral and Spatial Features. In: Proc. of 3rd International Airborne Remote Sensing Conference and Exhibition, vol. I, pp. 640–647 (1997)

2. Sabins, F.F.: Remote Sensing: Principles and Interpretation. John Wiley and Sons, Toronto (1994)
3. Campbell, J.: Introduction to Remote Sensing. The Guilford Press, New York (1996)
4. Sanders, L.C., Raqueno, R.V., Schott, J.R.: Atmospheric correction algorithm for hyperspectral imagery. In: Proceedings of SPIE, Image and Signal Processing for Remote Sensing, vol. 3871, pp. 2–9 (1999)
5. Guatieri, J.A., Bechdol, M., Chettri, S., Robinson, J.W., Garegnani, J., Vermeulen, A., Antonille, S.: From Spectra to Cassification. In: 1st International Symposium on Hyperspectral Analysis, Caceres (Spain) (2000)
6. Yamazaki, A., Imanaka, M., Shikada, M., Okumura, T., Kawata, Y.: An atmospheric correction algorithm for space remote sensing data and its validation. In: Proc. of IGARSS 1997, vol. 4, pp. 1899–1901 (1997), ISBN: 0-7803-3836-7
7. Thome, K., Palluconi, F., Takashima, T., Masuda, K.: Atmospheric Correction of ASTER. IEEE Transactions on geoscience and remote sensing 36(4), 1199–1211 (1998)
8. Liang, S., Fang, H.: An Improved Atmospheric Correction Algorithm for Hyperspectral Remotely Sensed Imagery. IEEE Geoscience and remote sensing letters 1(2), 112–117 (2004)
9. Rencz, A. (ed.): Manual of Remote Sensing. John Wiley and Sons Inc., New York (1999)
10. Van der Meer, F.D., de Jong, S.M.: Spectral Mapping Methods: many problems, some solutions. In: Habermeyer, M., Mueller, A., Holzwarth, S. (eds.) Proceedings of the 3rd EARSeL Workshop on Imaging Spectrometry, pp. 146–162 (2003), ISBN 2-90885-26-3
11. Hsu, P.H., Tseng, Y.H.: Feature Extraction for Hyperspectral Image. In: Proceedings of the 20th Asian Conference on Remote Sensing, vol. 1, pp. 405–410 (1999)
12. Schowengerdt, R.A.: Remote Sensing: Models and Methods for Image Processing. Academic Press, London (1994)
13. Ross, B.J., Gualtieri, A.G., Fueten, F., Budkewitsch, P.: Hyperspectral Image Analysis Using Genetic Programming Source. In: Proceedings of the Genetic and Evolutionary Computation Conference, pp. 1196–1203 (2002), ISBN:1-55860-878-8
14. Schouten, T.E., Klein, M.S.: Fuzzy classification of spectra. In: Neural networks: Best practice in Europe, pp. 198–201. World Scientific, Singapore (1997)
15. Plaza, J., Martínez, P., Pérez, R., Plaza, A.: Nonlinear neural network mixture models for fractional abundante estimation in AVIRIS hyperspectral images. In: Proc. of AVIRIS Airborne Earth Science Workshop, California (2004)
16. Nascimento, J.M.P., Bioucas Dias, J.M.: Vertex Component Analysis: A Fast Algorithm to Unmix Hyperspectral Data. IEEE Transactions on Geoscience and Remote Sensing 43(4), 898–910 (2005)
17. Bajorski, P.: Simplex projection methods for selection of endmembers in hyperspectral imagery. In: IGARSS 2004 Proceedings (2004), ISBN: 0-7803-8742-2
18. Bayliss, J., Gualtieri, J., Cromp, R.: Analyzing hyperspectral data with independent component analysis. In: Michael Selander, J. (ed.) Proc. SPIE AIPR Workshop, vol. 9 (1997)
19. Aguilar, P.L., Plaza, A., Martinez, P., Perez, R.M.: Endmember Extraction by a Self-Organizing Neural Network on Hyperspectral Images. In: VI International Conference on Automation, Robotics, Control and Vision (ICARCV), Singapore (2000)

20. Greiwe, A., Ehlers, M.: Combined Analysis of Hyperspectral and High resolution Image Data in an Object oriented Classification Approach. In: Proc. 5th International Symposium Remote Sensing of Urban Areas, Tempe, AZ, USA (2005)
21. Plaza, A., Martinez, P., Perez, R., Plaza, J.: Spatial spectral endmember extraction by multidimensional morphological operations. IEEE Transactions on Geoscience and Remote Sensing 40(9), 2025–2041 (2002)
22. Lee, C., Landgrebe, D.: Feature Extraction and Classification Algorithms for High Dimensional Data. Internal Report TR-EE 93-1. Purdue University (1993)
23. Merényi, E., Minor, T.B., Taranik, J.V., Farrand, W.H.: Quantitative Comparison of Neural Network and Conventional Classifiers for Hyperspectral Imagery. In: Summaries of the Sixth Annual JPL Airborne Earth Science Workshop, Pasadena, CA, vol. 1 (1996)
24. Ghosh, J.: Adaptive and neural methods for image segmentation. In: Bovik, A. (ed.) Handbook of Image and Video Processing, ch. 4.10, pp. 401–414. Academic Press, London (2000)
25. Tadjudin, S., Landgrebe, D.: Classification of High Dimensional Data with Limited Training Samples. Ph.D. Thesis and School of Electrical & Computer Engineering Technical Report TR-ECE 98-8, Purdue University
26. Merényi, E.: The Challenges in Spectral Image Analysis: an Introduction, and Review of ANN Approaches. In: Proceedings of the 7th European Symposium on Artificial Neural Networks, ESANN (1999)
27. Duro, R.J., Crespo, J.L., Santos, J.: Training Higher Order Gaussian Synapses. In: Mira, J. (ed.) IWANN 1999. LNCS, vol. 1606, pp. 537–545. Springer, Heidelberg (1999)
28. Prieto, A., Bellas, F., Duro, R.J., Lopez-Peña, F.: A Comparison of Gaussian Based ANNs for the Classification of Multidimensional Hyperspectral Signals. In: Cabestany, J., Prieto, A.G., Sandoval, F. (eds.) IWANN 2005. LNCS, vol. 3512, pp. 829–836. Springer, Heidelberg (2005)
29. Graña, M., Raducanu, B., Sussner, P., Ritter, G.: On Endmember Detection in Hyperspectral Images with Morphological Associative Memories. In: Garijo, F.J., Riquelme, J.-C., Toro, M. (eds.) IBERAMIA 2002. LNCS (LNAI), vol. 2527. Springer, Heidelberg (2002)
30. Landgrebe, D.: Indian Pines AVIRIS Hyperspectral Reflectance Data: 92av3c, `http://makalu.jpl.nasa.gov/`
31. Tadjudin, S., Landgrebe, D.: Covariance Estimation for Limited Training Samples. In: Proc. of Int. Geoscience and Remote Sensing Symp., Seattle (1998)
32. Tadjudin, S., Landgrebe, D.: Robust parameter estimation for mixture model. IEEE Trans. Geos. Rem. Sensing 38(1) (2000)
33. Boardman, J.W., Kruse, F.A., Green, R.O.: Mapping target signatures via partial unmixing of AVIRIS data. In: Fifth JPL Airborne Earth Science Workshop, vol. 1, pp. 23–26. JPL Publication 95-1 (1995)

15

Unsupervised Change Detection from Multichannel SAR Data by Markov Random Fields

Sebastiano B. Serpico and Gabriele Moser

University of Genoa, Dept. of Biophysical and Electronic Eng. (DIBE), I-16145 Genova, Italy, and Interuniversity Center of Research in Environmental Monitoring (CIMA), I-17100 Savona
sebastiano.serpico@unige.it

15.1 Introduction and Previous Work

In the contexts of environmental monitoring and disaster management, techniques allowing one to detect the changes that occurred in a given area between successive acquisition dates may serve as efficient data-analysis tools. Good discrimination between changed and unchanged areas can often be achieved by analyzing optical multispectral data in the related multidimensional feature space [20][43]. However, such data are affected by atmospheric and sun-illumination conditions. Synthetic aperture radar (SAR) [15][27][34] is insensitive to such issues, hence, multitemporal SAR imagery is expected to play an important role, for instance, in ecological applications [18] or in disaster assessment and prevention [11][42]. However, SAR usually allows only one amplitude/intensity observation, thus possibly resulting in poor discrimination. Multichannel (i.e., multipolarization and/or multifrequency) SAR represents an option with improved potential: as compared with single-channel SAR, it is expected to provide an increased discrimination capability, while maintaining its insensitivity to atmospheric and illumination issues. This potential is also reinforced by the availability of multichannel SAR data with short revisit times which are granted by future missions (e.g., COSMO/SkyMed and TerraSAR).

Due to the multiplicative nature of speckle, unsupervised change detection in SAR images is often addressed by image ratioing, i.e., by computing a ratio image obtained by dividing the pixel intensities in the image acquired at the first observation date by the ones in the image acquired at the second observation date (or vice-versa) [34]. The ratio image is then analyzed in order to quantitatively distinguish between changed and unchanged areas. Interactive thresholding procedures are often applied toward this end [34].

Recently, some efforts have been devoted to the automation of the change-detection process by single-channel SAR. The unsupervised change detection method proposed in [25] for optical multispectral data and based on the Kittler and Illingworth thresholding algorithm (K&I) is extended in [2] by introducing

M. Graña and R.J. Duro (Eds.): Comput. Intel. for Remote Sensing, SCI 133, pp. 363–388, 2008.
springerlink.com

a generalized Gaussian model for a logarithmic ratio image and by integrating it with K&I. Similar approaches are developed in [1] and [3] by combining generalized Gaussian distributions with the expectation-maximization (EM) technique for parametric probability density function (PDF) estimation [37] and with a multiple-threshold version of K&I, respectively. In [28], three SAR-specific parametric models for the PDFs of SAR amplitude ratios are developed and an unsupervised change-detection method is proposed by combining image ratioing with K&I and with the recently proposed "method of log-cumulants" (MoLC) for parameter estimation [33].

Other approaches, which have also been proposed in the literature to deal with SAR change detection, are multitemporal coherence analysis [38], integration of segmentation with multilayer perceptron and Kohonen neural networks [47], fuzzy-rule-based image analysis [21], multisource (optical and SAR) and multitemporal data fusion [4], spatio-temporal contextual classification [24][26][41], and likelihood ratio tests [6][23].

However, change detection in multichannel SAR images is still mostly an open issue, often addressed by using time-expensive "trial-and-error" procedures. In [9], image ratioing is applied to C- and L-band airborne SAR intensity images and compared with the use of the correlation coefficient and the phase difference between copolarized channels, with a particular focus on the detection of changes involving small objects and linear image features. Thresholds to be applied to the ratio channels are selected according to a Neyman-Pearson-type approach by manually choosing an acceptable value for the false-alarm probability [46]. In [6], the problem of change-detection in multilook fully polarimetric (complex) SAR images is formalized as a hypothesis testing problem, and case-specific likelihood-ratio tests are derived according to the well-known Wishart distribution for polarimetric covariance matrices [34]. Polarimetric basis transformations are used in [36] to optimize the similarity of the scattering mechanisms characterizing the images acquired at different observation dates and to define quantitative measures of change in multitemporal polarimetric SAR images, to be interactively analyzed to distinguish between changed and unchanged areas.

In the present chapter, two automatic unsupervised change-detection methods, both based on image-ratioing and Markov random fields (MRFs), are proposed for multichannel SAR images. The first technique extends to multichannel SAR images the method proposed in [25] for optical multispectral images. The EM algorithm with semilabeled samples developed by Landgrebe and Jackson (LJ-EM for short) [17] is combined with a SAR-specific version of the Fisher transform [12] to compute a scalar transformation of the multichannel ratio-image that optimally discriminates between "change" and "no-change." As is usual in SAR image classification, the use of spatial contextual information is crucial to reduce the impact of speckle on the change-detection results. An MRF-based approach is adopted toward this end: this is known as a flexible and powerful strategy to model the contextual information in a classification task by defining suitable "energy functions" and formulating the maximum-*a-posteriori* classification criterion as a minimum-energy decision rule [10]. This first proposed method (in the following denoted as "Fisher-transform-based multichannel change detection," FT-MCD) is iterative. Even though a good convergence behavior was experimentally observed (see Sec. 15.4), no theoretical proof of convergence is available yet. The second proposed technique overcomes this drawback by formalizing the problem of multichannel SAR change detection in terms of a multichannel data-fusion problem, again addressed by an MRF-based approach [45]. Each SAR channel is modeled as a separate "information source," and MRFs are used to define a decision rule based on multichannel fusion. In this case, MRFs are adopted thanks to their capability to integrate both the spatial contextual information and distinct information sources into the image classification process [45]. In order to address the task of the estimation of the model parameters, a case-specific formulation of LJ-EM is developed and combined with MoLC. This second approach will be denoted as "Data-fusion-based multichannel change detection" (DF-MCD). Both proposed techniques were experimentally validated on SIR-C/XSAR data.

The chapter is organized as follows. The two proposed techniques are described from the methodological viewpoint in Secs. 15.2 and 15.3, respectively. Experimental results of both approaches are presented in Sec. 15.4 and conclusions are drawn in Sec. 15.5. Analytical details and proofs concerning the parameter-estimation issues involved in the proposed techniques are reported in the Appendix.

15.2 The Fisher-Transform-Based Method

15.2.1 Overview of the FT-MCD Method

Let $\mathcal{I}_0$ and $\mathcal{I}_1$ be two coregistered SAR images, acquired over the same area at times t_0 and t_1, respectively ($t_1 > t_0$), and consisting of n SAR amplitude channels each. The change-detection problem is formulated as a binary hypothesis testing problem [46] by denoting the "change" and "no-change" hypotheses as H_0 and H_1, respectively. The image-ratioing approach is adopted, that generates a ratio image $\mathcal{R}$ by dividing pixel-by-pixel and channel-by-channel the amplitudes in $\mathcal{I}_0$ by the amplitudes in $\mathcal{I}_1$ [9]; hence $\mathcal{R}$ is an n-channel image as well.

The proposed FT-MCD technique represents a modified version of the LJ-EM technique developed in [25] for multispectral data. The key idea lies in the application of a SAR-specific version of the Fisher transform in order to compute a (scalar) feature that best discriminates between H_0 and H_1, and in the integration of LJ-EM and MRFs in order to perform a contextual classification in the Fisher-transformed space. The Fisher transform is intrinsically supervised [12], i.e., it requires training samples for H_0 and H_1, which are not available in the present unsupervised context. To overcome this difficulty, the proposed method employs an iterative approach. It is initialized with the change map generated by the (single-channel) "generalized Kittler and Illingworth thresholding" (GKIT) algorithm [28], applied to an automatically selected channel in

$\mathcal{R}$ (see Sec. 15.2.4). Then, at each q-th iteration ($q = 0, 1, 2, \ldots$), the following processing stages are performed:

- *feature-transformation stage:* apply the SAR-specific Fisher transform (see Sec. 15.2.2) to the ratio image $\mathcal{R}$ to generate a scalar transformed image $\mathcal{F}^q$ by using the current change map $\mathcal{M}^q$ as a training map in the computation of the transformation parameters;
- *EM-based classification stage:* apply an MRF-based formulation of LJ-EM (see Sec. 15.2.3) to $\mathcal{F}^q$ in order to update the change map (i.e., to generate $\mathcal{M}^{q+1}$).

The technique aims to iteratively improve the change map by generating a sequence of feature transformations, which progressively increase the separation between H_0 and H_1 in the transformed image. The above two stages are iterated until the percentage of pixels with different labels in $\mathcal{M}^q$ and $\mathcal{M}^{q+1}$ becomes smaller than a predefined threshold (here, equal to 0.1%).

15.2.2 Feature-Transformation Stage

The Fisher transform is a scalar feature mapping that linearly projects a (multidimensional) feature space in an optimal direction, computed by maximizing a functional related to the separation between the sets of samples drawn from two hypotheses (e.g., "change" and "no-change") and to the compactness of such sets [12]. This functional is defined in terms of the sample-means and sample-covariance matrices of the two sample sets; this intrinsically relates the transform to a Gaussian model for each hypothesis [12].

SAR amplitude ratio data are known to be strongly non-Gaussian [34]. By modelling $\mathcal{R}$ as a set $\{u_1, u_2, \ldots, u_N\}$ of n-variate identically distributed random vectors (N being the number of pixels), several parametric models can be used to describe the marginal statistics of the n features in u_k conditioned to H_i ($k = 1, 2, \ldots, N; i = 0, 1$), such as the log-normal (LN) [28], Nakagami-ratio (NR) [28][34], Weibull-ratio (WR) [28], and generalized Gaussian [2] models. We adopt a log-normal distribution, since it has already proven effective for single-channel SAR change-detection [28], and since, unlike the other mentioned models, it admits a natural extension to the multichannel case, i.e., the multivariate log-normal model [22][49]. Given H_i, such a model is assumed to hold for u_k ($k = 1, 2, \ldots, N; i = 0, 1$). Note that, by adopting this monomodal model for H_1, we implicitly focus on the case in which a single typology of change is present in the image.

A joint Gaussian distribution for the log-ratios $\ln u_{k1}, \ln u_{k2}, \ldots, \ln u_{kn}$ results ($k = 1, 2, \ldots, N$). Accordingly, at each q-th iteration of the proposed method, the transformed image $\mathcal{F}^q$ is computed by applying the Fisher transform not to $\mathcal{R}$ but to the corresponding log-ratio image, i.e., by mapping u_k into a scalar feature v_k such that:[1]

[1] Given $v \in \mathbb{R}^m$, we denote as v_j the j-th component of v ($j = 1, 2, \ldots, m$).

$$\ln v_k = \sum_{j=1}^{n} a_j \ln u_{kj} \quad \Longrightarrow \quad v_k = \prod_{j=1}^{n} u_{kj}^{a_j}, \tag{15.1}$$

where $a_1, a_2, \ldots, a_n$ are the transformation parameters. Given a training set for H_0 and H_1, a closed-form solution is known for such parameters [12]: as stated in Sec. 15.2.1, this training set is replaced here by the sets of pixels assigned to H_0 and H_1 in $\mathcal{M}^q$. Thanks to Eq. (15.1), an H_i-conditional log-normal model holds for v_k as well, i.e., the PDF of v_k, given H_i, is provided by:

$$p_i(v_k|\kappa_{1i}, \kappa_{2i}) = \frac{1}{v_k\sqrt{2\pi\kappa_{2i}}} \exp\left[-\frac{(\ln v_k - \kappa_{1i})^2}{2\kappa_{2i}}\right], \qquad v_k > 0, \tag{15.2}$$

where $\kappa_{1i} = E\{\ln v_k|\ell_k = H_i\}$ and $\kappa_{2i} = \mathrm{Var}\{\ln v_k|\ell_k = H_i\}$ are the logarithmic mean and variance of v_k, given H_i ($k = 1, 2, \ldots, N; i = 0, 1$).

15.2.3 EM-Based Classification Stage

Unlike [25], a contextual classification approach is adopted here in order to reduce the speckle impact. By denoting as $\ell_k \in \{H_0, H_1\}$ the hypothesis label of the k-th pixel ($k = 1, 2, \ldots, N$), the label configuration $\{\ell_1, \ell_2, \ldots, \ell_N\}$ is assumed to be an MRF, i.e., for each k-th pixel, a neighborhood is defined and the following properties are assumed to hold [10]:

- positivity: $P(\ell_1, \ell_2, \ldots, \ell_N) > 0$ for all $\ell_1, \ell_2, \ldots, \ell_N \in \{H_0, H_1\}$;
- Markovianity: $P\{\ell_k = H_i|\ell_h, h \neq k\} = P\{\ell_k = H_i|\mathcal{C}_k\}$, where $\mathcal{C}_k$ is the set of labels of the neighboring pixels (i.e., the context) of the k-th pixel ($k = 1, 2, \ldots, N; i = 0, 1$).

For instance, when adopting a second-order neighborhood, $\mathcal{C}_k$ is made up of the labels of the eight pixels surrounding the k-th pixel ($k = 1, 2, \ldots, N$). The Markovianity property allows modelling the spatial dependence among the labels of different pixels on a local basis. This approach is characterized by a dramatic reduction in the computational complexity, as compared with the use of a global model for the joint distribution $P(\ell_1, \ell_2, \ldots, \ell_N)$ of the pixel labels. Thanks to the so-called Hammersley-Clifford theorem [13], one may prove that, if the label configuration is an MRF, then the global maximum *a-posteriori* (MAP) classification rule (i.e., the maximization of the joint posterior probability distribution of all the N pixel labels, given all the N pixel intensities) is equivalent to the minimization of an "energy function" that is defined only locally, on the basis of the neighborhood. Hence, the intractable global maximization problem involved by MAP allows a tractable local formulation.

Specifically, under mild assumptions, the following expression results for the (local) posterior distribution of each pixel label ℓ_k, given v_k and $\mathcal{C}_k$ ($k = 1, 2, \ldots, N; i = 0, 1$) [10]:

$$P\{\ell_k = H_i|v_k, \mathcal{C}_k\} = \frac{\exp[-U(H_i|v_k, \mathcal{C}_k)]}{\sum_{j=0}^{1} \exp[-U(H_j|v_k, \mathcal{C}_k)]}, \tag{15.3}$$

where $U(H_i|v_k, \mathcal{C}_k)$ is the "energy function" characterizing the specific MRF model. Here, an isotropic second-order Potts MRF model is adopted [10]. According to this choice and to the LN distribution of v_k, given H_i, the energy function turns out to be [10]:

$$U(H_i|v_k, \mathcal{C}_k, \theta) = -\ln p_i(v_k|\kappa_{1i}, \kappa_{2i}) - \beta m_{ik}, \tag{15.4}$$

where m_{ik} is the number of labels equal to H_i in the neighborhood of the k-th pixel ($k = 1, 2, \ldots, N; i = 0, 1$), β is a positive parameter, and $\theta = (\beta, \kappa_{10}, \kappa_{20}, \kappa_{11}, \kappa_{21})$ is a vector collecting all the model parameters to be estimated (in Eq. (15.4) the dependence of the energy on θ has been stressed by including θ in the corresponding notation).

The energy function in Eq. (15.4) is a linear combination of two terms. The first is related to the information conveyed by the (pixel-wise) H_i-conditional distribution of v_k, which is in turn related, through the ratio u_k, to the SAR amplitude change in the k-th pixel ($k = 1, 2, \ldots, N; i = 0, 1$). The second term is related to the spatial contextual information, represented by m_{ik} thanks to the Potts model ($k = 1, 2, \ldots, N; i = 0, 1$). According to the minimum-energy MRF decision rule, the spatial term ($-m_{ik}$) encourages the generation of homogeneous regions including pixels assigned to the same hypothesis, thus reducing the impact of speckle on the classification results (further details can be found in [10] and [16]). The parameter β tunes the relative importance between the two energy terms. The model formalized by Eqs. (15.3) and (15.4) is intrinsically stationary. This condition is usually not fulfilled by real images, but is often accepted as a simplifying analytical assumption [16].

At the q-th step of FT-MCD, LJ-EM is used to estimate θ and to generate the updated map $\mathcal{M}^{q+1}$. LJ-EM is an iterative parameter estimation methodology that deals with problems of data incompleteness and aims to compute maximum-likelihood estimates [17]. The application of EM-type algorithms to contextual image models typically involves long computation times, due to the need to run a Gibbs sampler or to compute MonteCarlo approximations at each EM iteration [5]. To overcome this problem, we adopt the "mode-field" approach, which introduces an approximation in the log-likelihood function in order to reduce the computational burden, and which represents a good tradeoff between execution time and classification accuracy (a general description of this approach can be found in [5]; similar techniques have been applied to remote-sensing data in [7] and [16]).

Specifically, at the q-th step of FT-MCD, a full cycle of LJ-EM is performed; as a result, $\mathcal{M}^{q+1}$ is obtained. By focusing on the t-th LJ-EM iteration performed at the q-th step of the method and by marking with the superscript "t" the current parameter estimates and pixel labels ($q, t = 0, 1, 2, \ldots$; the index q is dropped for ease of notation), the following operations are performed [5]:

1. compute, for each k-th pixel, the current estimate of $P\{\ell_k = H_i|v_k, \mathcal{C}_k^t, \theta^t\}$ ($k = 1, 2, \ldots, N; i = 0, 1$) by using Eqs. (15.3) and (15.4);

2. update the label of each k-th pixel according to the MRF minimum-energy rule [10] by setting ℓ_k^{t+1} as the label H_i that corresponds to the smallest value of $U(H_i|v_k, C_k^t, \theta^t)$ $(k = 1, 2, \ldots, N; i = 0, 1)$;
3. update the parameter estimates as follows:

$$\begin{cases} \kappa_{1i}^{t+1} = \dfrac{\sum_{k=1}^{N} w_{ik}^t \ln v_k}{\sum_{k=1}^{N} w_{ik}^t}, & i = 0, 1 \\ \kappa_{2i}^{t+1} = \dfrac{\sum_{k=1}^{N} w_{ik}^t (\ln v_k - \kappa_{1i}^{t+1})^2}{\sum_{k=1}^{N} w_{ik}^t}, & i = 0, 1 \\ \beta^{t+1} = \arg\max_{\beta>0} \sum_{k=1}^{N} \left[\beta \sum_{i=0}^{1} w_{ik}^t m_{ik}^t - \ln \sum_{i=0}^{1} \exp(\beta m_{ik}^t) \right] \end{cases} \tag{15.5}$$

where $w_{ik}^t = P\{\ell_k = H_i | v_k, C_k^t, \theta^t\}$ if the k-th pixel has been assigned to H_i in step 2, and $w_{ik}^t = 0$ otherwise $(k = 1, 2, \ldots, N; i = 0, 1)$.

The formulation of the updated parameter estimates in step 3 is proven in Appendix A.1. The maximization problem to compute β^{t+1} in step 3 is solved by the Newton-Raphson method [35]. Thanks to the smooth differentiable behavior of the objective function, this numerical maximization process is not critical.

LJ-EM is a modified version of the EM algorithm. We recall that EM is known to converge, under mild assumptions, to a local maximum of the log-likelihood function (even though convergence to a global maximum is not ensured, usually a good solution is obtained) [37]; a similar convergent behavior is expected for LJ-EM as well [17]. LJ-EM is iterated until the differences between the parameter estimates at successive iterations are below a predefined threshold (here equal to 0.001). At each iteration, the parameters of H_i are updated only according to the samples assigned to H_i $(i = 0, 1)$. This aims to compute parameter estimates that reduce the overlapping between H_0 and H_1.

15.2.4 Initialization with the GKIT Algorithm

The initial change map $\mathcal{M}^0$ is generated on a single-channel basis. The GKIT technique [28] is adopted here; it can be based on an LN, NR, or WR conditional model for SAR ratio data. GKIT automatically computes the optimal threshold to be applied to a single-channel ratio image (in order to distinguish between "change" and "no-change") by minimizing a "criterion function" defined according to a Bayesian approach and related to the probability of error [28].

Here, according to the multivariate LN model adopted for the n-dimensional ratio vector and to the corresponding single-variate LN model for each ratio channel, the version of GKIT based on the LN distribution is applied. Let $J_r(\cdot)$ be the criterion function computed by GKIT when applied to the r-th channel in $\mathcal{R}$, $\tau_r^* = \arg\min_\tau J_r(\tau)$ the corresponding optimal threshold, and $\mathcal{M}_r^*$ the resulting change map $(r = 1, 2, \ldots, n)$ [28]. $\mathcal{M}^0$ is defined as the map $\mathcal{M}_r^*$ corresponding to the smallest optimal value $J_r(\tau_r^*)$ of the criterion function $(r = 1, 2, \ldots, n)$. According to the relationship between the criterion function and the probability of error [19][28], this change map is expected to be at least a suboptimal choice from among the single-channel maps.

For initialization purposes, this map is used to define the initial context $\mathcal{C}_k^0$ of each k-th pixel ($k = 1, 2, \ldots, N$) and the initial estimates of the PDF parameters. A sample mean estimate κ_{1i}^0 of κ_{1i} and a sample-variance estimate κ_{2i}^0 of κ_{2i} are computed according to the set of samples assigned to H_i in $\mathcal{M}_0$ ($i = 0, 1$).

15.3 The Data-Fusion-Based Method

15.3.1 The Proposed MRF Model

Adopting the same notations as in Sec. 15.2, the DF-MCD technique is still based on a two-hypothesis unsupervised classification of the ratio image $\mathcal{R}$. The key-idea of this method is to model each channel as a distinct information source and to formalize the change-detection problem as a multichannel data-fusion problem addressed by using MRFs. The label configuration $\{\ell_1, \ell_2, \ldots, \ell_N\}$ is assumed again to be an MRF. However, no vector-into-scalar transformations are adopted to map the multivariate ratio vector into a scalar feature. On the contrary, a multichannel fusion strategy is used in order to directly classify in the original multidimensional space. Accordingly, the following (local) posterior probability distribution is introduced ($k = 1, 2, \ldots, N; i = 0, 1$) [10]:

$$P\{\ell_k = H_i | u_k, \mathcal{C}_k\} = \frac{\exp[-U(H_i | u_k, \mathcal{C}_k)]}{\sum_{j=0}^{1} \exp[-U(H_j | u_k, \mathcal{C}_k)]}, \tag{15.6}$$

where $U(H_i | u_k, \mathcal{C}_k)$ is the "energy function" of the MRF model.

In accordance with the MRF-based approach to data fusion, this energy function is expressed as a linear combination of energy contributions, each related either to the label configuration or to the information conveyed by each channel of the ratio image [45]. The former is formulated again as a second-order isotropic Potts model and the latter is directly related to the single-variate PDF of the corresponding SAR ratio channel conditioned to "change" and "no-change." Hence, the proposed MRF model is defined by the following energy function ($k = 1, 2, \ldots, N; i = 0, 1$):

$$U(H_i | u_k, \mathcal{C}_k, \theta) = \sum_{r=1}^{n} \alpha_r [-\ln p_{ir}(u_{kr} | \xi_{ir})] - \beta m_{ik}, \tag{15.7}$$

where $p_{ir}(\cdot | \xi_{ir})$ is a parametric model for the PDF of the r-th amplitude ratio u_{kr} conditioned to H_i, ξ_{ir} is the vector of the parameters of this PDF, m_{ik} is again the number of labels equal to H_i in the neighborhood of the k-th pixel, α_r ($r = 1, 2, \ldots, n$) and β are positive parameters, and $\theta = (\beta, \alpha_1, \alpha_2, \ldots, \alpha_n, \xi_{01}, \xi_{11}, \xi_{02}, \xi_{12} \ldots, \xi_{0n}, \xi_{1n})$ is a vector collecting all the model parameters to be estimated (the dependence of the energy on θ has been included in the related notation). The same stationarity assumption, as mentioned with regard to the MRF model of FT-MCD is implicitly considered for DF-MCD, as well.

In the case of FT-MCD, the choice of an LN distribution is almost mandatory because the other parametric models proposed for amplitude ratio channels do not admit natural multivariate extensions. For DF-MCD, this constraint does not hold because only the marginal PDFs of the amplitude ratio channels are involved in the MRF model. The NR, WR, and LN distributions are adopted here. They are analytically derived (under mild assumptions) from the well-known Nagakami-Gamma, Weibull and log-normal distributions for the SAR amplitudes acquired at the two observation dates (further details on these derivations can be found in [28]) and are given by ($k = 1, 2, \ldots, N; r = 1, 2, \ldots, n; i = 0, 1$):

- LN distribution for u_{kr}, conditioned to H_i ($\xi_{ir} = (\mu_{ir}, \sigma_{ir})$, $\mu_{ir} \in \mathbb{R}$, $\sigma_{ir} > 0$):

$$p_{ir}(u_{kr}|\xi_{ir}) = \frac{1}{\sigma_{ir} u_{kr} \sqrt{2\pi}} \exp\left[-\frac{(\ln u_{kr} - \mu_{ir})^2}{2\sigma_{ir}^2}\right], \qquad u_{kr} > 0; \tag{15.8}$$

- NR distribution for u_{kr}, conditioned to H_i ($\xi_{ir} = (L_{ir}, \gamma_{ir}), L_{ir}, \gamma_{ir} > 0$):

$$p_{ir}(u_{kr}|\xi_{ir}) = \frac{2\Gamma(2L_{ir})}{\Gamma^2(L_{ir})} \frac{\gamma_{ir}^{L_{ir}} u_{kr}^{2L_{ir}-1}}{(\gamma_{ir} + u_{kr}^2)^{2L_{ir}}}, \qquad u_{kr} > 0, \tag{15.9}$$

  where $\Gamma(\cdot)$ is the usual Gamma function [39];
- WR distribution for u_{kr}, conditioned to H_i ($\xi_{ir} = (\eta_{ir}, \lambda_{ir}), \eta_{ir}, \lambda_{ir} > 0$):

$$p_{ir}(u_{kr}|\xi_{ir}) = \eta_{ir} \lambda_{ir}^{\eta_{ir}} \frac{u_{kr}^{\eta_{ir}-1}}{(\lambda_{ir}^{\eta_{ir}} + u_{kr}^{\eta_{ir}})^2}, \qquad u_{kr} > 0. \tag{15.10}$$

The parameters $\beta, \alpha_1, \alpha_2, \ldots, \alpha_n$ tune the relative importance of the energy terms. In particular, α_r plays the role of a "reliability factor" assigned to the information source represented by the r-th channel ($r = 1, 2, \ldots, n$) and is usually assumed to take on values in $[0, 1]$. This constraint can be analytically expressed as follows[2]:

$$||2\alpha - \mathbf{1}||_\infty \leq 1, \tag{15.11}$$

where $\mathbf{1}$ is an n-dimensional vector with n unitary components and $\alpha = (\alpha_1, \alpha_2, \ldots, \alpha_n)$. As detailed in Appendix A.2, a differentiable constraint is desirable for analytical convenience. Hence, we replace the nondifferentiable constraint in Eq. (15.11) with the following one:

$$||2\alpha - \mathbf{1}||_q \leq 1, \tag{15.12}$$

where $q \geq 2$ is an even integer number. For sufficiently large values of q (e.g., $q = 10$), Eq. (15.12) is basically a differentiable approximation of Eq. (15.11). Furthermore, it also allows undesired degenerate parameter values to be avoided (e.g., the undesired configuration $\alpha = 0$, which removes from the MRF model all the energy contributions related to the ratio channels, is permitted by Eq. (15.11) but forbidden by Eq. (15.12)).

[2] Given $v \in \mathbb{R}^m$ and $q \geq 1$, we denote as $||v||_\infty$ and $||v||_q$ the Tchebitchev and the q-th-order Minkowski norms (∞-norm and q-norm, for short) of v, i.e., $||v||_\infty = \max\{|v_1|, |v_2|, \ldots, |v_m|\}$ and $||v||_q = (|v_1|^q + |v_2|^q + \ldots + |v_m|^q)^{1/q}$ [40].

15.3.2 EM-Based Parameter-Estimation and Classification

The proposed DF-MCD change-detection method postulates the MRF model of Eqs. (15.6) and (15.7) to hold, and aims at generating a change map that minimizes the corresponding energy function. Hence, suitable parameter-estimation and classification strategies need to be defined. Toward this end, a case-specific variant of the LJ-EM method is developed by integrating this approach with MoLC.

The application of LJ-EM to the MRF model defined by Eqs. (15.6) and (15.7) gives rise to two difficulties. The first, already mentioned with regard to FT-MCD, lies in the computational burden involved by the combination of MRF and EM-type methods and is solved again by adopting the "mode-field" approach. A further critical point lies in the fact that the maximization problems involved by LJ-EM have no closed-form solutions, when one uses the NR and WR models [28][31]. In order to address this issue, we propose to combine LJ-EM with MoLC, a recently developed parameter estimation method that exploits the Mellin transform theory [44] to state a set of equations relating the unknown parameters to logarithmic moments or logarithmic cumulants of the gray-level distribution, thus expressing the parameter-estimation problem as the solution of a set of (typically nonlinear) equations [33].

MoLC has been developed in the context of SAR image analysis and has been proven to be numerically feasible and fast for many SAR-specific models and to exhibit good theoretical estimation properties [32][33]. Moreover, it turns out to be feasible for all the LN, NR, and WR distributions [28]. Denoting as $\kappa_{1ir} = E\{\ln u_{kr} | \ell_k = H_i\}$ and $\kappa_{2ir} = \mathrm{Var}\{\ln u_{kr} | \ell_k = H_i\}$ the logarithmic mean and variance of u_{kr}, given H_i ($k = 1, 2, \ldots, N; r = 1, 2, \ldots, n; i = 0, 1$), MoLC allows the following equations to be obtained for the LN, NR, and WR distributions [28]:

$$\begin{cases} \kappa_{1ir} = \mu_{ir} \\ \kappa_{2ir} = \sigma_{ir}^2 \end{cases} \qquad \begin{cases} 2\kappa_{1ir} = \ln \gamma_{ir} \\ 2\kappa_{2ir} = \Psi(1, L_{ir}) \end{cases} \qquad \begin{cases} \kappa_{1ir} = \ln \lambda_{ir} \\ \kappa_{2ir} = 2\Psi(1,1)\eta_{ir}^{-2} \end{cases} \tag{15.13}$$

where $\Psi(1, \cdot)$ is the first-order polygamma function (i.e., the second logarithmic derivative of $\Gamma(\cdot)$ [44]). The MoLC equations for LN simply state the identity between the distribution parameters and the logarithmic mean and variance. On the other hand, the ones for NR and WR represent a set of nontrivial equations relating the distribution parameters to κ_{1ir} and κ_{2ir} ($r = 1, 2, \ldots, n; i = 0, 1$).

LJ-EM is iterative and, as discussed in Sec. 15.2.3, when applied according to an LN model, it computes, at each iteration, estimates of the conditional logarithmic means and variances. In order to integrate LJ-EM with MoLC in the DF-MDC procedure, we still propose to update, at each iteration, estimates of the conditional logarithmic means and variances as in the LN version of LJ-EM, and to derive estimates of the distribution parameters by solving the related MoLC equations. A similar hybrid MoLC-EM approach was applied in [31] for probability modelling purposes.

Accordingly, the following operations are performed at the t-th iteration of the DF-MCD algorithm ($t = 0, 1, 2, \ldots$):

1. compute, for each k-th pixel, the current estimate of $P\{\ell_k = H_i|u_k, C_k^t, \theta^t\}$ $(k = 1, 2, \dots, N; i = 0, 1)$ by using Eqs. (15.6) and (15.7);
2. update the label of each k-th pixel according to the MRF minimum-energy rule [10] by setting ℓ_k^{t+1} as the label H_i that corresponds to the smallest value of $U(H_i|u_k, C_k^t, \theta^t)$ $(k = 1, 2, \dots, N; i = 0, 1)$;
3. update the estimates of the spatial parameter β and of the logarithmic means and variances as follows $(r = 1, 2, \dots, n; i = 0, 1)$:

$$\begin{cases} \kappa_{1ir}^{t+1} = \dfrac{\sum_{k=1}^N w_{ik}^t \ln u_{kr}}{\sum_{k=1}^N w_{ik}^t} \\ \kappa_{2ir}^{t+1} = \dfrac{\sum_{k=1}^N w_{ik}^t (\ln u_{kr} - \kappa_{1ir}^{t+1})^2}{\sum_{k=1}^N w_{ik}^t} \\ \beta^{t+1} = \arg\max_{\beta>0} \sum_{k=1}^N \left[\beta \sum_{i=0}^1 w_{ik}^t m_{ik}^t - \ln \sum_{i=0}^1 \exp(\beta m_{ik}^t) \right]; \end{cases} \tag{15.14}$$

   where $w_{ik}^t = P\{\ell_k = H_i|u_k, C_k^t, \theta^t\}$ if the k-th pixel has been assigned to H_i in step 2, and $w_{ik}^t = 0$ otherwise $(k = 1, 2, \dots, N; i = 0, 1)$;
4. compute, for each hypothesis H_i and each r-th channel, the updated estimate ξ_{ir}^{t+1} of the parameter vector ξ_{ir} of the adopted PDF model (i.e., NR, WR, or LN) by plugging the logarithmic mean and variance estimates κ_{1ir}^{t+1} and κ_{2ir}^{t+1} computed in step 3 in the corresponding MoLC equations (see Eq. (15.13)) and by solving for the unknown parameters $(r = 1, 2, \dots, n; i = 0, 1)$;
5. update the estimate of the reliability factor α_r of each r-th channel as follows $(r = 1, 2, \dots, n$; see Appendix A.2):

$$\alpha_r^{t+1} = \frac{1}{2} + \frac{1}{2} \sqrt[q-1]{\frac{c_r^t}{||c^t||_{q'}}}, \tag{15.15}$$

   where:

$$q' = \frac{q}{q-1}, \qquad c_r^t = \sum_{k=1}^N \sum_{i=0}^1 w_{ik}^t \ln p_{ir}(u_{kr}|\xi_{ir}^{t+1}), \tag{15.16}$$

   and $||c^t||_{q'}$ is the q'-norm of the vector $c^t = (c_1^t, c_2^t, \dots, c_n^t)$.

The formulations of the updated estimates of the spatial parameter (step 3) and of the reliability factors (step 5) are proven in Appendix A.2. The maximization problem to compute β^{t+1} is again numerically solved by the Newton-Raphson method. Step 4 requires solving the MoLC equations for the adopted PDF model. No actual solution process is required for LN, since the PDF parameters are exactly equal to the logarithmic mean and variance. The solution process for WR can be analytically carried out in closed form. A numerical procedure is needed for NR in order to solve the equation involving κ_{2ir} and L_{ir} $(r = 1, 2, \dots, n; i = 0, 1)$. However, thanks to the strict monotonicity of the involved polygamma function, the simple bisection method [35] can be effectively applied. The updates of the estimates of the reliability factors (step 5) are performed by a closed-form calculation.

As in the case with FT-MCD, the procedure is iterated until the differences between the parameter estimates at successive iterations are below a threshold (equal to 0.001). Since the iterative procedure performed by the proposed change-detection method is actually derived from EM (see Appendix A.2), a convergent behavior of the method can be theoretically expected [17][37][48].

A single-channel initialization strategy similar to the one employed for FT-MCD is used for DF-MCD, as well, by using GKIT. However, in the case of FT-MCD, GKIT was applied only in its LN version. Here, it is based on an NR, WR, or LN model, according to the distribution chosen in the energy function of Eq. (15.7). Then, the single-channel change map obtained by applying GKIT to each ratio channel is computed, and the best among such maps is chosen according to the minimum value of the GKIT criterion function and used to define the initial context and to compute the initial parameter estimates.

15.4 Experimental Results

15.4.1 Data Sets for Experiments

Two 700×280 pixel-sized coregistered multipolarization and multifrequency SAR images, acquired over an agricultural area near the city of Pavia (Italy) on April 16th and 18th, 1994, were used for experiments. At each date, a 4-look XSAR image (X band, VV polarization) and three 4-look SIR-C channels (C band, HH, HV polarizations and TP (total power) channels) were available (Figs. 1(a) and 1(b)). Ground changes occurred in the scene, as several rice fields were artificially flooded (for cultivation purpose). A test map, presenting 11287 "no-change" test pixels and 1870 "change" test pixels, was available (Fig. 1(c)).

As is usual in remote sensing, this map consisted of disjoint homogeneous regions with no test pixels at the interfaces between "change" and "no-change" areas. To assess the behavior of the Markovian approach at the border between different regions, a 130 × 100 pixel-sized test map with two classes was created (Fig. 2(a)), and a multichannel image was generated by filling the resulting regions with pixels drawn from the test areas of the SIR-C/XSAR data set (Figs. 2(b) and 2(c)). A semisimulated multichannel image was thus obtained at each date, such that every pixel was drawn from a real image and given a test label (in particular, 14.2% of the image area consisted of "change" pixels). We note that, with respect to a real data set, in this case no mixed pixels were present at the interfaces between changed and unchanged areas.

15.4.2 Experimental Results for FT-MCD

Let us focus on the results provided by the FT-MCD method. In order to focus on the role of the MRF model, two experimental comparisons were first performed. First, a noncontextual variant of the method was considered: in the "EM-based classification stage," LJ-EM was applied by assuming the labels not to be Markovian but rather independent and identically distributed (i.i.d.) and

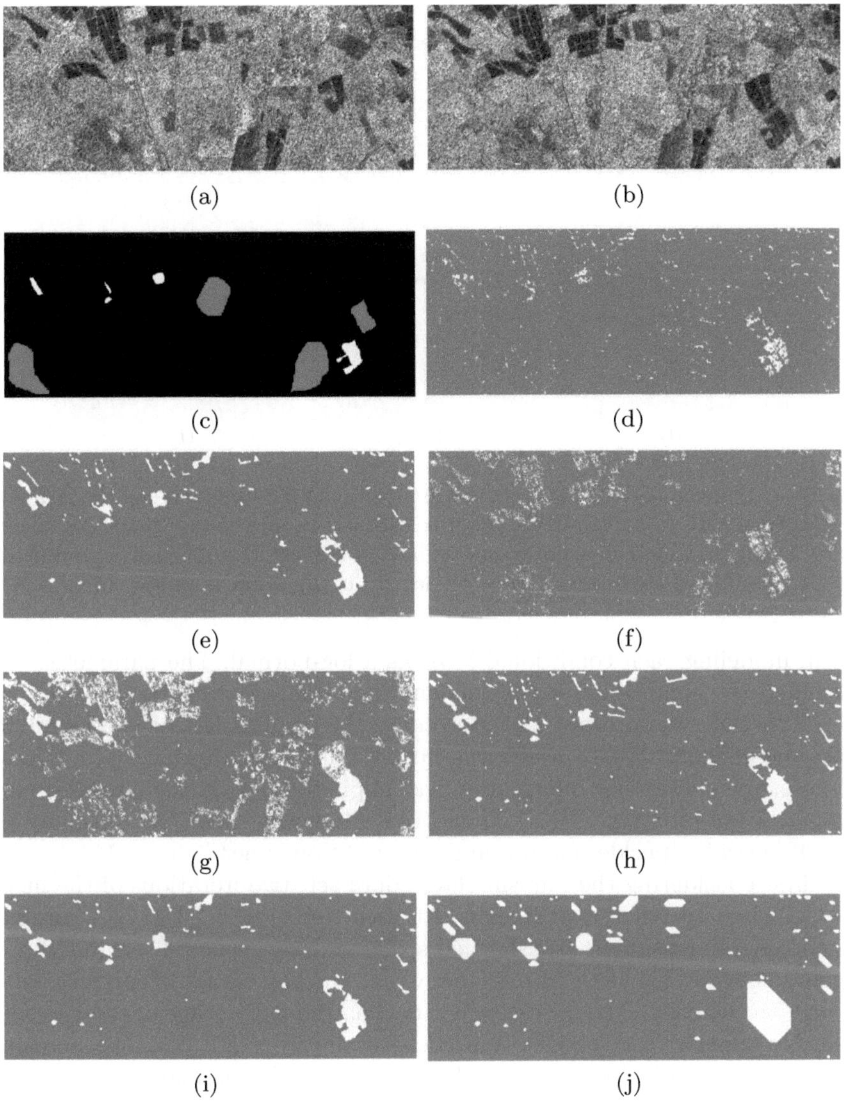

Fig. 15.1. Real data set: XSAR channels acquired on April 16th (a) and April 18th (b), 1994 (after histogram equalization), test map (c), and change maps generated by the initialization GKIT procedure (d), by FT-MCD (e), by the noncontextual variant of FT-MCD applied to the original images (f) and after despeckling (g), by DF-MCD, applied with $q = 2$ (h) and with $q = 10$ (i), and by the variant of DF-MCD endowed with the ∞-norm (j). All the methods were applied with the LN model. Legend to the change maps: white = "change", grey = "no-change."

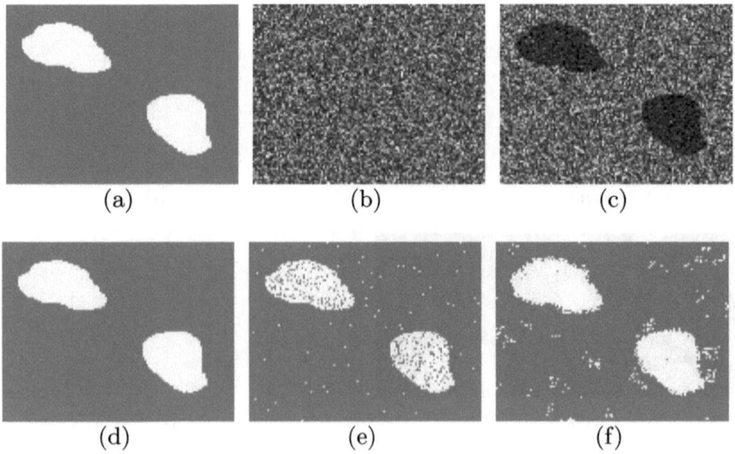

Fig. 15.2. Semisimulated data set: test map (a), XSAR channels for the April 16th (b) and April 18th (c), 1994, dates (after equalization), and change maps generated by FT-MCD (d) and by a noncontextual variant of FT-MCD applied to nonfiltered (e) and speckle-filtered (f) images. The color legend is the same as in Fig. 15.1.

by still modeling each conditional PDF as a log-normal. The parameter vector θ included the conditional log-means and log-variances and the prior probabilities of the two hypotheses [25]. Secondly, the application of speckle-filtering before the noncontextual variant was carried out, as an alternative to MRFs. In this case, the contextual information was exploited not by the classifier but by the spatial pre-filtering. The Gamma-MAP filter was chosen as a usually good tradeoff between speckle removal and computation time[3] [34].

By first considering the semisimulated data set, two iterations of the method were sufficient to reach convergence. At each iteration, LJ-EM was run in the related (one-dimensional) transformed space and, as expected, exhibited a convergent behavior. The detection accuracy (i.e., the percentage of correctly labeled "change" test pixels), the false-alarm rate (i.e., the percentage of erroneously labeled "no-change" test pixels), and the overall error rate (i.e., the percentage of erroneously labelled test pixels) of the resulting change maps were computed according to the exhaustive test map in Fig. 2(a) and are shown in Table 15.1. A very accurate map was obtained even though the initialization exhibited a lot of missed alarms. A visual analysis confirms this comment and also points out that, at least for this data set, the proposed contextual approach does not involve any loss of detail near the edges between "change" and "no-change" (see Fig. 2(d)).

The application of the noncontextual approach to nonfiltered data was not effective (Fig. 2(e)) and the resulting change map presented just a minor

[3] Two Gamma-MAP filtering iterations, with a 7×7 window, were adopted. Such values for the filter parameters were selected by a "trial-and-error" procedure, with a tradeoff between speckle removal and loss of detail.

Table 15.1. Change-detection performances of FT-MCD, of the initialization procedure, and of the noncontextual variant, applied both to the original (nonfiltered) data (NC-NF) and to speckle-filtered images (NC-SF).

Data set	Method	False-alarm rate	Detection accuracy	Overall error rate
semisimulated	FT-MCD	0.03%	99.73%	0.06%
	initialization	0.87%	82.27%	3.27%
	NC-NF	0.56%	81.95%	3.05%
	NC-SF	3.62%	99.73%	3.15%
real	FT-MCD	0%	93.10%	0.98%
	initialization	0.33%	49.09%	7.52%
	NC-NF	0.30%	31.55%	9.99%
	NC-SF	1.20%	99.89%	1.05%

improvement, as compared with the single-channel initialization (Table 15.1). This is interpreted as being due to the presence of speckle, and confirms the importance of contextual information. Better results were obtained by the noncontextual variant when applied to speckle-filtered images: a very high detection accuracy was achieved but along with many false alarms (Table 15.1). A numerical comparison between this contextual strategy and the one adopted by the proposed method suggests that the latter gives more accurate results (Table 15.1). Many false alarms can be noticed at the interfaces between "change" and "no-change" in the map generated by the noncontextual variant after despeckling (Fig. 2(f)), whereas the MRF-based method provided more accurate results in these critical areas as well.

Concerning the real data set, four iterations of the method were sufficient to reach convergence. At each iteration, LJ-EM was run in the transformed space and converged, in turn, in 14 or 15 iterations. A good detection accuracy (above 93%) was obtained by this method at very low error rates (below 1%; see Table 15.1). A visual analysis of the corresponding maps confirms these data (Fig. 1(e)). Note that no false alarm was observed in the test areas, although some false alarms can be noticed in other image areas, often due to small misregistration errors between the images acquired at the two dates. A considerable improvement was obtained by the proposed method, as compared with the initialization map. GKIT, applied to the single channel selected by the procedure described in Sec. 15.2.4 (which turned out to be SIR-C-HV), provided poor results due to the strong presence of speckle. Thanks to the joint use of multichannel and contextual information, the proposed method sharply increased the detection accuracy and decreased the error rate during each of the four steps.

Less accurate results were obtained by the noncontextual variant. When applied to nonfiltered data, it generated an even worse change map than the initialization one (see Table 15.1), with many missed and false alarms, the latter ones being particularly concentrated in image areas covered with wet soil at the second date (see Fig. 1(f)). A similar effect was also noticed when we applied the noncontextual variant after speckle filtering: even though the detection accuracy sharply improved, as compared with the case without despeckling, many false alarms were present in the areas covered with wet soil in the April-18th image (the corresponding false-alarm rate in Table 15.1 is still quite low, since, in this case, most of the false alarms were located outside the test areas). This can be interpreted as being due to the possible multimodality of "no-change," each mode being related to a specific land-cover subclass (a multimodal behavior is not expected for "change" because only one change typology is present in this data set). LJ-EM, formulated according to a two-component model ("change" vs "no-change"), can be sensitive to such multimodality because it can erroneously merge one or more "no-change" mode(s) with the "change" one. Since the map generated by the "EM-based classification stage" of each step is used as a training set in the "feature-transformation stage" of the next step, the merging is also expected to affect the update of the SAR-specific Fisher transform. This undesirable effect was not observed in the results of the proposed method, thanks to the fact that the spatial regularization energy term of the MRF model explicitly takes into account the spatial context in the iterative parameter update (see Eq. (15.5)), thus optimizing at each iteration the separation between the two hypotheses.

15.4.3 Experimental Results for DF-MCD

Three versions of DF-MCD can be defined according to the possible use of the three considered PDF models. In order to validate the method, the three versions were applied to the real data set by initializing the α_r ($r = 1, 2, 3, 4$) and β parameters with unitary values. Experiments on the semisimulated data set are not reported for brevity. Results very similar to the ones obtained by FT-MCD were yielded by DF-MCD as well, thus confirming the good behavior of this method, too, in the neighborhoods of the edges.

The initialization GKIT procedure, applied with either the LN, NR, or WR models, selected again the change map generated by the SIR-C-HV channel. As theoretically expected, the proposed method reached convergence in all the cases (fewer than 50 iterations were sufficient in all the experiments). Table 15.2 shows the accuracies of the resulting change maps computed according to the test map in Fig. 1(c).

Focusing first on the results obtained by $q = 2$ (i.e., Euclidean norm in the constraint of Eq. (15.12)), accurate change maps were generated by all the versions of the method, at very low false-alarm rates (always below 1% and often below 0.2%) and high detection accuracies (above 92% in all the cases), even though the initialization map showed a very low accuracy (see Table 15.1). This suggests a good capability of DF-MCD to jointly exploit both the information conveyed by the four ratio channels and the spatial information related to the correlations among neighboring pixels by Markovian data fusion. The three PDF models yielded results quite similar to one another. Specifically, the lowest overall error rate (0.84%) was obtained by LN, while the highest detection accuracy

Table 15.2. Change-detection performances of DF-MCD.

Norm index	Parametric model	False-alarm rate	Detection accuracy	Overall error rate
2	LN	0.16%	95.08%	0.84%
	NR	0.11%	92.24%	1.20%
	WR	0.62%	97.05%	0.95%
10	LN	0%	94.70%	0.75%
	NR	0%	91.01%	1.27%
	WR	0.62%	97.05%	0.95%
∞	LN	2.01%	99.14%	1.84%
	NR	1.06%	88.02%	2.58%
	WR	0.62%	97.05%	0.95%

(97.05%) was achieved by WR. A slightly worse result was obtained by NR. A visual analysis of the change maps (see, for instance, Fig. 1(h), which refers to the case of LN) further confirms both the limited impact of speckle and the correct detection of the changed areas. Several false alarms can be noticed in Fig. 1(h), as well, especially at the borders of the agricultural fields present in the imaged scene. Again, they are mainly due to small misregistration errors.

A remarkably stable behavior of the results was noticed when the Minkowski-norm index q was varied in the range $[2, 10]$ (the results obtained by $q = 10$ are given in Table 15.1; the ones obtained by $q = 4, 6, 8$ are similar so they are omitted for the sake of brevity). When using WR, the results were actually unaffected by the variation of $q \in [2, 10]$; when either LN or NR was employed, only small variations in the numerical accuracies were observed (Table 15.1). A visual analysis of the related change maps (see, for example, Fig. 1(i)) points out that such variations are actually related to a stronger spatial regularization effect obtained by using the 10-norm, as compared with the 2-norm. Accordingly, most of the false alarms, which were noticed when $q = 2$ due to misregistration errors, were not present when $q = 10$ (Fig. 1(i)). On the other hand, a slightly stronger smoothing of the borders between changed and unchanged areas can be noticed in the change map generated by using the 10-norm than in the map obtained by the 2-norm. This can be interpreted as a consequence of the different convergence values reached by the estimated reliability factors. Focusing, for instance, on the case of LN, $\alpha = (0.224, 0.238, 0.281, 0.260)$ and $\alpha = (0.060, 0.062, 0.071, 0.066)$ were reached at convergence in the 2- and 10-norm cases, respectively, i.e., much smaller values were obtained by $q = 10$ than by $q = 2$. On the other hand, the method provided almost unitary values for the estimate of β in both cases. Therefore, in the 10-norm case, the spatial energy contribution was assigned a comparatively larger relative weight than those of the energy terms related to the ratio channels, as compared with the 2-norm case, thus a stronger spatial regularization effect was obtained.

A specific experiment was performed to focus on the role of the constraints in Eqs. (15.11) and (15.12). A variant of the proposed method was run such that, at each LJ-EM iteration, the reliability factors were updated according to the ∞-norm constraint. As discussed in Appendix A.2, this choice results in a linear programming problem to compute, at each LJ-EM iteration, updated estimates of the reliability factors, and yields binary-valued solutions (i.e., the updated value of each α_r parameter is either 0 or 1; $r = 1, 2, 3, 4$). As suggested by Table 15.2, the same results as those obtained by the 2- and 10-norms were achieved when using WR. On the other hand, significantly higher error rates were obtained by LN and NR with the ∞-norm, as compared with the 2- and 10-norms. In particular, when using the LN model, LJ-EM even converged to parameter configurations with zero values for all the α_r factors ($r = 1, 2, 3, 4$). In such degenerate situations, the minimum-energy decision rule is affected only by the spatial energy contribution and simply performs a smoothing of the initial change map, thus giving a very poor result (Fig. 1(j)). These critical configurations are forbidden under the q-norm constraint employed by the proposed method; this further confirms the usefulness of the adopted formulation.

Finally, a comparison between Tables 15.1 and 15.2 shows that slightly higher accuracies were obtained by DF-MCD than by FT-MCD. In particular, the detection accuracy of DF-MCD, applied with WR, was almost 4% higher than the one provided by FT-MCD. This points out, at least for the adopted data set, a higher effectiveness of the data-fusion approach than the one of the multichannel transformation approach to solve the multichannel SAR change-detection problem.

15.5 Conclusions

Two unsupervised contextual change-detection techniques have been proposed for multichannel SAR images by combining the Markovian approach to classification, the LJ-EM parameter-estimation algorithm, and either a SAR-specific version of the Fisher transform or a data-fusion strategy. Accurate change maps were provided by both methods when applied to real multipolarization and multifrequency SAR data. No preliminary despeckling was needed, thanks to the capability of MRFs to exploit the spatial contextual information in the classification process.

As far as the FT-MCD technique is concerned, a specific experiment, carried out with semisimulated data endowed with an exhaustive test map, has also demonstrated the accurate behavior of the method at the interface between "change" and "no-change," with no significant loss of detail, at least, for this data set. A fully realistic experiment would need to take into account mixed pixels, which is a problem beyond the scope of this chapter. A noncontextual variant of FT-MCD, even when applied to despeckled images, provided less accurate results, thus proving to be strongly sensitive to the possible multimodality of the "no-change" hypothesis. This drawback has not been observed for the proposed approach, which further confirms the effectiveness of MRFs as contextual classification tools. Hence, at least as regards the considered data sets, the application of the proposed MRF-based contextual approach to change detection was proved to be more effective, as compared with the application of a noncontextual strategy after a standard speckle-filtering preliminary step.

Concerning DF-MCD, three versions of the method have been defined, according to three different parametric families (i.e., log-normal, Weibull-ratio, and Nakagami-ratio), derived from the corresponding well-known models for SAR amplitude statistics. They provided results quite similar to one another, with slightly lower overall error rates for the log-normal distribution and slightly higher detection accuracies for the Weibull-ratio family.

Both proposed techniques are iterative and the experiments have shown good convergent behaviors in small numbers of iterations, although the methods were initialized with (quite) low-accuracy change maps. Slightly longer execution times were required by DF-MCD with the Nakagami-ratio PDF, due to the need to numerically solve nonlinear equations at each iteration. Anyway, the computation time was small in all the cases, and no more than a few minutes were needed to reach convergence in any run of the two methods.

Concerning FT-MCD, the convergence properties have not yet been analytically proven and represent an interesting issue worth being investigated. On the other hand, the convergence of DF-MCD is a theoretically expected result, thanks to the well-known asymptotic properties of EM-like procedures. This represents a theoretical advantage of DF-MCD over FT-MCD. In addition, an experimental comparison of the two approaches also points out an improvement of DF-MCD over FT-MCD, especially in terms of detection accuracy.

On the other hand, DF-MCD presents an internal parameter whose value has to be preliminarily chosen, i.e., the index q of the Minkowski norm involved in the related MRF model, whereas FT-MCD does not require such a parameter setting and is fully automatic. Anyway, a remarkable stability of the change-detection results as functions of q has been pointed out by the experiments. This suggests that a preliminary choice of a proper value for q is not critical. In any case, the automation of the parameter-setting aspect for DF-MCD could be an interesting future extension of the present work.

The impact of the specific EM algorithm adopted in the proposed methods has also been investigated. Toward this end, different variants of EM have been developed and experimented using the same data sets. In particular, two further versions of the FT-MCD method (not considered here for the sake of brevity), adopting the standard EM [37] and the stochastic EM [5] algorithms, have been explored: they have provided very similar results (see [29] for details). A further version of DF-MCD, based on the standard EM technique, has also been developed. Slightly worse results were given by this approach, as compared with LJ-EM (details can be found in [30]).

For both proposed methods, according to the monomodal model for the "change" statistics, only one typology of change (either an increase or a decrease in the backscattering coefficient) is assumed to be present in the image. When both an increase and a decrease in the backscattering-coefficient occur, the proposed method can be separately applied to the two possible ratio images [28][34] (preliminary experiments, not reported here for brevity, have confirmed the effectiveness of this procedure). However, when many typologies of change are present, the adopted two-component model may lead to an incorrect merging of some "change" and "no-change" subclasses. A multiclass extension of the method could be developed to overcome this drawback.

Future extensions of this work could examine the integration of more sophisticated MRF models to further reduce the error rates (e.g., by taking into account the image nonstationarity [8] or by including edge information [13]), or the combination of the two methods with object-based approaches [14] to better adapt them to the application to high-resolution images. In particular, a further future generalization of DF-MCD could be the integration of multispectral data and/or of other (e.g., ancillary) information sources by suitably extending the related MRF-based data fusion approach.

Acknowledgments

The authors would like to thank Dr. Paolo Gamba from the University of Pavia (Italy) for providing the data employed in the experiments, and Mr. Daniele Chiapparoli and Mr. Andrea Sciarrone for their assistance with the implementation.

A Appendix

A.1 Proofs Related to FT-MCD

The mode-field EM algorithm expresses the parameter estimation problem as the iterative maximization of a modified log-likelihood function [5]. Adopting the notations in Sec. 15.2, the "mode-field" EM algorithm for the MRF model related to FT-MCD computes the updated parameter vector θ^{t+1} at the t-th iteration by maximizing, with respect to θ, the following pseudolikelihood function ($t = 0, 1, 2, \ldots$) [5]:

$$Q(\theta|\theta^t) = \sum_{k=1}^{N} \sum_{i=0}^{1} w_{ik}^t \left[\ln P\{\ell_k = H_i | C_k^t, \beta\} + \ln p_i(v_k | \kappa_{1i}, \kappa_{2i}) \right], \tag{15.17}$$

where $w_{ik}^t = P\{\ell_k = H_i | u_k, C_k^t, \theta^t\}$ ($k = 1, 2, \ldots, N; i = 0, 1$). Thanks to the Markovianity of the label configuration, the local prior probability distribution $P\{\ell_k = H_i | C_k^t, \beta\}$ of the k-th pixel label ℓ_k conditioned to its context is proved to be ($k = 1, 2, \ldots, N$) [10]:

$$P\{\ell_k = H_i | C_k^t, \beta\} = \frac{\exp(\beta m_{ik}^t)}{\sum_{j=0}^{1} \exp(\beta m_{jk}^t)}, \tag{15.18}$$

Plugging Eq. (15.18) into Eq. (15.17), one may obtain after standard algebraic manipulations:

$$Q(\theta|\theta^t) = \Phi^t(\beta) + \sum_{i=0}^{1} \Phi_i^t(\kappa_{1i}, \kappa_{2i}) \tag{15.19}$$

where:

$$\begin{cases} \Phi^t(\beta) = \sum_{k=1}^{N} \left[\beta \sum_{i=0}^{1} w_{ik}^t m_{ik}^t - \ln \sum_{i=0}^{1} \exp(\beta m_{ik}^t) \right] \\ \Phi_i^t(\kappa_{1i}, \kappa_{2i}) = \sum_{k=1}^{N} w_{ik}^t \ln p_i(v_k|\kappa_{1i}, \kappa_{2i}). \end{cases} \tag{15.20}$$

The function in Eq. (15.19) is maximized with respect to θ, subject to positivity constraints on β, κ_{20}, and κ_{21}. Since only the first term $\Phi^t(\beta)$ depends on β, the expression in Eq. (15.5) for the updated spatial parameter β^{t+1} immediately follows. Concerning the PDF parameters, since κ_{1i} and κ_{2i} are included only in the term $\Phi_i^t(\kappa_{1i}, \kappa_{2i})$, the updated estimates for such parameters are obtained by maximizing $\Phi_i^t(\kappa_{1i}, \kappa_{2i})$ $(i = 0, 1)$. This maximization problem is fairly standard in the EM literature and the solution is given in Eq. (15.5) [37].

The use of LJ-EM instead of the standard EM algorithm does not directly affect this analytical formulation. The parameter-update equations of LJ-EM are obtained by modifying the EM weights w_{ik}^t as defined in Sec. 15.2.3 (i.e., by setting $w_{ik}^t = 0$ if the k-th pixel is not assigned to H_i at the t-th iteration) and by letting the EM equations unchanged [17]. This modification is aimed at reducing the overlapping between the two hypotheses.

A.2 Proofs Related to DF-MCD

Concerning DF-MCD, according to the formulation of the related MRF model, the pseudo-likelihood to be maximized at the t-th EM iteration is $(t = 0, 1, 2, \ldots)$ [5]:

$$Q(\theta|\theta^t) = \sum_{k=1}^{N} \sum_{i=0}^{1} w_{ik}^t \left[\ln P\{\ell_k = H_i|\mathcal{C}_k^t, \beta\} + \sum_{r=1}^{n} \alpha_r \ln p_{ir}(u_{kr}|\xi_{ir}) \right], \tag{15.21}$$

where $w_{ik}^t = P\{\ell_k = H_i|u_k, \mathcal{C}_k^t, \theta^t\}$. The local prior probability distribution $P\{\ell_k = H_i|\mathcal{C}_k^t, \beta\}$ is still given by Eq. (15.18) $(k = 1, 2, \ldots, N)$ [10]. Hence, we have:

$$Q(\theta|\theta^t) = \Phi^t(\beta) + \sum_{r=1}^{n} \alpha_r \sum_{i=0}^{1} \Phi_{ir}^t(\xi_{ir}) \tag{15.22}$$

where $\Phi^t(\beta)$ has the same expression as in Eq. (15.20) and:

$$\Phi_{ir}^t(\xi_{ir}) = \sum_{k=1}^{N} w_{ik}^t \ln p_{ir}(u_{kr}|\xi_{ir}). \tag{15.23}$$

The updated parameter vector θ^{t+1} is obtained by maximizing the function in Eq. (15.22) subject to the constraint in Eq. (15.12) on $\alpha_1, \alpha_2, \ldots, \alpha_n$, to a positivity constraint on β, and to possible further constraints on each parameter vector ξ_{ir} (e.g., in the case of NR, both components of $\xi_{ir} = (L_{ir}, \gamma_{ir})$ are constrained to be positive). The value of β^{t+1} is computed as in the case of FT-MCD. Concerning the PDF parameters, since ξ_{ir} is included only in the term $\Phi_{ir}^t(\xi_{ir})$, which is weighted in Eq. (15.22) by the positive factor α_r, the mode-field EM algorithm updates ξ_{ir} as follows ($r = 1, 2, \ldots, n; i = 0, 1$):

$$\xi_{ir}^{t+1} = \arg\max_{\xi_{ir}} \Phi_{ir}^t(\xi_{ir}). \tag{15.24}$$

If the LN model is used, then $\xi_{ir} = (\mu_{ir}, \sigma_{ir})$ and this maximization process yields, as in the case of FT-MCD ($r = 1, 2, \ldots, n; i = 0, 1$):

$$\begin{cases} \mu_{ir}^{t+1} = \dfrac{\sum_{k=1}^N w_{ik}^t \ln u_{kr}}{\sum_{k=1}^N w_{ik}^t} \\ \sigma_{ir}^{t+1} = \left[\dfrac{\sum_{k=1}^N w_{ik}^t (\ln u_{kr} - \mu_{1ir}^{t+1})^2}{\sum_{k=1}^N w_{ik}^t} \right]^{1/2} \end{cases}. \tag{15.25}$$

Since the parameters of LN are exactly the logarithmic mean and variance of the distribution, this actually proves Eq. (15.14) in the LN case. On the other hand, if either NR or WR is adopted, the maximization process in Eq. (15.24) does not yield a closed-form solution [28]. On the contrary, the MoLC approach is feasible for such parametric families; hence, we plug this approach in the EM iterative process, as described in Sec. 15.3.2.

The vector $\alpha = (\alpha_1, \alpha_2, \ldots, \alpha_n)$ of the reliability factors is involved only in the double summation term in Eq. (15.22). Hence, plugging the updated PDF parameter estimates ξ_{ir}^{t+1} ($r = 1, 2, \ldots, n; i = 0, 1$) into Eq. (15.22) and taking into account the definition of the c_r^t coefficients ($r = 1, 2, \ldots, n$; see Eq. (15.16)), the vector α^{t+1} is obtained by maximizing the following function:

$$F^t(\alpha) = \sum_{r=1}^n \alpha_r \sum_{i=0}^1 \Phi_{ir}^t(\xi_{ir}^{t+1}) = \sum_{r=1}^n c_r^t \alpha_r, \tag{15.26}$$

subject to the following constraint:

$$G(\alpha) = ||2\alpha - \mathbf{1}||_q^q - 1 = \sum_{r=1}^n (2\alpha_r - 1)^q - 1 \leq 0. \tag{15.27}$$

The domain $\Omega = \{\alpha \in \mathbb{R}^n : G(\alpha) \leq 0\}$ defined by the constraint is a compact subset of $\mathbb{R}^n$ and F^t is continuous on Ω: hence, a global constrained maximum point of F^t exists, thanks to the Weierstrass theorem [39]. Both F^t and G are differentiable and F^t has a constant and (in general) nonzero gradient. Hence, the constrained maximum point will lie at the boundary of Ω, i.e., it will satisfy the equality condition $G(\alpha) = 0$. According to the Lagrange multiplier theorem, a necessary condition for α to be a solution of this constrained maximization problem is expressed as:

$$\frac{\partial F^t(\alpha)}{\partial \alpha_r} + \lambda \frac{\partial G(\alpha)}{\partial \alpha_r} = 0, \qquad r = 1, 2, \ldots, n, \tag{15.28}$$

where λ is a Lagrange multiplier. Therefore:

$$c_r^t + \lambda \cdot 2q(2\alpha_r - 1)^{q-1} = 0, \qquad r = 1, 2, \ldots, n, \tag{15.29}$$

and, since $q \geq 2$ is an even integer number (and so $(q-1)$ is odd), we obtain:

$$2\alpha_r - 1 = -\sqrt[q-1]{\frac{c_r^t}{2\lambda q}}, \qquad r = 1, 2, \ldots, n. \tag{15.30}$$

Plugging this result into the equality constraint, we have:

$$G(\alpha) = \frac{1}{|2\lambda q|^{q/(q-1)}} \sum_{r=1}^{n} |c_r^t|^{q/(q-1)} - 1 = \left(\frac{||c^t||_{q'}}{|2\lambda q|} \right)^{q'} - 1 = 0 \Longrightarrow |2\lambda q| = ||c^t||_{q'}. \tag{15.31}$$

This yields two solutions α^+ and α^-, i.e.:

$$\alpha_r^{\pm} = \frac{1}{2} \pm \frac{1}{2} \sqrt[q-1]{\frac{c_r^t}{||c^t||_{q'}}}, \qquad r = 1, 2, \ldots, n. \tag{15.32}$$

They correspond to local extrema of F^t subject to the constraint defined by G. In order to choose the maximum point, we substitute into Eq. (15.26), which yields, after standard algebraic calculations:

$$F^t(\alpha^{\pm}) = \frac{1}{2} \left(\sum_{r=1}^{n} c_r^t \pm ||c^t||_{q'} \right). \tag{15.33}$$

Hence, α^+ is the solution corresponding to the maximum constrained value of F^t, and the reliability factors are updated as in Eq. (15.15).

Using LJ-EM, the weigths are modified as discussed with regard to FT-MCD. Note that the above calculations were feasible thanks to the differentiability of both the objective function F^t and the constraint G. The latter holds after the replacement of the ∞-norm with the q-norm. Furthermore, the ∞-norm constraint in Eq. (15.11) means that $0 \leq \alpha_r \leq 1$ for $r = 1, 2, \ldots, n$, i.e., α belongs to the n-dimensional unitary hypercube $[0, 1]^n$. Since F^t is linear, if the ∞-norm constraint were used, a linear programming problem would result, and the corresponding solution would be, in general, one of the vertices of the hypercube. Hence, binary values for the reliability factors would be obtained. This is not formally forbidden but is intuitively quite undesired from a data-fusion viewpoint because this solution does not actually fuse the information conveyed by the n ratio channels but simply selects a subset of the channels, while discarding the remaining ones.

References

1. Bazi, Y., Bruzzone, L., Melgani, F.: Change detection in multitemporal SAR images based on generalized Gaussian distribution and EM algorithm. In: Proceedings of the SPIE Conference on Image and Signal Processing for Remote Sensing X, Gran Canaria, Spain, pp. 364–375 (2004)
2. Bazi, Y., Bruzzone, L., Melgani, F.: An unsupervised approach based on the generalized Gaussian model to automatic change detection in multitemporal SAR images. IEEE Trans. Geosci. Remote Sensing 43(4), 874–887 (2005)
3. Bazi, Y., Bruzzone, L., Melgani, F.: Automatic identification of the number and values of decision thresholds in the log-ratio image for change detection in SAR images. IEEE Geosci. Remote Sensing Letters 3(3), 349–353 (2006)
4. Bruzzone, L., Prieto, D.F., Serpico, S.B.: A neural-statistical approach to multitemporal and multisource remote-sensing image classification. IEEE Trans. Geosci. Remote Sensing 37, 1350–1359 (1999)
5. Celeux, G., Forbes, F., Peyrand, N.: EM procedures using mean field-like approximations for Markov model-based image segmentation. Pattern Recogn. 36, 131–144 (2003)
6. Conradsen, K., Nielsen, A.A., Schou, J., Skriver, H.: A test statistic in the complex Wishart distribution and its application to change detection in polarimetric SAR data. IEEE Trans. Geosci. Remote Sensing 41(1), 4–19 (2003)
7. Cossu, R., Chaudhuri, S., Bruzzone, L.: A context-sensitive bayesian technique for the partially supervised classification of multitemporal images. IEEE Geosci. Remote Sensing Lett. 2(2), 352–356 (2005)
8. Descombes, X., Sigelle, M., Preteux, F.: Estimating Gaussian Markov Random Field parameters in a nonstationary framework: Application to remote sensing imaging. IEEE Trans. Image Process. 8(4), 490–503 (1999)
9. Dierking, W., Skriver, H.: Change detection for thematic mapping by means of airborne multitemporal polarimetric SAR imagery. IEEE Trans. Geosci. Remote Sensing 40(3), 618–636 (2002)
10. Dubes, R.C., Jain, A.K.: Random Field models in image analysis. J. Appl. Stat. 16, 131–163 (1989)
11. Fransson, J.E.S., Walter, F., Blennow, K., Gustavsson, A., Ulander, L.M.H.: Detection of storm-damaged forested areas using airborne CARABAS-II VHF SAR image data. IEEE Trans. Geosci. Remote Sensing 40(10), 2170–2175 (2002)
12. Fukunaga, K.: Introduction to statistical pattern recognition, 2nd edn. Academic Press, London (1990)
13. Geman, S., Geman, D.: Stochastic relaxation, Gibbs distributions, and the Bayesian restoration. IEEE Trans. Pattern Anal. Machine Intell. 6, 721–741 (1984)
14. Hazel, G.G.: Object-level change detection in spectral imagery. IEEE Trans. Geosci. Remote Sensing 39(3), 553–561 (2001)
15. Inglada, J.: Change detection on SAR images by using a parametric estimation of the Kullback-Leibler divergence. In: Proc. of IGARSS-2003, Toulouse, France, July 21-25, 2003, vol. 6, pp. 4104–4106 (2003)
16. Jackson, Q., Landgrebe, D.: Adaptive Bayesian contextual classification based on Markov Random Fields. IEEE Trans. Geosci. Remote Sensing 40(11), 2454–2463 (2002)
17. Jackson, Q., Landgrebe, D.A.: An adaptive classifier design for high-dimensional data analysis with a limited training data set. IEEE Trans. Geosci. Remote Sensing 39(12), 2664–2679 (2001)

18. Kasischke, E.S., Melack, J.M., Dobson, M.C.: The use of imaging radars for ecological applications - a review. Remote Sens. Environ. 59, 141–156 (1997)
19. Kittler, J., Illingworth, J.: Minimum error thresholding. Pattern Recogn. 19, 41–47 (1986)
20. Landgrebe, D.A.: Signal Theory Methods in Multispectral Remote Sensing. Wiley-InterScience, Chichester (2003)
21. Li, X., Yeh, A.G.: Multitemporal SAR images for monitoring cultivation systems using case-based reasoning. Remote Sens. Environ. 90, 524–534 (2004)
22. Lombardo, P., Oliver, C.J., Pellizzeri, T.M., Meloni, M.: A new maximum-likelihood joint segmentation technique for multitemporal SAR and multiband optical images. IEEE Trans. Geosci. Remote Sensing 11(11), 2500–2518 (2003)
23. Lombardo, P., Pellizzeri, T.M.: Maximum likelihood signal processing techniques to detect a step pattern of change in multitemporal SAR images. IEEE Trans. Geosci. Remote Sensing 40(4), 853–870 (2002)
24. Melgani, F.: Classification of multitemporal remote-sensing images by a fuzzy fusion of spectral and spatio-temporal contextual information. Int. J. Pattern. Recogn. Artificial Intell. 18(2), 143–156 (2002)
25. Melgani, F., Moser, G., Serpico, S.B.: Unsupervised change detection methods for remote sensing images. Opt. Eng. 41(12), 3288–3297 (2002)
26. Melgani, F., Serpico, S.B.: A statistical approach to the fusion of the spectral and spatio-temporal contextual information for the classification of remote sensing images. Pattern Recogn. Lett. 23, 1053–1061 (2002)
27. Mercier, G., Derrode, S.: SAR image change detection using distance between distributions of classes. In: Proc. of IGARSS-2004, Anchorage, USA, vol. 6, pp. 3872–3875 (2004)
28. Moser, G., Serpico, S.B.: Generalized minimum-error thresholding for unsupervised change detection from SAR amplitude imagery. IEEE Trans. Geosci. Remote Sensing 40(10), 2972–2982 (2006)
29. Moser, G., Serpico, S.B.: Unsupervised change-detection from multi-channel SAR data. In: Proc. of NORSIG-2005, Reykjavík, Iceland, June 7-9 (2006)
30. Moser, G., Serpico, S.B.: Unsupervised change detection by multichannel sar data fusion. In: Proc. of IGARSS-2007, Barcelona, Spain, July 23-27 (in print, 2007)
31. Moser, G., Zerubia, J., Serpico, S.B.: Dictionary-based Stochastic Expectation-Maximization for SAR amplitude probability density function estimation. IEEE Trans. Geosci. Remote Sensing 44(1), 188–200 (2005)
32. Moser, G., Zerubia, J., Serpico, S.B.: SAR amplitude probability density function estimation based on a generalized Gaussian model. IEEE Trans. Image Process. 15(6), 1429–1442 (2005)
33. Nicolas, J.-M.: Introduction aux statistiques de deuxième espéce: applications des log-moments et des log-cumulants a l'analyse des lois d'images radar. Traitement du Signal (in French) 19, 139–167 (2002)
34. Oliver, C., Quegan, S.: Understanding Synthetic Aperture Radar Images, Artech House, Norwood (MA) (1998)
35. Press, W.H., Teukolsky, S.A., Wetterling, W.T., Flannery, B.P.: Numerical recipes in C. Cambridge University Press, Cambridge (2002)
36. Qong, M.: Polarization state conformation and its application to change detection in polarimetric SAR data. IEEE Geosci. Remote Sensing Lett. 1(4), 304–308 (2004)
37. Redner, R.A., Walker, H.F.: Mixture densities, maximum likelihood, and the EM algorithm. SIAM Review 26(2), 195–239 (1984)
38. Rignot, E.J.M., Van Zyl, J.J.: Change detection techniques for ERS-1 SAR data. IEEE Trans. Geosci. Remote Sensing 31(4), 896–906 (1993)

39. Rudin, W.: Principles of Mathematical Analysis, 2nd edn. McGraw-Hill, New York (1976)
40. Rudin, W.: Functional Analysis, 2nd edn. McGraw-Hill, New York (1990)
41. Serpico, S.B., Moser, G.: MRF-based remote-sensing image classification with automatic model-parameter estimation. In: Chen, C.H., Taylor, Francis (eds.) Signal and Image Processing for Remote Sensing, pp. 305–326. CRC Press, Boca Raton (2006)
42. Siegert, F., Hoffmann, A.A.: The 1998 forest fires in East Kalimantan (Indonesia): a quantitative evaluation using high resolution ERS-2 SAR images and NOAA AVHRR Hotspot data. Remote Sens. Environ. 72, 64–67 (1999)
43. Singh, A.: Digital change detection techniques using remotely-sensed data. Int. J. Remote Sens. 10, 989–1003 (1989)
44. Sneddon, I.: The use of integral transforms. McGraw-Hill, New York (1972)
45. Solberg, A.H.S., Taxt, T., Jain, A.K.: A Markov Random Field model for classification of multisource satellite imagery. IEEE Trans. Geosci. Remote Sensing 34(1), 100–113 (1996)
46. Trees, H.L.V.: Detection, estimation and modulation theory, vol. 1. John Wiley & Sons, New York (1968)
47. White, R.G.: Change detection in SAR imagery. Int. J. Remote Sens. 12(2), 339–360 (1990)
48. Wu, C.F.J.: On the convergence properties of the EM algorithm. The Ann. Stat. 11, 95–103 (1983)
49. Zografos, K., Nadarajah, S.: Expressions for Renyi and Shannon entropies for multivariate distributions. Statistics and Probability Letters 71, 71–84 (2005)

Index

Author Index

MIX
Papier aus verantwortungsvollen Quellen
Paper from responsible sources
FSC® C105338

If you have any concerns about our products,
you can contact us on
ProductSafety@springernature.com

In case Publisher is established outside the EU,
the EU authorized representative is:
Springer Nature Customer Service Center GmbH
Europaplatz 3, 69115 Heidelberg, Germany

Printed by Libri Plureos GmbH
in Hamburg, Germany